FOURIER TRANSFORMS
An Introduction for Engineers

THE KLUWER INTERNATIONAL SERIES
IN ENGINEERING AND COMPUTER SCIENCE

FOURIER TRANSFORMS
An Introduction for Engineers

by

Robert M. Gray
Joseph W. Goodman

Information Systems Laboratory
Department of Electrical Engineering
Stanford University

KLUWER ACADEMIC PUBLISHERS
Boston / Dordrecht / London

Distributors for North America:
Kluwer Academic Publishers
101 Philip Drive
Assinippi Park
Norwell, Massachusetts 02061 USA

Distributors for all other countries:
Kluwer Academic Publishers Group
Distribution Centre
Post Office Box 322
3300 AH Dordrecht, THE NETHERLANDS

Library of Congress Cataloging-in-Publication Data

A C.I.P. Catalogue record for this book is available
from the Library of Congress.

to Ron Bracewell
whose teaching and research on Fourier transforms
and their applications have been an inspiration
to us and to generations of students

Contents

Preface

The Fourier transform is one of the most important mathematical tools in a wide variety of fields in science and engineering. In the abstract it can be viewed as the transformation of a signal in one domain (typically time or space) into another domain, the frequency domain. Applications of Fourier transforms, often called *Fourier analysis* or *harmonic analysis*, provide useful decompositions of signals into fundamental or "primitive" components, provide shortcuts to the computation of complicated sums and integrals, and often reveal hidden structure in data. Fourier analysis lies at the base of many theories of science and plays a fundamental role in practical engineering design.

The origins of Fourier analysis in science can be found in Ptolemy's decomposing celestial orbits into cycles and epicycles and Pythagorus' decomposing music into consonances. Its modern history began with the eighteenth century work of Bernoulli, Euler, and Gauss on what later came to be known as Fourier series. J. Fourier in his 1822 *Théorie analytique de la Chaleur* [16] (still available as a Dover reprint) was the first to claim that arbitrary periodic functions could be expanded in a trigonometric (later called a Fourier) series, a claim that was eventually shown to be incorrect, although not too far from the truth. It is an amusing historical sidelight that this work won a prize from the French Academy, in spite of serious concerns expressed by the judges (Laplace, Lagrange, and Legendre) regarding Fourier's lack of rigor. Fourier was apparently a better engineer than mathematician. (Unhappily for France, he subsequently proved to be an even worse politician than mathematician.) Dirichlet later made rigorous the basic results for Fourier series and gave precise conditions under which they applied. The rigorous theoretical development of general Fourier transforms did not follow until about one hundred years later with the development of the Lebesgue integral.

The current extent of the influence of Fourier analysis is indicated by a partial list of scientists and engineers who use it:

- Circuit designers, from audio to microwave, characterize circuits in terms of their frequency response.

- Systems engineers use Fourier techniques in signal processing and communications algorithms for applications such as speech and image processing and coding (or compression), and for estimation and system identification. In addition to its widespread use for the analysis of linear systems, it also plays a fundamental role in the analysis of nonlinear systems, especially memoryless nonlinearities such as quantizers, hard limiters, and rectifiers.

- Audio engineers use Fourier techniques, partially because the ear seems to be sensitive to frequency domain behavior.

- Statisticians and probabilists characterize and compute probability distributions using Fourier transforms (called *characteristic functions* or *operational transforms*). Fourier transforms of covariance functions are used to characterize and estimate the properties of random processes.

- Error control code designers use Fourier techniques to characterize cyclic codes for error correction and detection.

- Radio astronomers use the Fourier transform to form images from interferometric data gathered from antenna arrays.

- Antenna designers evaluate beam patterns of periodic arrays using z-transforms, a form of Fourier transform, and evaluate beam patterns for more general arrays using Fourier transforms.

- Spectroscopists use the Fourier transform to obtain high resolution spectra in the infrared from interferograms (Fourier spectroscopy).

- Crystallographers find crystal structure using Fourier transforms of X-ray diffraction patterns.

- Lens designers specify camera performance in terms of spatial frequency response.

- Psychologists use the Fourier transform to study perception.

- Biomedical engineers use Fourier transforms for medical imaging, as with magnetic resonance imaging (MRI) wherein data collected in the frequency domain is inverse Fourier transformed to obtain images.

- Mathematicians and engineers use Fourier transforms in the solution of differential, integral, and other equations.

This book is devoted to a development of the basic definitions, properties, and applications of Fourier analysis. The emphasis is on techniques important for applications to linear systems, but other applications are occasionally described as well. The book is intended for engineers, especially for electrical engineers, but it attempts to provide a careful treatment of the fundamental mathematical ideas wherever possible. The assumed prerequisite is familiarity with complex variables and basic calculus, especially sums and Riemann integration. Some familiarity with linear algebra is also assumed when vector and matrix ideas are used. Since knowledge of real analysis and Lebesgue integration is not assumed, many of the mathematical details are not within the scope of this book. Proofs are provided in simple cases when they can be accomplished within the assumed background, but for more general cases we content ourselves with traditional engineering heuristic arguments. These arguments can always be made rigorous, however, and such details can be found in the cited mathematical texts.

This book is intended to serve both as a reference text and as a teaching text for a one quarter or one semester course on the fundamentals of Fourier analysis for a variety of types of signals, including discrete time (or parameter), continuous time (or parameter), finite duration, and infinite duration. By "finite duration" we mean a signal with a finite domain of definition; that is, the signal is only defined for a finite range of its independent variable. The principal types of infinite duration signals considered are absolutely summable (or integrable), finite energy, impulsive, and periodic. All of these signal types commonly arise in applications, although sometimes only as idealizations of physical signals. Many of the basic ideas are the same for each type, but the details often differ significantly. The intent of this book is to highlight the common aspects in these cases and thereby build intuition from the simple examples, which will be useful in the more complicated examples where careful proofs are not included.

Traditional treatments tend to focus on infinite duration signals, either beginning with the older notion of a Fourier series of a periodic function and then developing the Fourier integral transform as the limit of a Fourier series as the period approaches infinity, or beginning with the integral transform and defining the Fourier series as a special case of a suitably generalized transform using generalized functions (Dirac delta functions). Most texts emphasize the continuous time case, with the notable exception of treatments in the digital signal processing literature. Finite duration signals are usually considered late in the game when the discrete Fourier transform (DFT) is introduced prior to discussing the fast Fourier transform (FFT). We here take a less common approach of introducing all of the basic types of Fourier transform at the beginning: discrete and continuous time (or parameter) and finite and infinite duration. The DFT is emphasized early because it is the easiest to work with and its properties are the easiest to demonstrate without cumbersome mathematical details, the so-called "delta-epsilontics" of real analysis. Its importance is enhanced by the fact that virtually all digital computer implementations of the Fourier transform eventually reduce to a DFT. Furthermore, a slight modification of the DFT provides Fourier series for infinite duration periodic signals and thereby generalized Fourier transforms for such signals.

This approach has several advantages. Treating the basic signal types in parallel emphasizes the common aspects of these signal types and avoids repetitive proofs of similar properties. It allows the basic properties to be proved in the simplest possible context. The general results are then believable as simply the appropriate extensions of the simple ones even though the detailed proofs are omitted. This approach should provide more insight than the common engineering approach of quoting a result such as the basic Fourier integral inversion formula without proof. Lastly, this approach emphasizes the interrelations among the various signal types, for example, the production of discrete time signals by sampling continuous time signals or the production of a finite duration signal by windowing an infinite duration signal. These connections help in understanding the corresponding different types of Fourier transforms.

No approach is without its drawbacks, however, and an obvious problem here is that students interested primarily in one particular signal type such as the classical infinite duration continuous time case or the finite duration discrete time case dominating digital signal processing texts may find the variety of signal types annoying and regret the loss of additional detailed results peculiar to their favorite signal type. Although a sequential treatment of the signal types would solve this problem, we found the cited advantages of the parallel approach more persuasive. An additional factor in favor of the parallel approach is historical. The Stanford University course (EE261) in which these notes were developed has served as a general survey and review for a diverse group of students from many departments. Most students taking this course had had bits and pieces of Fourier theory scattered throughout their undergraduate courses. Few had had any form of overview relating the apparently different theories and taking advantage of the common ideas. Such a parallel treatment allows readers to build on their intuition developed for a particular signal class in order to understand a collection of different applications. A sequential treatment would not have accomplished this as effectively.

Synopsis

The topics covered in this book are:

1. Signals and Systems. This chapter develops the basic definitions and examples of signals, the mathematical objects on which Fourier transforms operate, the inputs to the Fourier transform. Included are continuous time and discrete time signals, two-dimensional signals, infinite and finite duration signals, time-limited signals, and periodic signals. Combinations of signals to produce new signals and *systems* which operate on an *input signal* to produce an *output* signal are defined and basic examples considered.

2. The Fourier Transform. The basic definitions of Fourier transforms are introduced and exemplified by simple examples. The analytical and numerical evaluation of transforms is considered. Several transforms closely related to the Fourier transform are described, including the cosine and sine transforms, the Hartley transform, the Laplace transform, and z-transforms.

3. Fourier Inversion. Basic results on the recovery of signals from their transform are developed and used to describe signals by Fourier integrals and Fourier series.

4. Basic Properties. This chapter is the heart of the book, developing the basic properties of Fourier transforms that make the transform useful in applications and theory. Included are linearity, shifts, modulation, Parseval's theorem, sampling, the Poisson summation formula, aliasing, pulse amplitude modulation, stretching, downsampling and upsampling, differentiating and differencing, moment generating, bandwidth and pulse width, and symmetry properties.

5. Generalized Transforms and Functions. Here the Fourier transform is extended to general signals for which the strict original definitions do not work. Included are Dirac delta functions, periodic signals, and impulse trains.

6. Convolution. Fourier methods are applied to the analysis of linear time invariant systems. Topics include impulse responses, superposition and convolution, the convolution theorem, transfer functions, integration, sampling revisited, correlation, the correlation theorem, pulsewidth and bandwidth in terms of autocorrelation, the uncertainty relation, and the central limit theorem.

7. Two-Dimensional Fourier Transforms. Methods of Fourier analysis particularly useful for two-dimensional signals such as images are considered. Topics include two-dimensional linear systems and reconstruction of 2D signals from projections.

8. Memoryless Nonlinearities. Fourier methods are shown to be useful in certain nonlinear problems. The so-called "transform method" of nonlinear analysis is described and applied to phase modulation, frequency modulation, uniform quantization, and transform coding.

The final two topics are relatively advanced and may be tackled in any order as time and interest permit. The construction of Fourier transform tables is treated in the Appendix and can be referred to as appropriate throughout the course.

Instructional Use

This book is intended as an introduction and survey of Fourier analysis for engineering students and practitioners. It is a mezzanine level course in the sense that it is aimed at senior engineering students or beginning Master's level students. The basic core of the course consists of the unstarred sections of Chapters 1 through 6. The starred sections contain additional details and proofs that can be skipped or left for background reading without classroom presentation in a one quarter course. This core plus one of the topics from the final chapters constitutes a one quarter course. The entire book, including many of the starred sections, can be covered in a semester.

Many of the figures were generated using Matlab ^{TM} on both unix ^{TM} and Apple Macintosh ^{TM} systems and the public domain NIH Image program (written by Wayne Rasband at the U.S. National Institutes of Health and available from the Internet by anonymous ftp from zippy.nimh.nih.gov or on floppy disk from NITS, 5285 Port Royal Rd., Springfield, VA 22161, part number PB93-504568) on Apple Macintosh ^{TM} systems.

The problems in each chapter are intended to test both fundamentals and the mechanics of the algebra and calculus necessary to find transforms. Many of these are old exam problems and hence often cover material from previous chapters as well as the current chapter. Whenever yes/no answers are called for, the answer should be justified, e.g., by a proof for a positive answer or a counterexample for a negative one.

Recommended Texts

Fourier analysis has been the subject of numerous texts and monographs, ranging from books of tables for practical use to advanced mathematical treatises. A few are mentioned here for reference. Some of the classic texts still make good reading.

The two most popular texts for engineers are *The Fourier Transform and its Applications* by R. Bracewell [6] and *The Fourier Integral and its Applications* by A. Papoulis [24]. Both books are aimed at engineers and emphasize the infinite duration, continuous time Fourier transform. *Circuits, Signals, and Systems* by W. McC. Siebert [30] is an excellent (and enormous) treatment of all forms of Fourier analysis applied to basic circuit and linear system theory. It is full of detailed examples and emphasizes applications. A detailed treatment of the fast Fourier transform may be found in O. Brigham's *The Fast Fourier Transform and its Applications* [9]. Treatments of two-dimensional Fourier transforms can be found in Goodman [18] and Bracewell [8] as well as in books on image processing or digital image processing. For example, Gonzales and Wintz [17] contains a variety of applications of Fourier techniques to image enhancement, restoration, edge detection, and filtering.

Mathematical treatments include Wiener's classic text *The Fourier Integral and Certain of its Applications* [36], Carslaw's *An Introduction to the Theory of Fourier's Series and Integrals* [10], Bochner's classic *Lectures on Fourier Integrals* [3], Walker's *Fourier Analysis* [33], and Titchmarsh's *Introduction to the Theory of Fourier Integrals* [32]. An advanced and modern (and inexpensive) mathematical treatment can also be found in *An Introduction to Harmonic Analysis* by Y. Katznelson [21]. An elementary and entertaining introduction to Fourier analysis applied to music may be found in *The Science of Musical Sound*, by John R. Pierce [25].

Discrete time Fourier transforms are treated in depth in several books devoted to digital signal processing such as the popular text by Oppenheim and Schafer [23].

Some Notation

We will deal with a variety of functions of real variables. Let $\mathcal{R}$ denote the real line. Given a subset $\mathcal{T}$ of the real line $\mathcal{R}$, a real-valued *function g* of a real variable t with domain of definition $\mathcal{T}$ is an assignment of a real number $g(t)$ to every point in $\mathcal{T}$. Thus denoting a function g is shorthand for the more careful and complete notation $\{g(t); \ t \in \mathcal{T}\}$ which specifies the name of the function (g) and the collection of values of its argument for which it is defined. The most common cases of interest for a domain of definition are intervals of the various forms defined below.

- $\mathcal{T} = \mathcal{R}$, the entire real line.

- $\mathcal{T} = (a, b) = \{r : a < r < b\}$, an *open interval* consisting of the points between a and b but not a and b themselves. The real line itself is often written in this form as $\mathcal{R} = (-\infty, \infty)$.

- $\mathcal{T} = [a, b] = \{r : a \le r \le b\}$, a *closed interval* consisting of the points between a and b together with the endpoints a and b (a and b both finite).

- $\mathcal{T} = [a, b) = \{r : a \le r < b\}$, a *half open* (or half closed) interval consisting of the points between a and b together with the lower endpoint a (a finite).

- $\mathcal{T} = (a, b] = \{r : a < r \le b\}$, a *half open* (or half closed) interval consisting of the points between a and b together with the upper endpoint b (b finite).

- $\mathcal{T} = \mathcal{Z} = \{\cdots, -1, 0, 1, \cdots\}$, the collection of integers.

- $\mathcal{T} = \mathcal{Z}_N = \{0, 1, \cdots, N-1\}$, the collection of integers from 0 through $N-1$.

Complex numbers can be expressed as

$$z = x + iy,$$

where $x = \Re(z)$ is the *real part* of z, $y = \Im(z)$ is the *imaginary part* of z, and

$$i = \sqrt{-1}$$

(often denoted by j in the engineering literature). Complex numbers can also be represented in polar coordinates or magnitude/phase form as

$$z = Ae^{i\theta}.$$

The *magnitude* or *modulus* A is given by

$$A = |z| = \sqrt{x^2 + y^2}.$$

If we restrict the phase angle θ to be within $[-\frac{\pi}{2}, \frac{\pi}{2})$ (or, equivalently, in $[0, \pi)$), then θ is given by the principal value of the inverse tangent (or arctangent): $\theta = \arctan(\frac{y}{x})$ or $\theta = \tan^{-1}(\frac{y}{x})$ in radians. The quantity $\theta/2\pi$ is the phase in units of cycles. Figure 0.1 illustrates these quantities. Here,

Imaginary

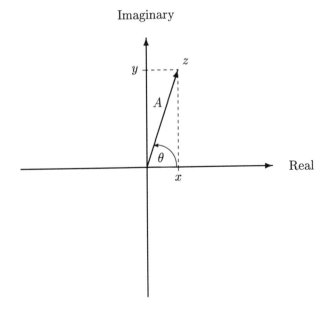

Figure 0.1: Complex Numbers

the complex number is denoted by the vector z, with magnitude A and phase θ. It has a real component x and imaginary component y.

The *complex conjugate* z^* is defined as $x - iy$. The real and imaginary parts of z are easily found by

$$x = \Re(z) = \frac{1}{2}(z + z^*),\ y = \Im(z) = \frac{1}{2}(z - z^*)$$

Sinusoids and complex exponentials play a basic role throughout this book. They are related by Euler's formulas: For all real θ

$$e^{i\theta} = \cos\theta + i\sin\theta$$

$$\cos\theta = \frac{e^{i\theta} + e^{-i\theta}}{2}$$

$$\sin\theta = \frac{e^{i\theta} - e^{-i\theta}}{2i}.$$

A complex-valued function $g = \{g(t);\ t \in \mathcal{T}\}$ is an assignment of complex-valued numbers $g(t)$ to every $t \in \mathcal{T}$.

A signal $\{g(t);\ t \in \mathcal{R}\}$ is said to be *even* if

$$g(-t) = g(t);\ t \in \mathcal{R}.$$

It is said to be *odd* if

$$g(-t) = -g(t); \ t \in \mathcal{R}.$$

A slight variation of this definition is common: strictly speaking, an odd signal must satisfy $g(0) = 0$ since $-g(0) = g(0)$. This condition is sometimes dropped so that the definition becomes $g(-t) = -g(t)$ for all $t \neq 0$. For example, the usual definition of the sign function meets the strict definition, but the alternative definition (which is $+1$ for $t \geq 0$ and -1 for $t < 0$) does not. The alternative definition is, however, an odd function if one ignores the behavior at $t = 0$. A signal is *Hermitian* if

$$g(-t) = g^*(t); \ t \in \mathcal{R}.$$

For example, a complex exponential $g(t) = e^{i2\pi f_0 t}$ is Hermitian. A signal is *anti-Hermitian* if

$$g(-t) = -g^*(t); \ t \in \mathcal{R}.$$

As examples, $\sin t$ and $te^{-\lambda |t|}$ are odd functions of $t \in \mathcal{R}$, while $\cos t$ and $e^{-\lambda |t|}$ are even.

We shall have occasion to deal with modular arithmetic. Given a positive real number $T > 0$, any real number a can be written uniquely in the form $a = kT + r$ where the "remainder" term is in the interval $[0, T)$. This formula defines $a \bmod T = r$, that is, $a \bmod T$ is what is left of a when the largest possible number of integer multiples of T is subtracted from a. This is often stated as "a modulo T." The definition can be summarized by

$$a \bmod T = r \text{ if } a = kT + r \text{ where } r \in [0, T) \text{ and } k \in \mathcal{Z}. \qquad (0.1)$$

More generally we can define modular arithmetic on any interval $[a, b)$, $b > a$ by

$$a \bmod [a, b) = r \text{ if } a = kT + r \text{ where } r \in [a, b) \text{ and } k \in \mathcal{Z}. \qquad (0.2)$$

Thus the important special case $a \bmod T$ is an abbreviation for $a \bmod [0, T)$. By "modular arithmetic" is meant doing addition and subtraction within an interval. For example, $(0.5 + 0.9) \bmod 1 = 1.14 \bmod 1 = .14 \bmod 1$ and $(-0.3) \bmod 1 = 0.7$.

Acknowledgements

We gratefully acknowledge our debt to the many students who suffered through early versions of this book and who considerably improved it by their corrections, comments, suggestions, and questions. We also acknowlege the Industrial Affiliates Program of the Information Systems Laboratory, Stanford University, whose continued generous support provided the computer facilities used to write and design this book.

FOURIER TRANSFORMS
An Introduction for Engineers

Chapter 1

Signals and Systems

1.1 Waveforms and Sequences

The basic entity on which all Fourier transforms operate is called a *signal*.
The model for a signal will be quite general, although we will focus almost
entirely on a few special types. Intuitively, the definition of a signal should
include things like sinusoids of the form $\sin t$ or, more precisely, $\{\sin t; \ t \in
(-\infty, \infty)\}$, as well as more general waveforms $\{g(t); \ t \in \mathcal{T}\}$, where $\mathcal{T}$ could
be the real line or the positive real line $[0, \infty)$ or perhaps some smaller
interval such as $[0, T)$ or $[-T/2, T/2)$. In each of these cases the signal is
simply a function of an independent variable (or parameter), here called t
for "time." In general, however, the independent variable could correspond
to other physical quantities such as "space."

$\mathcal{T}$ provides the allowable values of the time index or parameter and is
called the *domain of definition* or, simply, *domain* of the signal. It is also
called the *index set* of the signal since one can think of the signal as an
indexed set of values, one value for each choice of the index t. This set can
be *infinite duration*, i.e., infinite in extent, as in the case of the real line
$\mathcal{R} = (-\infty, \infty)$, or *finite duration*, i.e., finite in extent as in the case $[0, T)$.
There is potential confusion in this use of the phrase "finite duration" since
it could mean either a finite extent domain of definition or a signal with
an infinite extent index set with the property that the signal is 0 except
on a finite region. We adopt the first meaning, however, and hence "finite
duration" is simply a short substitute for the more precise but clumsy
"finite extent domain of definition." The infinite duration signal with the
property that it is 0 except for a finite region will be called a *time-limited*
signal. For example, the signal $= \{\sin t; \ t \in [0, 2\pi)\}$ has finite duration,

while the signal $\{h(t);\ t \in (-\infty, \infty)\}$ defined by

$$h(t) = \begin{cases} \sin t & t \in [0, 2\pi) \\ 0 & t \in \mathcal{R},\ t \notin [0, 2\pi) \end{cases}$$

is infinite duration but time-limited. Clearly the properties of these two signals will strongly resemble each other, but there will be subtle and important differences in their analysis.

When the set $\mathcal{T}$ is continuous as in these examples, we say that the signal is a *continuous time signal,* or *continuous parameter signal,* or simply a *waveform.*

The shorter notation $g(t)$ is often used for the signal $\{g(t);\ t \in \mathcal{T}\}$, but this can cause confusion as $g(t)$ could either represent the value of the signal at a specific time t or the entire waveform $g(t)$ for all $t \in \mathcal{T}$. To lessen this confusion it is common to denote the entire signal by simply the name of the function g; that is,

$$g = \{g(t);\ t \in \mathcal{T}\}$$

when the index set $\mathcal{T}$ is clear from context. It is also fairly common practice to use boldface to denote the entire signal; that is, $\mathbf{g} = \{g(t);\ t \in \mathcal{T}\}$.

Signals can also be sequences, such as sampled sinusoids $\{\sin(nT);\ n \in \mathcal{Z}\}$, where $\mathcal{Z}$ is the set of all integers $\{\ldots, -2, -1, 0, 1, 2, \ldots\}$, a geometric progression $\{r^n;\ n = 0, 1, 2, \ldots\}$, or a sequence of binary data $\{u_n; n \in \mathcal{Z}\}$, where all of the u_n are either 1 or 0. Analogous to the waveform case we can denote such a sequence as $\{g(t);\ t \in \mathcal{T}\}$ with the index set $\mathcal{T}$ now being a set of integers. It is more common, however, to use subscripts rather than functional notation and to use indexes like k, l, n, m instead of t for the index for sequences. Thus a sequence will often be denoted by $\{g_n; n \in \mathcal{T}\}$. We still use the generic notation g for a signal of this type. The only difference between the first and second types is the nature of the index set $\mathcal{T}$. When $\mathcal{T}$ is a discrete set such as the integers or the nonnegative integers, the signal g is called a *discrete time signal, discrete parameter signal, sequence,* or *time series.* As in the waveform case, $\mathcal{T}$ may have infinite duration (e.g., all integers) or finite duration (e.g., the integers from 0 to $N - 1$).

In the above examples the index set $\mathcal{T}$ is *one-dimensional,* that is, consists of some collection of real numbers. Some signals are best modeled as having multidimensional index sets. For example, a two-dimensional square sampled image intensity raster could be written as $\{g_{n,k};\ n = 1, \ldots, K;\ k = 1, \ldots, K\}$, where each $g_{n,k}$ represents the intensity (a nonnegative number) of a single picture element or pixel in the image, the pixel located in the nth column and kth row of the square image. Note that in this case the dummy arguments correspond to space rather than time. There is nothing magic about the name of a dummy variable, however, and we could still use t here if we wished.

For example, a typical magnetic resonance (MR) image consists of a 256×256 square array of pixel intensities, each represented by an integer from 0 (black, no light) to $2^9 - 1 = 511$ (white, fully illuminated). As we shall wish to display such images on screens which support only 8 and not 9 bits in order to generate examples, however, we shall consider MR images to consist of 256×256 square array of pixel intensities, each represented by an integer from 0 to $2^8 - 1 = 255$. We note that, in fact, the raw data used to generate MR images constitute (approximately) the Fourier transform of the MR image. Thus when MR images are rendered for display, the basic operation is an inverse Fourier transform.

A square continuous image raster might be represented by a waveform depending on two arguments $\{g(x,y); \ x \in [0,a], \ y \in [0,a]\}$. A three-dimensional sequence of image rasters could be expressed by a signal $\{g_{n,k,l}; \ n = 0,1,2,\ldots, \ k = 1,2,\ldots,K, \ l = 1,2,\ldots,K\}$, where now n is the time index (any nonnegative integer) and k and l are the spatial indices. Here the index set $\mathcal{T}$ is three-dimensional and includes both time and space.

In all of these different signal types, the signal has the general form

$$g = \{g(t); \ t \in \mathcal{T}\},$$

where $\mathcal{T}$ is the *domain of definition* or *index set* of the signal, and where $g(t)$ denotes the value of the signal at "time" or parameter or dummy variable t. In general, $\mathcal{T}$ can be finite, infinite, continuous, discrete, or even vector valued. Similarly $g(t)$ can take on vector values, that is, values in Euclidean space. We shall, however, usually focus on signals that are *real* or *complex* valued, that is, signals for which $g(t)$ is either a real number or a complex number for all $t \in \mathcal{T}$. As mentioned before, when $\mathcal{T}$ is discrete we will often write g_t or g_n or something similar instead of $g(t)$.

In summary, *a signal is just a function whose domain of definition is $\mathcal{T}$ and whose range is the space of real or complex numbers.* The nature of $\mathcal{T}$ determines whether the signal is continuous time or discrete time and finite duration or infinite duration. The signal is *real-valued* or *complex-valued* depending on the possible values of $g(t)$.

Although $\mathcal{T}$ appears to be quite general, we will need to impose some structure on it to get useful results and we will focus on a few special cases that are the most important examples for engineering applications. The most common index sets for the four basic types of signals are listed in Table 1.1. The subscripts of the domains inherit their meaning from their place in the table, that is,

DTFD = discrete time, finite duration
CTFD = continuous time, finite duration
DTID = discrete time, infinite duration
CTID = continuous time, infinite duration.

The superscripts will be explained shortly.

Duration	Time	
	Discrete	Continuous
Finite	$\mathcal{T}_{DTFD}^{(1)} = \mathcal{Z}_N \triangleq \{0, 1, \cdots, N-1\}$	$\mathcal{T}_{CTFD}^{(1)} = [0, T)$
Infinite	$\mathcal{T}_{DTID}^{(2)} = \mathcal{Z} \triangleq \{\cdots, -1, 0, 1, \cdots\}$	$\mathcal{T}_{CTID}^{(2)} = \mathcal{R}$

Table 1.1: Common Index Sets for Basic Signal Types

It should be pointed out that the index sets of Table 1.1 are not the only possibilities for the given signal types; they are simply the most common. The two finite duration examples are said to be *one-sided* since only nonnegative indices are considered. The two infinite duration examples are *two-sided* in that negative and nonnegative indices are considered. Common alternatives are to use two-sided sets for finite duration and one-sided sets for infinite duration as in Table 1.2. Superscripts are used to distinguish between one- and two-sided time domains when convenient. They will be dropped when the choice is clear from context. The modifications in

Duration	Time	
	Discrete	Continuous
Finite	$\mathcal{T}_{DTFD}^{(2)} = \{-N, \cdots, -1, 0, 1, \cdots, N\}$	$\mathcal{T}_{CTFD}^{(2)} = [-\frac{T}{2}, \frac{T}{2})$
Infinite	$\mathcal{T}_{DTID}^{(1)} = \{0, 1, \cdots\}$	$\mathcal{T}_{CTID}^{(1)} = [0, \infty)$

Table 1.2: Alternative Index Sets for Basic Signal Types

transform theory for these alternative choices are usually straightforward.

We shall emphasize the choices of Table 1.1, but we shall often encounter examples from Table 1.2. The careful reader may have noticed the use of half open intervals in the definitions of finite duration continuous time signals and be puzzled as to why the apparently simpler open or closed intervals were not used. This was done for later convenience when we construct periodic signals by repeating finite duration signals. In this case one endpoint of the domain is not included in the domain as it will be provided by another domain that will be concatenated.

1.2 Basic Signal Examples

One of the most important signals in Fourier analysis is the sinusoid. Figure 1.1 shows several periods of a continuous time sinusoid $\{g(t) = \sin(2\pi t); \ t \in \mathcal{R}\}$. A *period* of a signal g is a $T \in \mathcal{T}$ such that $g(t+T) = g(t)$ for all t satisfying $t + T \in \mathcal{T}$. When we refer to *the* period rather than *a* period of a signal, we mean the smallest period $T > 0$. The period of the sinusoidal g under consideration is 1. Since figures obviously cannot depict infinite duration signals for the entire duration, it is not clear from the figure alone if it is a piece of an infinite duration signal or the entire finite duration signal. This detail must always be provided.

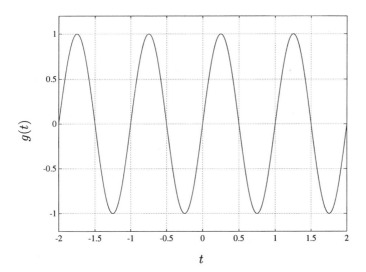

Figure 1.1: Continuous Time Sine Signal: Period 1

A sinusoid can also be considered as a discrete time signal by sampling the continuous time version. For example, consider the discrete time signal g defined by

$$g(n) = \sin(\frac{2\pi}{10}n); \; n \in \mathcal{Z}. \tag{1.1}$$

A portion of this signal is plotted in Figure 1.2. Note that the period has

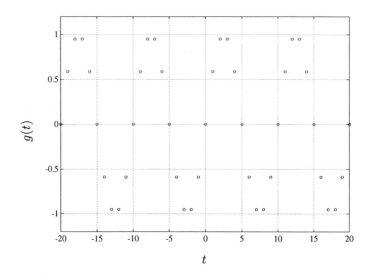

Figure 1.2: Discrete Time Sine Signal: Period 10

changed because of the change in the definition of the variable corresponding to "time." Not only has time been made discrete, it has also been scaled.

We can consider yet another signal type using the same sinusoid by both sampling and truncating. For example, form the one-sided discrete time signal defined by

$$g(n) = \sin(\frac{2\pi}{8}n); \; n = \{0, 1, \cdots, 31\} \tag{1.2}$$

as shown in Figure 1.3. Here the figure shows the entire signal, which is not possible for infinite duration signals. For convenience we have chosen the number of time values to be a power of 2; this will lead to simplifications when we consider numerical evaluation of Fourier transforms. Also for convenience we have chosen the signal to contain an integral number of

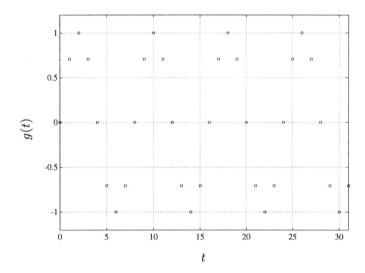

Figure 1.3: One-Sided Finite Duration Discrete Time Sine Signal: Period 8

periods. This simplifies some of the manipulations we will perform on this signal.

The two previous figures are, of course, different in a trivial way. It is worth understanding, however, that they depict distinct signal types although they originate with the same function. For historical reasons, the most common finite duration signals are one-sided. This often makes indexing and numerical techniques a little simpler.

Another basic signal is the continuous time exponential

$$g(t) = ae^{-\lambda t}; \ t \in [0, \infty), \tag{1.3}$$

where $\lambda > 0$ and a is some real constant. The signal is depicted in Figure 1.4 for the case $a = 1$ and $\lambda = .9$. This signal is commonly considered as a two-sided signal $\{g(t); \ t \in \mathcal{R}\}$ by defining

$$g(t) = \begin{cases} ae^{-\lambda t} & t \geq 0 \\ 0 & \text{otherwise} \end{cases}. \tag{1.4}$$

The discrete time analog of the exponential signal is the geometric signal. For example, consider the signal given by the finite sequence

$$g = \{r^n; n = 0, 1, \ldots, N - 1\} = \{1, r, \ldots, r^{N-1}\}. \tag{1.5}$$

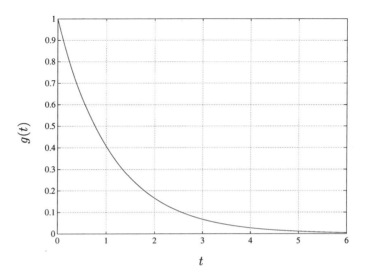

Figure 1.4: Continuous Time Exponential Signal

This signal has discrete time and finite duration and is called the finite duration *geometric signal* because it is a finite length piece of a geometric progression. It is sometimes also called a discrete time exponential. The signal is plotted in Figure 1.5 for the case of $r = .9$ and $N = 32$.

Given any real $T > 0$, the box function $\Box_T(t)$ is defined for any real t by

$$\Box_T(t) = \begin{cases} 1 & |t| \leq T \\ 0 & \text{otherwise} \end{cases}. \tag{1.6}$$

A variation of the box function is the rectangle function considered by Bracewell [6]:

$$\Pi(t) = \begin{cases} 1 & |t| < \frac{1}{2} \\ \frac{1}{2} & |t| = \frac{1}{2} \\ 0 & \text{otherwise} \end{cases}.$$

Note that $\Pi(t/2T) = \Box_T(t)$ except at the edges $t = T$. The reason for this difference in edge values is that, as we shall see later, signals with discontinuities require special care when doing Fourier inversion. We shall see that the rectangle function is chosen so as to ensure that it is reproduced exactly after a sequence of a Fourier transform and an inverse Fourier transform. Both box functions, when considered as continuous time signals, will give the same transform, but Fourier inversion in the continuous time case yields Bracewell's definition. On the other hand, the extra simplicity of

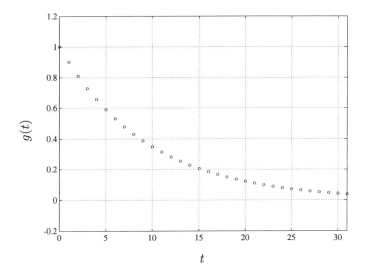

Figure 1.5: Finite Duration Geometric Signal

$\Box_T(t)$ will prove useful, especially when used to define a discrete time signal, since then discontinuities do not pose problems. The notation $\text{rect}(t)$ is also used for $\Box(t)$.

As an example, a portion of the two-sided infinite duration discrete time box signal $\{\Box_5(n); \ n \in \mathcal{Z}\}$ is depicted in Figure 1.6 and a finite duration one-sided box signal $\{\Box_5(n); \ n \in \{0, 1, \cdots, 15\}$ is depicted in Figure 1.7. The corresponding continuous time signal $\{\Box_5(t); \ t \in \mathcal{R}\}$ is depicted in Figure 1.8.

The Kronecker delta function δ_t is defined for any real t by

$$\delta_t = \begin{cases} 1 & t = 0 \\ 0 & t \neq 0 \end{cases}.$$

This should not be confused with the Dirac delta or unit impulse which will be introduced later when generalized functions are considered. The Kronecker delta is primarily useful as a discrete time signal, as exemplified in Figure 1.9. The Kronecker delta is sometimes referred to as a "unit sample" or as the "discrete time impulse" or "discrete impulse" in the literature, but the latter terms should be used with care as "impulse" is most associated with the Dirac delta and the two deltas have several radically different properties. The Kronecker and Dirac delta functions will play similar roles in discrete time and continuous time systems, respectively, but the

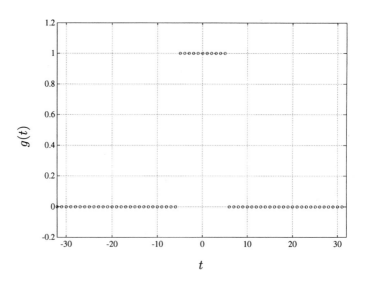

Figure 1.6: Discrete Time Box Signal

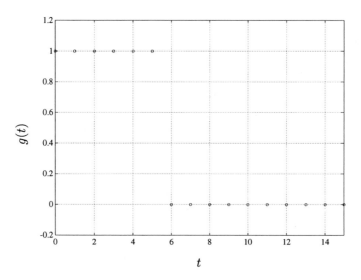

Figure 1.7: One-sided Discrete Time Box Signal

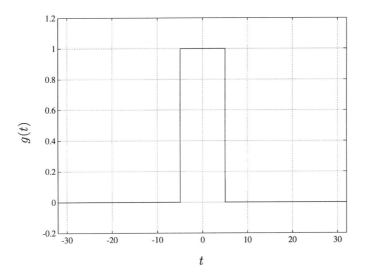

Figure 1.8: Continuous Time Box Signal

Kronecker delta function is an ordinary signal, while a Dirac delta is not; the Dirac delta is an example of a *generalized function* and it will require special treatment.

Note that

$$\delta_n = \square_0(n); \ n \in \mathcal{Z}.$$

The sinc function sinc t is defined for real t by

$$\text{sinc}\, t = \frac{\sin \pi t}{\pi t}$$

and is illustrated in Figure 1.10. The sinc function will be seen to be of fundamental importance to sampling theory and to the theory of Dirac delta functions and continuous time Fourier transform inversion.

The unit step function is defined for all real t by

$$u_{-1}(t) = \begin{cases} 1 & t \geq 0 \\ 0 & \text{otherwise} \end{cases} \tag{1.7}$$

The notation indicates that the unit step function is one of a class of special functions $u_k(t)$ related to each other by integration and differentiation. See, e.g., Siebert [30]. The continuous time step function and the discrete time step function are depicted in Figures 1.11 and 1.12.

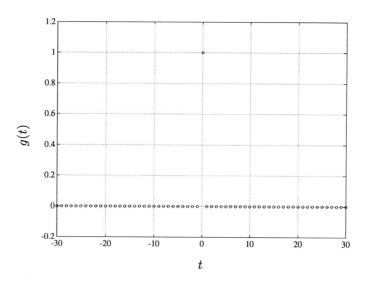

Figure 1.9: Discrete Time Kronecker Delta

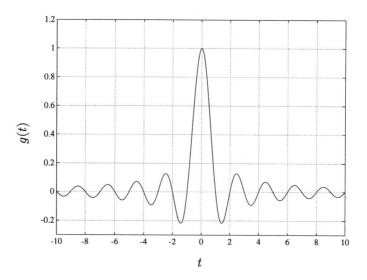

Figure 1.10: Sinc Function

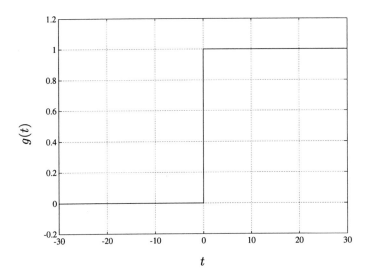

Figure 1.11: Continuous Time Step Function

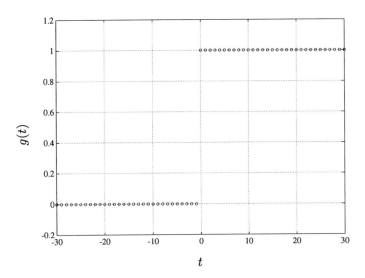

Figure 1.12: Discrete Time Step Function

As with the box functions, we consider two slightly different forms of the unit step function. Following Bracewell [6], define the *Heaviside unit step function* $H(t)$ for all real t by

$$H(t) = \begin{cases} 1 & t > 0 \\ \frac{1}{2} & t = 0 \\ 0 & t < 0 \end{cases}. \qquad (1.8)$$

The Heaviside step function is used for the same purpose as the rectangle function; the definition of its value at discontinuities as the midpoint between the values above and below the discontinuity will be useful when forming Fourier transform pairs. Both step functions share a common Fourier transform, but the inverse continuous time Fourier transform yields the Heaviside step function.

The signum or sign function also has two common forms: The most common (especially for continuous time) is

$$\text{sgn}(t) = \begin{cases} +1 & \text{if } t > 0 \\ 0 & \text{if } t = 0 \\ -1 & \text{if } t < 0. \end{cases} \qquad (1.9)$$

The most popular alternative is to replace the 0 at the origin by $+1$. The principal difference is that the first definition has three possible values while the second has only two. The second is useful, for example, when modeling the action of a hard limiter (or binary quantizer) which has two possible outputs depending on whether the input is smaller than a given threshold or not. Rather than add further clutter to the list of names of special functions, we simply point out that both definitions are used. Unless otherwise stated, the first definition will be used. We will explicitly point out when the second definition is being used.

The continuous time and discrete time $\text{sgn}(t)$ signals are illustrated in Figures 1.13–1.14.

Another common signal is the triangle or wedge $\wedge(t)$ defined for all real t by

$$\wedge(t) = \begin{cases} 1 - |t| & \text{if } |t| < 1 \\ 0 & \text{otherwise.} \end{cases} \qquad (1.10)$$

The continuous time triangle signal is depicted in Figure 1.15. In order to simplify the definition of the discrete time triangle signal, we introduce first the time scaled triangle function $\wedge_T(t)$ defined for all real t and any $T > 0$:

$$\wedge_T(t) = \wedge(\frac{t}{T}). \qquad (1.11)$$

Thus $\wedge(t) = \wedge_1(t)$. The discrete time triangle is defined as $\wedge_T(n)$; $n \in \mathcal{Z}$ for any positive integer T. Figure 1.16 shows the discrete time triangle signal $\wedge_5(n)$.

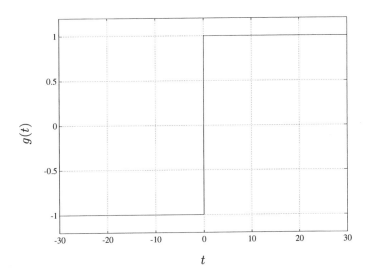

Figure 1.13: Continuous Time Sign Function

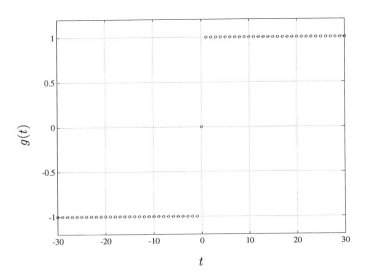

Figure 1.14: Discrete Time Sign Function

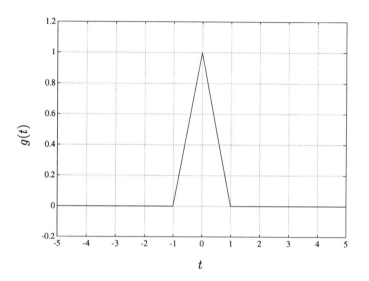

Figure 1.15: Continuous Time Triangle Signal

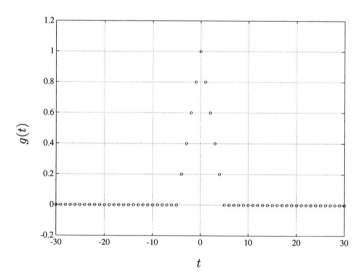

Figure 1.16: Discrete Time Triangle Signal

As a final special function that will be encountered on occasion, the nth order ordinary Bessel function of the first kind is defined for real t and any integer n by

$$J_n(t) = \frac{1}{2\pi} \int_{-\pi}^{\pi} e^{it\sin\phi - in\phi}\, d\phi. \tag{1.12}$$

Bessel functions arise as solutions to a variety of applied mathematical problems, especially in nonlinear systems such as frequency modulation (FM) and quantization, as will be seen in Chapter 8. Figure 1.17 shows a plot of the $J_n(t)$ for various indices n.

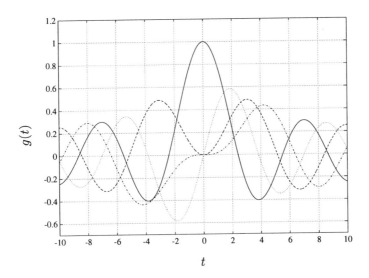

Figure 1.17: Bessel Functions J_n: Solid line $n = 0$, dotted line $n = 1$, dashed line $n = 2$, dash-dot line $n = 3$.

The basic properties and origins of Bessel functions may be found in standard texts on the subject such as Watson [35] or Bowman [5]. For example, Bessel functions can be expressed in a series form as

$$J_n(t) = (\frac{t}{2})^n \sum_{k=0}^{\infty} \frac{(-\frac{t^2}{4})^k}{k!(n+k)!}. \tag{1.13}$$

Table 1.3 summarizes several examples of signals and their index sets along with their signal type. ω, $\lambda > 0$, and x are fixed real parameters, and m is a fixed integer parameter.

Signal	Time Domain	Duration	Comments
$\{\sin \pi t; \, t \in \mathcal{R}\}$	continuous	infinite	sinusoid, periodic
$\{\sin \pi t; \, t \in [0,1]\}$	continuous	finite	sinusoid
$\{e^{i\omega t}; \, t \in \mathcal{R}\}$	continuous	infinite	complex exponential
$\{e^{-\lambda t}; \, t \in [0,\infty)\}$	continuous	infinite	real exponential
$\{1; \, t \in \mathcal{R}\}$	continuous	infinite	constant (dc)
$\{e^{-\pi t^2}; \, t \in \mathcal{R}\}$	continuous	infinite	Gaussian
$\{e^{i\pi t^2}; \, t \in \mathcal{R}\}$	continuous	infinite	chirp
$\{H(t); \, t \in \mathcal{R}\}$	continuous	infinite	Heaviside step function
$\{u_{-1}(t); \, t \in \mathcal{R}\}$	continuous	infinite	unit step function
$\{\text{sgn}(t); \, t \in \mathcal{R}\}$	continuous	infinite	signum (sign) function
$\{2B \, \text{sinc}(2Bt); \, t \in \mathcal{R}\}$	continuous	infinite	sinc function
$\{\Box_T(t); \, t \in \mathcal{R}\}$	continuous	infinite	box, time-limited
$\{t; \, t \in [-1,1]\}$	continuous	finite	ramp
$\{\wedge(t); \, t \in \mathcal{R}\}$	continuous	infinite	triangle, time-limited
$\{t\Box_1(t); \, t \in \mathcal{R}\}$	continuous	infinite	time-limited ramp
$\{J_m(2\pi t); \, t \in \mathcal{R}\}$	continuous	infinite	Bessel function
$\{J_n(x); \, n \in \mathcal{Z}\}$	discrete	infinite	Bessel function
$\{r^n; \, n \in \{0,1,\ldots\}\}$	discrete	infinite	geometric progression
$\{r^n; \, n \in \mathcal{Z}_N\}$	discrete	finite	
$\{e^{i\omega n}; \, n \in \mathcal{Z}\}$	discrete	infinite	
$\{u_{-1}(n); \, n \in \mathcal{Z}\}$	discrete	infinite	unit step
$\{\delta_n; \, n \in \mathcal{Z}\}$	discrete	infinite	Kronecker delta

Table 1.3: Common Signal Examples

1.3   Random Signals

All of the signals seen so far are well defined functions and knowing the signal means knowing exactly its form and structure. Obviously in real life one will meet signals that are not so well understood and are not constructible in any useful and simple form from the above building blocks. One way of modeling unknown signals is to consider them as having been produced by a random process. Although probability and random processes are not prerequisite for this book and we do not intend to teach such a general subject here, it is fair to use important examples of such processes as a means of producing interesting signals for illustration and analysis. We shall confine attention to discrete time when considering randomly generated signals to avoid the mathematical and notational complexity involved in dealing with continuous time random processes. Figure 1.18 shows an example of a discrete time signal produced by a Gaussian random process produced by Matlab's random number generator. The values are produced as a sequence of independent, zero mean, unit variance Gaussian random variables. (Don't be concerned if these terms are unfamiliar to you; the point is the signal is produced randomly and has no nice description.) This

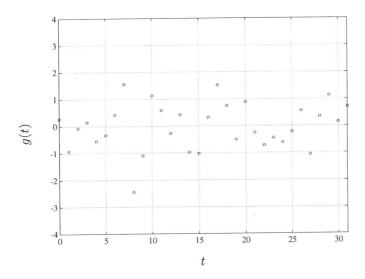

Figure 1.18: Sequence Produced by an Independent, Identically Distributed Gaussian Random Process.

is an example of what is often called discrete time white Gaussian noise and it is a common and often good model of random phenomena in signal processing and statistical applications.

1.4 Systems

A common focus of many of the application areas mentioned in the introduction is the action of *systems* on signals to produce new signals. A system is simply a mapping which takes one signal, often called the input, and produces a new signal, often called the output. A particularly trivial system is the identity system which simply passes the input signal through to the output without change (an ideal "wire"). Another trivial system is one which sets the output signal equal to 0 regardless of the input (an ideal "ground"). More complicated systems can perform a variety of linear or nonlinear operations on an input signal to produce an output signal. Of particular interest to us will be how systems affect Fourier transforms of signals.

Mathematically, a *system* (also called a *filter*) is a mapping $\mathcal{L}$ of an input signal $v = \{v(t); t \in \mathcal{T}_i\}$ into an output signal $w = \{w(t); t \in \mathcal{T}_o\} = \mathcal{L}(v)$. When we wish to emphasize the output at a particular time, we write

$$w(t) = \mathcal{L}_t(v).$$

Note that the output of a system at a particular time can, in principle, depend on the entire past and future of the input signal (if indeed t corresponds to "time"). While this may seem unphysical, it is a useful abstraction for introducing properties of systems in their most general form. We shall later explore several physically motivated constraints on system structure.

The ideal wire mentioned previously is modeled as a system by $\mathcal{L}(v) = v$. An ideal ground is defined simply by $\mathcal{L}(v) = 0$, where here 0 denotes a signal that is 0 for all time.

In many applications the input and output signals are of the same type, that is, $\mathcal{T}_i$ and $\mathcal{T}_o$ are the same; but they need not always be. Several examples of systems with different input and output signal types will be encountered in section 1.8. As the case of identical signal types for input and output is the most common, it is usually safe to assume that this is the case unless explicitly stated otherwise (or implied by the use of different symbols for the input and output time domains of definition).

The key thing to remember when dealing with systems is that they map an entire input signal v into a complete output signal w.

A particularly simple type of system is a *memoryless system*. A memoryless system is one which maps an input signal $v = \{v(t); t \in \mathcal{T}\}$ into an output signal $w = \{w(t); t \in \mathcal{T}\}$ via a mapping of the form

$$w(t) = \alpha_t(v(t)); \ t \in \mathcal{T}$$

so that the output at time t depends only on the current input and not on any past or future inputs (or outputs).

1.5   Linear Combinations

By scaling, shifting, summing, or combining in any other way the previously treated signals a wide variety of new signals can be constructed. Scaling a signal by a constant is the simplest such operation. Given a signal $g = \{g(t); t \in \mathcal{T}\}$ and a complex number a, define the new signal ag by

$\{ag(t); t \in \mathcal{T}\}$; that is, the new signal formed by multiplying all values of the original signal by a. This production of a new signal from an old one provides another simple example of a system, where here the system $\mathcal{L}$ is defined by $\mathcal{L}_t(g) = ag(t); t \in \mathcal{T}$.

Similarly, given two signals g and h and two complex numbers a and b, define a linear combination of signals $ag + bh$ as the signal $\{ag(t) + bh(t); t \in \mathcal{T}\}$. We have effectively defined an algebra on the space of signals. This linear combination can also be considered as a system if we extend the definition to include multiple inputs; that is, here we have a system $\mathcal{L}$ with two input signals and an output signal defined by $\mathcal{L}_t(g, h) = ag(t) + bh(t); t \in \mathcal{T}$.

As a first step toward what will become Fourier analysis, consider the specific example of the signal shown in Figure 1.19 obtained by adding two sines together as follows:

$$g(n) = \frac{\sin(\frac{2\pi}{8}n) + \sin(\frac{\pi}{8}n)}{2}.$$

The resulting signal is clearly not a sine, but it is equally clearly quite well

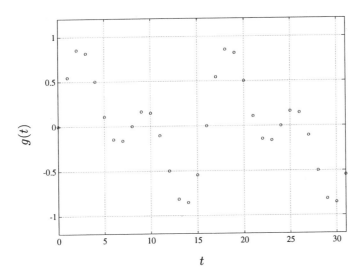

Figure 1.19: Sum of Sinusoids.

behaved and periodic. One might guess that given such a signal one should be able to decompose it into its sinusoidal components. Furthermore, one might suspect that by using general linear combinations of sinusoids one should be able to approximate most reasonable discrete time signals. The general form of these two goals constitute a large part of Fourier analysis.

Suppose that instead of combining two sinusoids, however, we combine a finite duration one-sided sinusoid and a random signal. If the signals of Figures 1.3 and 1.18 are summed and divided by two, the result is that of Figure 1.20. Unlike the previous case, it is not clear if one can recover

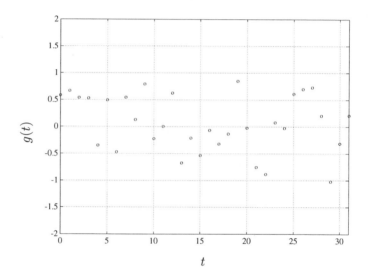

Figure 1.20: Sum of Random Signal and Sine.

the original sinusoid from the sum. In fact, several classical problems in detection theory involve such sums of sinusoids and noise. For example: given a signal that is known to be either noise alone or a sinusoid plus noise, how does one intelligently decide if the sinusoid is present or not? Given a signal that is known to be noise plus a sinusoid, how does one estimate the amplitude or period or the phase of the sinusoid? Fourier methods are crucial to the solutions of such problems. Although such applications are beyond the scope of this book, we will later suggest how they are approached by simply computing and looking at some transforms.

Linear Systems

A system is *linear* if linear combinations of input signals yield the corresponding linear combination of outputs; that is, if given input signals $v^{(1)}$

and $v^{(2)}$ and complex numbers a and b, then

$$\mathcal{L}(av^{(1)} + bv^{(2)}) = a\mathcal{L}(v^{(1)}) + b\mathcal{L}(v^{(2)}).$$

Linearity is sometimes referred to as additivity or as superposition. A system is *nonlinear* if it is not linear. As a special case linearity implies that a mapping is *homogeneous* in the sense that

$$\mathcal{L}(av) = a\mathcal{L}(v)$$

for any complex constant a.

Common examples of linear systems include systems that produce an output by adding (in discrete time) or integrating (in continuous time) the input signal times a weighting function. Since integrals and sums are linear operations, using them to define systems result in linear systems. For example, the systems with output w defined in terms of the input v by

$$w(t) = \int_{-\infty}^{\infty} v(\tau)h_t(\tau)\,d\tau$$

in the infinite duration continuous time case or the analogous

$$w_n = \sum_{k=-\infty}^{\infty} v_k h_{n,k}$$

in the discrete time case yield linear systems. In both cases $h_t(\tau)$ is a weighting which depends on the output time t and is summed or integrated over the input times τ. We shall see in Chapter 6 that these weighted integrals and sums are sufficiently general to describe all linear systems.

A good way to get a feel for linear systems is to consider some nonlinear systems. The following systems are easily seen to be nonlinear:

$$
\begin{aligned}
\mathcal{L}_t(v) &= v^2(t) \\
\mathcal{L}_t(v) &= a + bv(t) \\
\mathcal{L}_t(v) &= \text{sgn}(v(t)) \\
\mathcal{L}_t(v) &= \sin(v(t)) \\
\mathcal{L}_t(v) &= e^{-i2\pi v(t)}.
\end{aligned}
$$

(Note that all of the above systems are also memoryless.) Thus a square law device, a hard limiter (or binary quantizer), a sinusoidal mapping, and a phase-modulator are all nonlinear systems.

1.6   Shifts

The notion of a shift is fundamental to the association of the argument of a
signal, the independent variable t called "time," with the physical notion of
time. Shifting a signal means starting the signal sooner or later, but leaving
its basic shape unchanged. Alternatively, shifting a signal means redefining
the time origin. If the independent variable corresponds to space instead of
time, shifting the signal corresponds to moving the signal in space without
changing its shape. In order to define a shift, we need to confine interest
to certain index sets $\mathcal{T}$. Suppose that we have a signal $g = \{g(t); t \in \mathcal{T}\}$
and suppose also that $\mathcal{T}$ has the property that if $t \in \mathcal{T}$ and $\tau \in \mathcal{T}$, then
also $t - \tau \in \mathcal{T}$. This is obviously the case when $\mathcal{T} = \mathcal{R}$ or $\mathcal{T} = \mathcal{Z}$, but it
is not immediately true in the finite duration case (which we shall remedy
shortly). We can define a new signal

$$g^{(\tau)} = \{g^{(\tau)}(t); t \in \mathcal{T}\} = \{g(t - \tau); t \in \mathcal{T}\}$$

as the original signal *shifted* or *delayed* by τ. Since by assumption $t - \tau \in \mathcal{T}$
for all $t \in \mathcal{T}$, the values $g(t - \tau)$ are well-defined. The shifted signal can be
thought of as a signal that starts τ seconds after $g(t)$ does and then mimics
it. The property that the difference (or sum) of any two members of $\mathcal{T}$ is
also a member of $\mathcal{T}$ is equivalent to saying that $\mathcal{T}$ is a *group* in mathematical
terms. For the two-sided infinite cases the shift has the normal meaning.
For example, a continuous time wedge signal with width $2T$ is depicted in
Figure 1.21 and the same signal shifted by $2T$ is shown in Figure1.22.

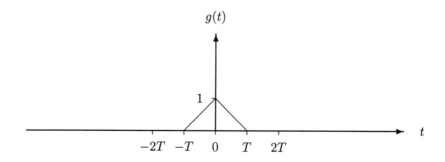

$$g(t)$$

Figure 1.21: The Wedge or Triangle Signal $\wedge_T(t)$

The operation of shifting or delaying a signal by τ can be viewed as a
system: the original signal put into the system produces an output that is
a delayed version of the original signal. This is a mathematical model for
an idealized delay line with delay τ, which can be expressed as a system

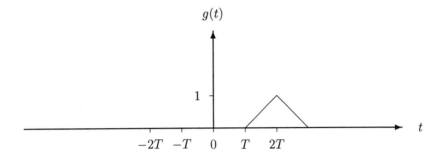

Figure 1.22: Shifted Triangle Signal $\wedge_T(t - 2T)$

$\mathcal{L}$ defined by $\mathcal{L}_t(v) = v(t - \tau)$. It should be obvious that this system is a linear system.

The idea of a shift extends in a natural fashion to finite duration signals by taking advantage of the relation between a finite duration signal and an infinite duration signal formed by replicating the finite duration signal forever: its *periodic extension*. The continuous time case is considered first. To define carefully a periodic extension of a finite duration signal, we use the modulo operation of (0.1).

Define the periodic extension $\tilde{g} = \{\tilde{g}(t); \ t \in \mathcal{R}\}$ of a finite duration signal $g = \{g(t); \ t \in [0, T)\}$ by

$$\tilde{g}(t) = g(t \bmod T); \ t \in \mathcal{R}. \tag{1.14}$$

In the discrete time case, the natural analog is used. The periodic extension $\tilde{g} = \{\tilde{g}_n; \ n \in \mathcal{Z}\}$ of a finite duration signal $g = \{g_n; \ n \in \mathcal{Z}_N\}$ is defined by

$$\tilde{g}(t) = g(n \bmod N); \ t \in \mathcal{Z}. \tag{1.15}$$

Periodic extensions for more complicated finite duration time domains can be similarly defined using the general definition of modulo of (0.2).

Roughly speaking, given a finite duration signal $g = \{g(t); t \in [0, T)\}$ (discrete or continuous time) then we will define the shifted signal $g^{(\tau)} = \{g^{(\tau)}(t); t \in \mathcal{T}\}$ as one period of the shifted periodic extension of g; that is, if $\tilde{g}(t) = g(t \bmod T)$, then

$$g^{(\tau)}(t) = \tilde{g}(t - \tau) = g((t - \tau) \bmod T); \ t \in [0, T).$$

This is called a *cyclic shift* and can be thought of as wrapping the original finite duration signal around so that what is shifted out one end is shifted back into the other. The finite duration cyclic shift is depicted in Figure 1.23

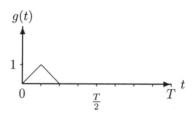

(a) $g(t) = \wedge_{T/8}(t - \frac{T}{8})$

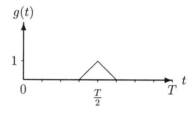

(b) $g(t) = \wedge_{T/8}(t - \frac{T}{2})$

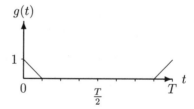

(c) $g(t) = \wedge_{T/8}(t - T) = \wedge_{T/8}(t)$

Figure 1.23: Shifted Finite Duration Triangle Signal

for the triangle signal. The choice of index set $[0, T) = \{t : 0 \le t < T\}$ does not include the endpoint T because the periodic extension starts its replication of the signal at T, that is, the signal at time T is the same as the signal at time 0.

An alternative and equivalent definition of a cyclic shift is to simply redefine our "time arithmetic" $t - \tau$ to mean difference modulo T (thereby again making $\mathcal{T}$ a group) and hence we are defining the shift of $\{g(t); t \in [0, T)\}$ to be the signal $\{g((t - \tau) \bmod T); t \in [0, T)\}$. Since $(t - \tau) \bmod T \in [0, T)$, the shifted signal is well defined.

Time-Invariant Systems

We have seen that linear systems handle linear combinations of signals in a particularly simple way. This fact will make Fourier methods particulary amenable to the analysis of linear systems. In an analogous manner, some systems handle shifts of inputs in a particularly simple way and this will result in further simplifying the application of Fourier methods.

A system $\mathcal{L}$ is said to be *time invariant* or *shift invariant* or *stationary* if shifting the input results in a corresponding shift of the output. To be precise, a system is time invariant if for any input signal v and any shift τ, the shifted input signal $v^{(\tau)} = \{v(t - \tau); t \in \mathcal{T}_i\}$ yields the shifted output signal

$$\mathcal{L}_t(v^{(\tau)}) = \mathcal{L}_{t-\tau}(v); \ t \in \mathcal{T}_o. \tag{1.16}$$

In other words, if $w(t) = \mathcal{L}_t(\{v(t); t \in \mathcal{T}_i\})$ is the output at time t when v is the input, then $w(t - \tau)$ is the output at time t when the shifted signal v^τ is the input.

One can think of a time-invariant system as one which behaves in the same way at any time. If you apply a signal to the system next week at this time the effect will be the same as if you apply a signal to the system now except that the results will occur a week later.

Examples of time-invariant systems include the ideal wire, the ideal ground, a simple scaling, and an ideal delay. A memoryless system defined by $w(t) = \alpha_t(v(t))$ is time-invariant if α_t does not depend on t, in which case we drop the subscript.

As an example of a system that is linear but not time invariant, consider the infinite duration continuous time system defined by

$$w(t) = \mathcal{L}_t(v) = v(t) \cos 2\pi f_0 t.$$

This is a double sideband suppressed carrier (DSB-SC or, simply, DSB) modulation system. It is easily seen to be linear by direct substitution. It is not time invariant since, for example, if $v(t) = 1$, all t, then shifting the input by $\pi/2$ does not shift the output by $\pi/2$. Alternatively, the system is time-varying because it always produces an output of 0 when $2\pi f_0 t$ is an odd multiple of $\pi/2$. Thus the action of the system at such times is different from that at other times.

Another example of a time varying system is given by the infinite duration continuous time system

$$\mathcal{L}_t(v) = v(t) \sqcap (t).$$

This system can be viewed as one which closes a switch and passes the input during the interval $[-\frac{1}{2}, \frac{1}{2}]$, but leaves the switch open (producing a zero output) otherwise. This system is easily seen to be linear by direct substitution, but it is clearly not time invariant. For example, shifting an input of $v(t) = \sqcap(t)$ by 1 time unit produces an output of 0, not a shifted square pulse. Another way of thinking about a time-invariant system is that its action is independent of the definition of the time origin $t = 0$.

A more subtle example of a time varying system is given by the continuous time "stretch" system, a system which compresses or expands the time scale of the input signal. Consider the system which maps a signal $\{v(t); t \in \mathcal{R}\}$ into a stretched signal defined by $\{v(at); t \in \mathcal{R}\}$, i.e., we have a system mapping $\mathcal{L}$ that maps an input signal $\{v(t); t \in \mathcal{R}\}$ into an output signal $\{w(t); t \in \mathcal{R}\}$ where $w(t) = v(at)$. Assume for simplicity that $a > 0$ so that no time reversal is involved.

Shift the input signal to form a new input signal $\{v^\tau(t); t \in \mathcal{R}\}$ defined by $v^\tau(t) = v(t - \tau)$. If this signal is put into the system, the output signal, say $w_0(t)$, is defined by $w_0(t) = v^\tau(at) = v(at - \tau)$.

On the other hand, if the unshifted v is put into the system to get $w(t) = v(at)$, and then the output signal is delayed by τ, then $w^\tau(t) = w(t - \tau) = v(a(t - \tau)) = v(at - a\tau)$, since now w directly plugs $t - \tau$ into the functional form defining w.

Since $w_0(t)$ and $w(t-\tau)$ are not equal, the system is not time invariant. The above shows that it makes a difference in which order the stretch and shift are done.

1.7   Two-Dimensional Signals

Recall that a two-dimensional or 2D signal is taken to mean a signal of the form $\{g(x,y); \ x \in \mathcal{T}_X, \ y \in \mathcal{T}_Y\}$, that is, a signal with a two-dimensional domain of definition. Two dimensional signal processing is growing in importance. Application areas include image processing, seismology, radio astronomy, and computerized tomography. In addition, signals depending on two independent variables are important in applied probability and random processes for representing two dimensional probability density functions and probability mass functions which provide probabilistic descriptions of two dimensional random vectors.

A particularly simple class of 2D signals is constructed by taking products of one-dimensional (or 1D) signals. For example, suppose that we have two 1D signals $\{g_X(x);\ x \in \mathcal{T}_X\}$ and $\{g_Y(y);\ y \in \mathcal{T}_Y\}$. For simplicity we usually consider both 1D signals to have the same domain, i.e., $\mathcal{T}_X = \mathcal{T}_Y$. We now define a two-dimensional domain of definition as the cartesian product space $\mathcal{T} = \mathcal{T}_X \times \mathcal{T}_Y$ consisting of all (x, y) such that $x \in \mathcal{T}_X$ and $y \in \mathcal{T}_Y$. An example of a 2D function defined on this domain is

$$g(x, y) = g_X(x)g_Y(y);\ x \in \mathcal{T}_X,\ y \in \mathcal{T}_Y,$$

or, equivalently,

$$g(x, y) = g_X(x)g_Y(y);\ (x, y) \in \mathcal{T}_X \times \mathcal{T}_Y.$$

A 2D signal of this form is said to be *separable* in rectangular coordinates.

A continuous parameter example of such a separable signal is the 2D box signal

$$g(x, y) = \Box_T(x)\Box_T(y);\ (x, y) \in \mathcal{R}^2, \qquad (1.17)$$

where $\mathcal{R}^2 = \mathcal{R} \times \mathcal{R}$ is the real plane. For display we often focus on a discrete parameter version of the basic 2D signals. Consider the discrete time box function of Figure 1.6. If $g_1(n) = g_2(n) = \Box_5(n),\ n \in \mathcal{T}_1 = \{-32, \cdots, 31\}$, then

$$g(n, k) = \Box_5(n)\Box_5(k)$$

is a 2D box signal. A question now arises as to how to display such a 2D signal to help the reader visualize it. There are two methods, both of which we will use. The first method depicts the signal in three dimensions, that is, as a surface above the 2D domain. This is accomplished in Matlab by using the *mesh* command and is illustrated in Figure 1.24. The mesh figure plots the signal values at the collection of times in the time domain and collects the points by straight lines to create a "wire figure" of the signal. Here the signal appears as a skyscraper on a flat field. We choose a fairly small domain of definition, $64 \times 64 = 4096$ index pairs so that the Matlab mesh graphics can clearly show the shape and the individual pixels. The 64×64 points in the grid at which the signal is defined are called *pixels* or *pels* as an abbreviation for "picture elements." The real images we will consider as examples will have a larger domain of definition of $256 \times 256 = 65,536$ pixels.

An alternative method, which will be more useful when dealing with nonnegative signals such as image intensity rasters, is to plot the intensities

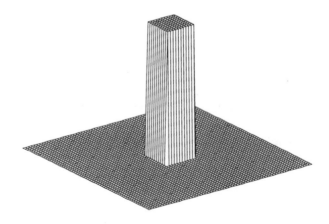

Figure 1.24: 2D Box: Three Dimensional Representation

in two dimensions. This we do using the public domain NIH Image program as in Figure 1.25. Here the light intensity at each pixel is proportional to the signal value, i.e., the larger the signal value, the whiter the pixel appears. Image rescales the pixel values of a signal to run from the smallest value to the largest value and hence the image appears as a light square in a dark background.

Both mesh and image representations provide depictions of the same 2D signal.

The above 2D signal was easy to describe because it could be written as a product, in two "separable" pieces. It is a product of separate signals in each of the two rectangular coordinates. Another way to construct simple signals that separate into product terms is to use polar coordinates. To convert rectangular coordinates (x, y) into polar coordinates (r, θ) so that

$$x = r\cos\theta, \quad y = r\sin\theta. \tag{1.18}$$

The radius r is given by

$$r = r(x, y) = \sqrt{x^2 + y^2}. \tag{1.19}$$

If we restrict the phase angle θ to $[-\pi/2, \pi/2)$, then θ is given by the

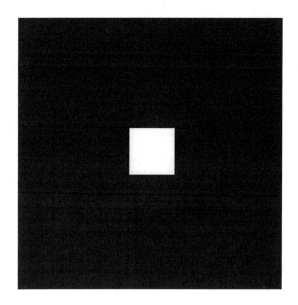

Figure 1.25: 2D Box: Image Representation

principal value of the inverse tangent:

$$\theta(x,y) = \tan^{-1}\frac{y}{x}. \tag{1.20}$$

Consider for example the one-dimensional signals $g_R(r) = \text{sinc}\, r$ for all positive real r and $g_\Theta(\theta) = 1$ for all $\theta \in [-\pi, \pi)$. Form the 2D signal from these two 1D signals by

$$g(x,y) = g_R(r)g_\Theta(\theta)$$

for all real x and y. Once again the signal is separable, but this time in polar coordinates.

A simple and common special case of separable signals in polar coordinates is obtained by setting

$$g_\Theta(\theta) = 1; \text{ for all } \theta$$

so that

$$g(x,y) = g_R(r). \tag{1.21}$$

A 2D signal having this form is said to be *circularly symmetric*.

Perhaps the simplest example of a circularly symmetric signal separable in polar coordinates is the 2D disc

$$g(x, y) = \Box_r(\sqrt{x^2 + y^2}); \ (x, y) \in \mathcal{R}^2. \qquad (1.22)$$

This signal is a disc of radius r and height 1. For purposes of illustration, we focus on a discrete parameter analog and consider the 2D signal having a 64×64 pixel domain of definition defined by

$$g(n, k) = \Box_{10}(\sqrt{n^2 + k^2}); \ k, n \in \{-32, \cdots, 31\}.$$

The mesh representation of this signal is presented in Figure 1.26 and it is seen to be approximately a circle or disk above a flat plane. The circle is not exact because the domain of definition is discrete and not continuous. If more pixels had been used, the approximation to the continuous parameter 2D disk would be better. At this lower resolution the individual pixels are seen as small squares. These blocky artifacts are not visible in the higher resolution images such as the 256×256 images to be considered shortly. The image representation is seen in Figure 1.27, where it is seen as a white circle on a black background.

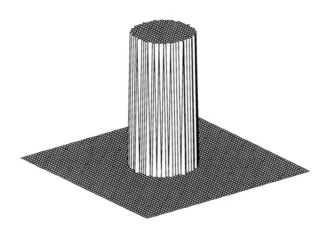

Figure 1.26: Disk: Three Dimensional Representation

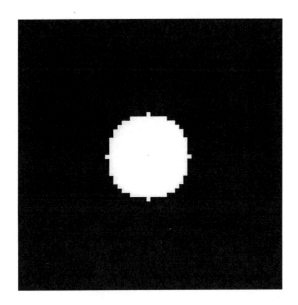

Figure 1.27: Disk: Image Representation

Another polar separable circularly symmetric signal is

$$g(x,y) = \text{sinc}(\sqrt{x^2 + y^2}); \ (x,y) \in \mathcal{R}^2. \qquad (1.23)$$

For illustration we again emphasize a discrete parameter analog, a 2D signal having a 64 × 64 pixel domain of definition defined by

$$g(n,k) = \text{sinc}(\sqrt{n^2 + k^2}); \ k,n \in \{-32, \cdots, 31\}.$$

The mesh representation of this signal is presented in Figure 1.28. The plotting program makes the discrete signal appear continuous. The signal does not have an obvious image representation because it has negative values and hence does not correspond to image intensities at all points. We can make it into an image by taking the absolute value, resulting in the mesh representation of Figure 1.29 and the image representation of Figure 1.30. The discrete nature of the signal can be seen in the blockiness of the image representation.

As alternatives to the previous artificial images representing examples of 2D signals, we include two real world images as examples. The first is a

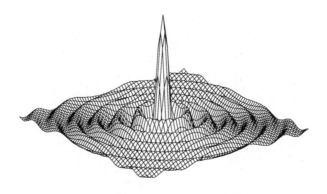

Figure 1.28: 2D Sinc: Three Dimensional Representation

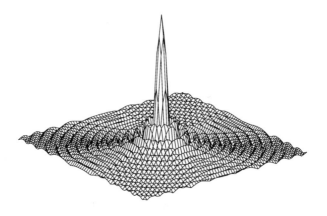

Figure 1.29: Magnitude 2D Sinc: Three Dimensional Representation

Figure 1.30: 2D Sinc: Image Representation

256×256 section of a digitized version of the Mona Lisa taken from the NIH collection of image examples and depicted in Figure 1.31. The second image is a magnetic resonance (MR) brain scan image, which we shall refer to as "Eve." This image is 256×256 pixels and is 8-bit gray scale as previously discussed. The printed version is, however, half-toned.

1.8 Sampling, Windowing, and Extending

The various types of signals are clearly related to each other. We constructed some of the examples of discrete time signals by "sampling" continuous time signals, that is, by only defining the signal at a discrete collection of regularly spaced time values. This suggests that more generally signals of a given type might naturally produce signals of another type. Another reason for considering the relations among different types of signals is that often a physical signal can be well modeled by more than one of the given types and it is useful to know which, if any, of the models is best. For example, suppose that one observes a continuous time sine wave

Figure 1.31: Mona Lisa

produced by a oscillator for T seconds. This could be modelled by a finite
duration waveform $\{\sin(\omega t);\ t \in [0,T)\}$, but it might be useful to consider
it as a piece of an infinite duration sine wave $\{\sin(\omega t);\ t \in \mathcal{R}\}$ or even as a
time-limited waveform that lasts forever and assumes the value 0 for t not
in $[0,T)$. Which model is more "correct"? None; the appropriate choice
for a particular problem depends on convenience and the goal of the analysis. If one only cares about system behavior during $[0,T)$, then the finite
duration model is simpler and leads to finite limits of integrals and sums.
If, however, the signal is to be used in a system whose behavior outside
this time range is important, then the infinite (or at least larger) duration
model may be better. Knowing only the output during time $[0,T)$ may
force one to guess the behavior for the rest of time, and that this can be
done in more than one way. If we know the oscillator behaves identically
for a long time, then repeating the sinusoid is a good idea. If we do not
know what mechanism produced the sinusoid, however, it may make more
sense to set unknown values to zero or something else. The only general

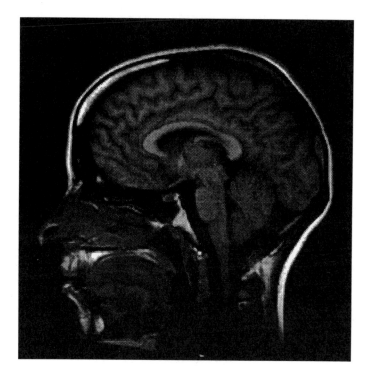

Figure 1.32: Magnetic Resonance Image

rule is to use the simplest useful model.

All of the signal conversions considered in this section are examples of systems where a signal of one type is converted into a signal of a different type.

Continuous Time Infinite Duration to Discrete Time Infinite Duration

Given an infinite duration continuous time signal, say $\{g(t);\ t \in \mathcal{R}\}$, we can form a discrete time infinite duration signal $\{g_n; n \in \mathcal{Z}\}$ by *sampling*: given a fixed positive real number T (called the *sampling period*) define

$$g_n = g(nT);\ n \in \mathcal{Z};\qquad(1.24)$$

that is, g_n is formed as the sequence of successive values of the waveform $g(t)$ each spaced T units apart. Note that the new signal is *different* from the original and the original may or may not be reconstructible from its sampled version. In other words, the sampling operation is not necessarily invertible. One of the astonishing results in Fourier analysis (which we will prove in a subsequent chapter) is the Whittaker-Shannon-Kotelnikov sampling theorem which states that under certain conditions having to do with the shape of the Fourier transform of g and the sampling period T, the original waveform can (in theory) be perfectly reconstructed from its samples. This result is fundamental to sampled-data systems and digital signal processing of continuous waveforms. The sampling idea also can be used if the duration of the original signal is finite.

Figure 1.1 shows a continuous time sinusoid having a frequency of one Hz; that is, $g(t) = \sin(2\pi t)$. Figure 1.33 shows the resulting discrete time signal formed by sampling the continuous time signal using a sampling period of $T = .1$, that is, $\sin(2\pi t)$ for $t = n/10$ and integer n. Note that the sampled waveform looks different in shape, but it is still periodic and resembles the sinusoid. Figure 1.2 shows the resulting discrete time signal $g_n = \sin(2\pi n/10)$, where we have effectively scaled the time axis so that there is one time unit between each sample.

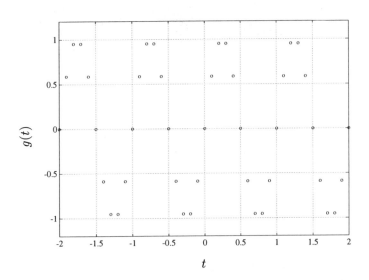

Figure 1.33: Sampled Sinusoid: Sampling Period .1

If instead we take a sampling period of $1/3\pi \approx .1061$, the resulting signal $\sin(2\pi t)$ for $t = n/3\pi$ is shown in Figure 1.34. Rescaling the time axis so as to have one time unit between consecutive samples yields the discrete time signal $g_n = \sin(2n/3)$ shown in Figure 1.35. This discrete time signal is not periodic in n (for example, it never returns to the value $0 = \sin 0$) and it bears less resemblance to the original continuous time signal. This

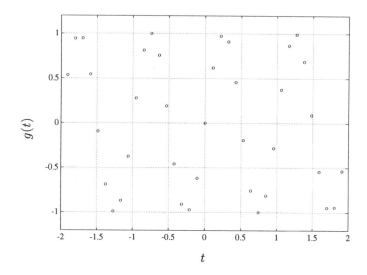

Figure 1.34: Sampled Sinusoid: Sampling Period $1/3\pi$

simple example shows that discrete time signals obtained from continuous time signals can be quite different in appearance and behavior even if the original signal being sampled is fixed.

Discrete Time Infinite Duration to Continuous Time Infinite Duration

Given a discrete time signal, we can construct a continuous time signal by using the discrete time signal to "modulate" (i.e., modify) shifted versions of a waveform. For example, suppose that g_n is a discrete time signal and $\{p(t); \ t \in \mathcal{R}\}$ is a continuous time signal such as an ideal pulse

$$p(t) = \begin{cases} 1 & \text{if } 0 \le t < T \\ 0 & \text{otherwise} \end{cases} \tag{1.25}$$

as depicted in Figure 1.36 for $T = .1$. This pulse is an example of a time-limited signal, an infinite duration signal that is nonzero only on an interval

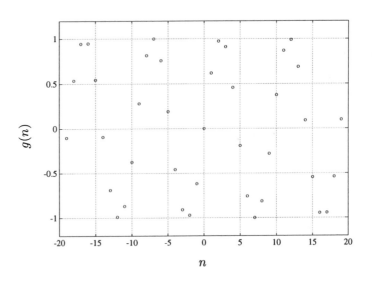

Figure 1.35: Discrete Time Sinusoid: Sampling Period $1/3\pi$

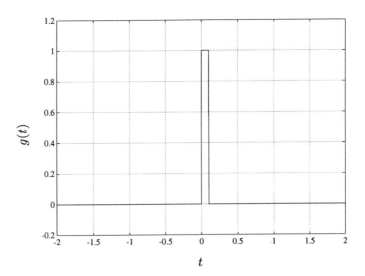

Figure 1.36: Ideal Rectangular Pulse

of finite length. Note that except for the point $t = T$

$$p(t) = \Box_{T/2}(t - \frac{T}{2}), \tag{1.26}$$

a shifted box function. If we had defined $p(t)$ to be 1 at $t = T$ then (1.26) would hold everywhere, but that would cause trouble with the application we consider next. Alternatively,

$$p(t) = u_{-1}(t) - u_{-1}(t - T). \tag{1.27}$$

For a fixed T, $p(t - T)$ is a delayed version of $p(t)$, which here means it is 1 for $T \leq t < 2T$ and 0 otherwise. Similarly, for any integer n $p(t - nT)$ is a delayed version of $p(t)$ that is 1 for $nT \leq t < (n + 1)T$ and 0 otherwise. For each n the signal $\{p(t - nT); t \in \mathcal{R}\}$ is thus a pulse that is 1 for $nT \leq t < (n + 1)T$ and 0 otherwise and hence the pulses in this collection have no overlap with each other. Form a continuous time signal $\{\tilde{g}(t); t \in \mathcal{R}\}$ as

$$\tilde{g}(t) = \sum_{n=-\infty}^{\infty} g_n p(t - nT). \tag{1.28}$$

This is an example of a pulse amplitude modulation (PAM) system where the "pulses" $p(t - nT)$ are modulated by the values of g_n. For the flat pulse under consideration, during the time interval $[nT, (n + 1)T)$ the output of the PAM signal is simply g_n.

If the sequence g_n is the sampled sinusoid of Figure 1.2, then the resulting PAM signal is as depicted in Figure 1.37. If the waveforms $p(t)$ are the ideal pulses of (1.25) and if the g_n are the samples $g(nT)$ of a waveform, then the overall operation is called a *sample-and-hold* system. Here we have chosen the pulse width to match the sampling period used originally to sample the sinusoid so that the reconstructed signal resembles the original signal. The pulse width and the original sampling period need not be the same, however. For example, doubling the pulse width yields the stretched out sample-and-hold waveform of Figure 1.38.

Note that there is no guarantee that in a sample-and-hold system $\tilde{g}(t)$ and $g(t)$ will be the same; in fact they will almost always be different waveforms as above. (If the pulse $p(t)$ is 0 for $t < 0$ and $t \geq T$, then at least the waveforms $\tilde{g}(t)$ and $g(t)$ will agree at the sample times nT.)

An alternative means of generating the continuous time signal would be to interpolate between successive values of the discrete time signal.

Infinite Duration to Finite Duration

A finite duration signal can be obtained from an infinite duration signal by simply truncating a sequence $g = \{g_n; n \in \mathcal{Z}\}$ to form a finite duration

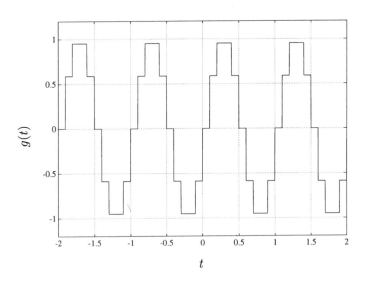

Figure 1.37: Pulse Amplitude Modulated (PAM) Sequence

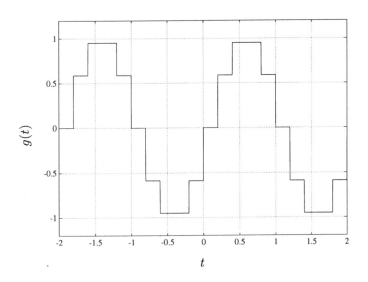

Figure 1.38: Pulse Amplitude Modulated Sequence: Doubled Pulse Width

sequence $\hat{g} = \{g_n; n \in \{0, 1, \ldots, N-1\}\}$. More generally, the truncation can also include *windowing*. Define a *window function* w_n of length N so that $w_n = 0$ if n is not in some finite subset W, say $\mathcal{Z}_N = \{0, 1, \ldots, N-1\}$, of $\mathcal{Z}$. The truncated and windowed

$$\hat{g} = \{\hat{g}_n; \ n = 0, 1, \ldots, N-1\} = \{w_n g_n; \ n = 0, 1, \ldots, N-1\}. \quad (1.29)$$

A common choice of w_n is to set $w_n = 1$ for all $n \in \mathcal{Z}_N$ and 0 otherwise. This is called the *boxcar* window function. Again observe that something has been lost in forming the new signal from the old and that the original infinite duration signal is in general not perfectly recoverable from the finite duration signal.

The same idea can be used to construct finite duration continuous time signals from infinite duration continuous time signals. For example, given a continuous time signal $\{g(t); \ t \in \mathcal{R}\}$, we can define a continuous time window function $w(t)$ with $w(t) = 0$ for t not in $[0, T)$ and then define the windowed and truncated signal $\hat{g} = \{\hat{g}(t) = g(t)w(t); \ t \in [0, T)\}$. Once again, the constant window is called a "boxcar" window. If the continuous 1 Hz sine wave (a portion of which is shown in Figure 1.1) is truncated to the time interval $[-.5, .5]$, then the resulting finite duration signal is as shown in Figure 1.39.

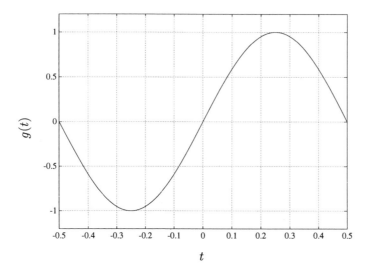

Figure 1.39: Finite Duration Signal by Truncation

Finite Duration to Infinite Duration

Zero Filling

An infinite duration signal can be constructed from a finite duration signal by "filling in" the missing signal values according to some rule. Suppose that $g = \{g_n; \ n = 0, 1, \ldots, N-1\}$ is a finite duration signal. We can define an infinite duration signal $\bar{g} = \{\bar{g}_n; \ n \in \mathcal{Z}\}$ by "zero filling" as follows:

$$\bar{g}_n = \begin{cases} g_n & \text{if } n \in \mathcal{Z}_N \\ 0 & \text{otherwise} \end{cases} \tag{1.30}$$

The infinite duration signal has simply taken the finite duration signal and inserted zeros for all other times. Observe that if the finite duration signal was originally obtained by windowing an infinite duration signal, then it is likely that the infinite duration signal constructed as above from the finite duration signal will differ from the original infinite duration signal. The one notable exception will be if the original infinite duration signal was in fact 0 outside of the window $\mathcal{Z}_N$, in which case the original signal will be perfectly reconstructed. Extending a finite duration signal by zero filling always produces a time-limited signal.

In a similar fashion, a continuous time finite duration signal can be extended by zero filling. The signal $\{g(t); \ t \in [0, T)\}$ can be extended to the infinite duration signal

$$\bar{g} = \begin{cases} g(t) & t \in [0, T) \\ 0 & \text{otherwise} \end{cases}$$

As an example, extending the finite duration sinusoid of Figure 1.39 by zero filling produces an infinite duration signal which has one period of a sinusoid and is zero elsewhere, as illustrated in Figure 1.40. Another example is given by noting that the two-sided continuous time ideal pulse can be viewed as a one-sided box function extended by zero filling.

Periodic Extension

Another approach to constructing an infinite duration signal from a finite duration signal is to replicate the finite duration signal rather than just insert zeros. For example, given a discrete time finite duration signal $g = \{g_n; \ n = 0, 1, \cdots, N-1\}$ we can form an infinite duration signal $\tilde{g} = \{\tilde{g}_n; \ n \in \mathcal{Z}\}$ by defining

$$\tilde{g}_n = g_{n \bmod N}, \tag{1.31}$$

where the mod operation was defined in (0.1). Note that the infinite duration signal $\tilde{g}$ has the property that $\tilde{g}_{n+N} = \tilde{g}_n$ for all integers n; that is, it is

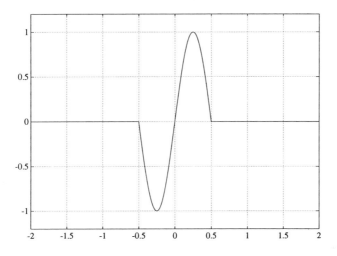

Figure 1.40: Signal Extension by Zero Filling

periodic with period N. We refer to $\tilde{g}$ as the *periodic extension* of the finite duration signal. Observe that if the finite duration signal were obtained via a boxcar window function of length N, then the periodic extension of the finite duration signal will not equal the original signal unless the original signal was itself periodic with period N.

Given a continuous time finite duration signal $g = \{g(t); t \in [0, T)\}$ we can similarly form a periodic extension $\tilde{g}(t)$ as in the discrete time case:

$$\tilde{g}(t) = g(t \bmod T); \ t \in \mathcal{R}. \tag{1.32}$$

As an example, consider the continuous time finite duration ramp $g = \{g(t); t \in [0, 1)\}$ defined by

$$g(t) = t \tag{1.33}$$

and depicted in Figure 1.41. The periodic extension of this having period 1 is the sawtooth signal $t \bmod (1)$ depicted in Figure 1.42.

A convenient form for the periodic extension that relates the zero filled and periodic extension is given by

$$\tilde{g}(t) = \sum_{n=-\infty}^{\infty} \bar{g}(t - nT), \tag{1.34}$$

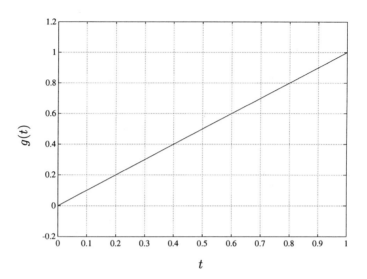

Figure 1.41: Continuous Time Ramp

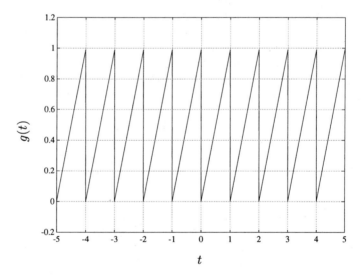

Figure 1.42: Sawtooth Signal

where $\bar{g}$ is the zero filled extension.

We shall see that the theory of finite duration signals is intimately connected with that of infinite duration periodic signals and both periodic and zero filled extensions will prove useful for different applications.

Although each transformation of one signal type can be considered to be a system, more complex systems may have many such transformations. For example, a sampled data system may start with continuous time signals and then sample to form discrete time signals for use in digital signal processing. Digital audio is a popular example. A digital-to-analog converter, such as is used in a compact disc player, begins with discrete time signals (possibly stored in a sequence of memory locations) and produces a continuous time signal.

1.9   Probability Functions

It is important to keep in mind that a signal is just a function of a dummy variable or parameter. The idea of a waveform or data sequence is perhaps the most common example in engineering, but transforms are important for a variety of other examples as well. We close this section with two examples of important special cases that do not immediately fit in the category of waveforms and sequences, but which can be profitably modeled as signals. A discrete parameter signal $\{p_n;\ n \in \mathcal{T}\}$ such that $p_n \geq 0$ for all n and

$$\sum_{n \in \mathcal{T}} p_n = 1 \tag{1.35}$$

is called a *probability mass function* or *pmf* for short. (Here $\mathcal{T}$ is called the *alphabet*.) A continuous parameter signal $\{p(x); x \in \mathcal{T}\}$ with the properties that $p(x) \geq 0$ all x and

$$\int_{x \in \mathcal{T}} p(x)dx = 1 \tag{1.36}$$

is called a *probability density function* or *pdf*. These quantities play a fundamentally important role in probability theory. Here we simply note that they are a source of examples for demonstrating and practicing Fourier analysis.

Probability functions are simply nonnegative signals with integrals or sums normalized to 1. Nonnegative signals having bounded sums or integrals will also arise in other applications, e.g., power spectral densities.

Some of the more common named probability functions are listed below. The parameters p, A, λ, and σ are positive real numbers, with $0 < p < 1$. The parameters m and $b > a$ are real numbers. n is a positive integer.

The Bernoulli pmf: $p(k) = p^k(1-p)^{1-k}$; $k \in \mathcal{T} = \{0, 1\}$.

The binomial pmf: $p(k) = \begin{pmatrix} n \\ k \end{pmatrix} p^k(1-p)^{n-k}$; $k \in \mathcal{T} = \{0, 1, 2, \cdots, n\}$.

The geometric pmf: $p(k) = p(1-p)^{k-1}$; $k \in \mathcal{T} = \{1, 2, \cdots\}$.

The Poisson pmf: $p(k) = \lambda^k e^{-\lambda}/k!$; $k \in \mathcal{T} = \{0, 1, 2, \cdots\}$.

The uniform pdf: $\mathcal{T} = \mathcal{R}$. $f(r) = \frac{1}{A}\Box_{\frac{A}{2}}(r-m)$; $r \in \mathcal{T} = \mathcal{R}$.

The exponential pdf: $f(r) = \lambda e^{-\lambda r} u_{-1}(t)$; $r \in \mathcal{T} = \mathcal{R}$.

The Laplacian pdf: $f(r) = \frac{\lambda}{2} e^{-\lambda|r|}$; $r \in \mathcal{T} = \mathcal{R}$.

The Gaussian pdf: $f(r) = (2\pi\sigma^2)^{-1/2} e^{-(r-m)^2/2\sigma^2}$; $r \in \mathcal{T} = \mathcal{R}$.

Probability functions can also be two-dimensional, e.g., $p = \{p_{n,k}; n \in \mathcal{Z}_N, k \in \mathcal{Z}_N\}$. Because of the requirement of nonnegativity, a 2D probability function is mathematically equivalent to a 2D intensity image that has been normalized.

1.10   Problems

1.1. Which of the following signals are periodic and, if so, what is the period?

    (a) $\{\sin(2\pi f t); t \in (-\infty, \infty)\}$

    (b) $\{\sin(2\pi f n); n \in \{\ldots, -1, 0, 1, \ldots\}\}$ with f a rational number.

    (c) Same as the previous example except that f is an irrational number.

    (d) $\{\sum_{n=1}^{N} \sin(2\pi f_0 t n); t \in (-\infty, \infty)\}$

    (e) $\{\sin(2\pi f_0 t) + \sin(2\pi f_1 t); t \in (-\infty, \infty)\}$ with f_0 and f_1 relatively prime; that is, their only common divisor is 1.

1.2. Is the sum of two continuous time periodic signals having different periods itself periodic? Is your conclusion the same for discrete time signals? *Here and hereafter questions requiring a yes/no answer also require a justification of the answer, e.g., a proof for a positive answer or a counter example for a negative answer.*

1.3. Suppose that $g = \{g(t); \ t \in \mathcal{R}\}$ is an arbitrary signal. Prove that the signal

$$h(t) = \sum_{n=-\infty}^{\infty} g(t - nT)$$

is periodic with period T. Sketch h for the case where

$$g(t) = \begin{cases} 1 - \frac{2|t|}{T} & |t| \le \frac{T}{2} \\ 0 & \text{else} \end{cases}$$

and
$$g(t) = \begin{cases} 1 - \frac{|t|}{T} & |t| \leq T \\ 0 & \text{else.} \end{cases}$$

1.4. Prove the basic geometric progression formulas:

$$\sum_{n=0}^{N-1} r^n = \frac{1 - r^N}{1 - r} \tag{1.37}$$

If $|r| < 1$,

$$\sum_{n=0}^{\infty} r^n = \frac{1}{1 - r}.$$

Evaluate these sums for the case $r = \frac{1}{2}$.

1.5. Evaluate the following sums using the geometric progression formulas and Euler's relations:

(a) $\sum_{n=0}^{N} \cos n\omega$

(b) $\sum_{n=0}^{N} \sin n\omega$

(c) $\sum_{n=0}^{N} r^n \cos n\omega$

(d) $\sum_{n=0}^{N} r^n \sin n\omega$

1.6. Evaluate the sum

$$\sum_{n=0}^{N} n r^n.$$

Hint: What happens if you differentiate the geometric progression formula of Eq. 1.37 with respect to r? Use a similar trick to relate the integrals $\int_0^T t e^{-ft} \, dt$ and $\int_0^T e^{-ft} \, dt$. Verify your answer by the usual method of integration by parts.

1.7. The following input/output relations describe infinite duration discrete time systems with input v and output w. Are the systems linear? time invariant? (Justify your answers!)

(a) $w_n = v_n - v_{n-1}$.

(b) $w_n = \text{sgn}(v_n)$.

(c) $w_n = r^{v_n}$, $|r| < 1$.

(d) $w_n = \sum_{k=-\infty}^{n} v_k$.

(e) $w_n = a v_n + b$, a, b real constants.

(f) $w_n = n^2 v_{n+2}$.

1.8. The following systems describe infinite duration continuous time systems with input $v(t)$ and output $w(t)$. Are the systems linear? time invariant? (Justify your answers!)

(a) $w(t) = v(t)\cos(2\pi f_0 t + \phi)$, where f_0 and ϕ are constants.

(b) $w(t) = A\cos(2\pi f_0 t + \Delta v(t))$ where Δ is a constant (this is phase modulation).

(c) $w(t) = \int_{-\infty}^{t} v(\tau)\, d\tau$.

(d) $w(t) = \int_{-\infty}^{\infty} v(\tau) e^{-i2\pi t\tau}\, d\tau$.

(e) $w(t) = \frac{d}{dt} v(t)$.

(f) $w(t) = av(t) + b$ (a and b are constants).

(g) $w(t) = t^3 v(t)$.

(h) $w(t) = v(t - \pi)$.

(i) $w(t) = v(-t)$.

(j) $w(t) = \Pi(v(t))$.

1.9. The operations of sampling, windowing, and extending can all be viewed as systems. Are these systems linear? (Explain for each operation.)

1.10. Define the infinite duration continuous time signal $g(t) = e^{-t}u_{-1}(t)$ for all $t \in \mathcal{R}$.

(a) Sketch the discrete time signal $\{g_n = g(n);\ n \in \mathcal{Z}\}$ formed by sampling.

(b) Let p be the pulse defined by (1.25) with $T = 1$ and sketch the PAM signal $\hat{g}$ defined by

$$\hat{g}(t) = \sum_{n \in \mathcal{Z}} g_n p(t - n).$$

(c) Evaluate the time-average magnitude error between g and the approximation $\hat{g}$ defined by

$$\int_{-\infty}^{\infty} |g(t) - \hat{g}(t)|\, dt.$$

1.11. Define a continuous time finite duration waveform $g = \{g(t);\ t \in [-1, 1)\}$ by
$$g(t) = \begin{cases} +1 & t \in [0, 1) \\ -1 & t \in [-1, 0) \end{cases}.$$

  (a) Express g as a sum of box functions. Is the expression valid for all t?

  (b) Sketch the zero filled extension $\{\bar{g};\ t \in \mathcal{R}\}$ of g.

  (c) Sketch a periodic extension $\{\tilde{g};\ t \in \mathcal{R}\}$ having period 2 of g.

  (d) Write an expression for $\tilde{g}$ in terms of g using the modular notation.

  (e) Write an expression for $\tilde{g}$ in terms of $\bar{g}$ as an infinite sum.

  (f) Sketch a periodic extension $\{g';\ t \in \mathcal{R}\}$ having period 4 of g.

1.12. Prove that the Poisson pmf sums to 1. *Hint:* You may find the Taylor series expansion of an exponential useful.

1.13. Prove that the Gaussian pdf is indeed a pdf, i.e., that it is nonnegative and integrates to 1. (Do the integral, do not just quote a table and change variables.)

1.14. Express the uniform pdf in terms of the box function and in terms of unit step functions.

1.15. Define the circle function
$$\mathrm{circ}(x, y) = \begin{cases} 1 & x^2 + y^2 \leq 1 \\ 0 & \text{otherwise} \end{cases} \qquad (1.38)$$

for all real x, y. Provide a labeled sketch of the mesh and image forms of the signal
$$g(x, y) = \mathrm{circ}(\frac{x}{10}, \frac{y}{10});\ x \in \mathcal{R}, y \in \mathcal{R}.$$

1.16. Define a 2-dimensional signal g
$$g(x, y) = \begin{cases} 1 & |x| + |y| \leq 1 \\ 0 & \text{else} \end{cases}$$

for all real x and y. Provide a labeled sketch of the mesh and image forms of the signal g. Repeat for the signal h defined by
$$h(x, y) = g(\frac{x - 2}{2}, \frac{y + 1}{2}).\ x \in \mathcal{R}, y \in \mathcal{R}.$$

Chapter 2

The Fourier Transform

2.1 Basic Definitions

The definition of a Fourier transform will differ depending on the signal type. The definitions all have a common form, however, and all can be thought of as a means of mapping a signal $g = \{g(t); t \in \mathcal{T}\}$, which depends on a parameter t in some time index set $\mathcal{T}$, into another signal $G = \{G(f); f \in \mathcal{S}\}$, which depends on a new parameter or independent variable f, which we shall call *frequency*. As does t, the independent variable f takes values in a domain of definition or index set, denoted $\mathcal{S}$. We will eventually show that there are natural frequency domains of definitions for use with each of the basic signal types, as summarized in Table 2.1.

There are two basic forms of the Fourier transform and several special cases. Unless specifically stated otherwise, we assume one-dimensional index sets; that is, we assume that $\mathcal{T}$ is a subset of the real line $\mathcal{R}$. The basic form of the transform depends on whether the index set $\mathcal{T}$ is continuous (e.g., the real line itself) or discrete (e.g., the integers). Given a signal $g = \{g(t); \ t \in \mathcal{T}\}$, its Fourier transform, Fourier spectrum, or frequency spectrum is defined by $G = \{G(f); \ f \in \mathcal{S}\}$, where $\mathcal{S}$ is a set of real numbers, and

$$G(f) = \mathcal{F}_f(g) = \begin{cases} \displaystyle\sum_{n \in \mathcal{T}} g(n)e^{-i2\pi fn} & \mathcal{T} \text{ discrete} \\[2em] \displaystyle\int_{t \in \mathcal{T}} g(t)e^{-i2\pi ft} dt & \mathcal{T} \text{ continuous} \end{cases} . \qquad (2.1)$$

The frequency parameter f is required to be real, although later we shall briefly consider generalizations which permit it to be complex (the z transform of discrete time signals and the Laplace transform of continuous time

signals). For reasons that will be made clear in the next section, we do not
always need to consider the Fourier transform of a signal to be defined for
all real f; each signal type will have a corresponding domain of definition
for f. There is nothing magic about the sign of the exponential, but the
choice of the negative sign is the most common for the Fourier transform.
The *inverse* Fourier transform will later be seen to take similar form except
that the sign of the exponent will be reversed.

We sometimes refer to the original signal g as a *time domain signal* and
the second signal G a *frequency domain signal* or *spectrum*. We will denote
the general mapping by $\mathcal{F}$; that is,

$$G = \mathcal{F}(g). \tag{2.2}$$

When we wish to emphasize the value of the transform for a particular
frequency f we write

$$G(f) = \mathcal{F}_f(g).$$

Another popular notation is

$$\{g(t); \ t \in \mathcal{T}\} \supset \{G(f); \ f \in \mathcal{S}\} \tag{2.3}$$

or, more simply,

$$g \supset G.$$

When the index sets are clear from context, (2.3) is commonly written

$$g(t) \supset G(f)$$

to denote that $G(f)$, considered as a function of the independent variable
f, is the Fourier transform of $g(t)$, which is considered as a function of the
independent variable t.

The Fourier transform is a mapping or transformation of a signal, in
the time or space domain, into its spectrum or transform, in the frequency
domain. If we do not distinguish between the independent variables as being
associated with time and frequency, then the Fourier transform operation
can be considered to be a system.

In this section we consider the details of the definition of the Fourier
transform for the signal types introduced. The remainder of the book is
devoted to developing the properties and applications of such transforms.
Some of the basic questions to be explored include:

- How do we compute transforms either analytically or numerically?

- Does the transform exist; i.e. does the sum or integral defining the
  transform exist? Under what conditions is existence of the transform
  guaranteed?

- Have we lost information by taking the transform; that is, can the original signal be recovered from its spectrum? Is the Fourier transform *invertible*?

- What are the basic properties of the mapping, e.g., linearity and symmetry?

- What happens to the spectrum if we do something to the original signal such as scale it, shift it, filter it, scale its argument, or modulate it? By filtering we include, for example, integrating or differentiating continuous time signals and summing or differencing discrete time signals.

- Suppose that we are given two signals and their transforms. If we combine the signals to form a new signal, e.g., using addition, multiplication, or convolution, how does the transform of the new signal relate to those of the old signals?

- What happens to the spectrum if we change the signal type, e.g., sample a continuous signal or reconstruct a continuous signal from a discrete one?

Before specializing the basic definitions to the most important cases, it is useful to make several observations regarding the definitions and the quantities involved. The basic definitions require that the sum or integral exists, e.g., the limits defining the Riemann integrals converge. If the sum or integral exists, we say the Fourier transform exists. To distinguish the two cases of discrete and continuous $\mathcal{T}$ we often speak of the sum form as a *discrete time (or parameter) Fourier transform* or *DTFT*, and the integral form as the *continuous time (or parameter) Fourier transform* or *CTFT* or *integral Fourier transform*. Note that even if the original signal is real, its transform is in general complex valued because of the multiplicative complex exponential $e^{-i2\pi ft}$.

The dimensions of the frequency variable f are inverse to those of t. Thus if t has seconds as units, f has cycles per second or hertz as units. Often frequency is measured as $\omega = 2\pi f$ with radians per second as units. If t has meters as units, f has cycles/meter as units. If t uses the dimensionless spatial units of distance/wavelength, then f has cycles as units. The symbol $\omega/2\pi$ is also commonly used as the frequency variable, the units of ω being radians per second (or radians per meter, etc.).

One fundamental difference between the discrete and continuous time cases follows from the fact that the exponential $e^{-i2\pi fn}$ is a periodic function in f with period one for *every* fixed integer n; that is,

$$e^{-i2\pi(f+1)n} = e^{-i2\pi fn}e^{-i2\pi n} = e^{-i2\pi fn}, \text{ all } f \in \mathcal{R}.$$

This means that if we consider a DTFT $G(f)$ to be defined for all real f, it is a periodic function with period 1 (since sums of periodic functions of a common period are also periodic). Thus

$$G(f + 1) = G(f) \qquad (2.4)$$

for the DTFT discrete time signals. $G(f)$ does not exhibit this behavior in the CTFT case; that is, $e^{-i2\pi ft}$ is not periodic in f with a fixed period for all values of $t \in \mathcal{T}$ when $\mathcal{T}$ is continuous. The periodicity of the spectrum in the discrete time case means that we can restrict consideration of the spectrum to only a single period of its argument when performing our analysis.

In addition to distinguishing the Fourier transforms of discrete time signals and continuous time signals, the transforms will exhibit different behavior depending on whether or not the index set $\mathcal{T}$ is finite or infinite, that is whether or not the signal has *finite duration* or *infinite duration*. For example, if g has a finite duration index set $\mathcal{T} = \mathcal{Z}_N = \{0, 1, \ldots, N - 1\}$, then

$$G(f) = \sum_{n=0}^{N-1} g(n)e^{-i2\pi fn}. \qquad (2.5)$$

To define a Fourier transform completely we need to specify the domain of definition of the frequency variable f. While the transforms appear to be defined for all real f, in many cases only a subset of real frequencies will be needed in order to recover the original signal and have a useful theory. We have already seen, for example, that all the information in the spectrum of a discrete time signal can be found in a single period and hence if $\mathcal{T} = \mathcal{Z}$, we could take the frequency domain to be $\mathcal{S} = [0, 1)$ or $[-1/2, 1/2)$, for example, since knowing $G(f)$ for $f \in [0, 1)$ gives us $G(f)$ for *all* real f by taking the periodic extension of $G(f)$ of period 1. We introduce the appropriate frequency domains at this point so as to complete the definitions of the Fourier transforms and to permit a more detailed solution of the examples. The reasons for these choices, however, will not become clear until the next chapter. The four basic types of Fourier transform are presented together with their most common choice of frequency domain of definition in Table 2.1. When evaluating Fourier transforms it will often be convenient first to find the functional form for arbitrary real f and then to specialize to the appropriate set of frequencies for the given signal type. This is particularly true when we may be considering differing signal types having a common functional form.

Common alternatives are to use a two-sided finite duration DTFT

$$G(f) \quad = \quad \sum_{n=-N}^{N} g_n e^{-i2\pi fn};$$

Duration	Time	
	Discrete	Continuous
Finite	Finite Duration DTFT (DFT) $$G(f) = \sum_{n=0}^{N-1} g_n e^{-i2\pi fn};$$ $f \in \{0, 1/N, \cdots, (N-1)/N\}$	Finite Duration CTFT $$G(f) = \int_0^T g(t)e^{-i2\pi ft}\, dt;$$ $f \in \{k/T; \; k \in \mathcal{Z}\}$
Infinite	Infinite Duration DTFT $$G(f) = \sum_{n=-\infty}^{\infty} g_n e^{-i2\pi fn};$$ $f \in [-1/2, 1/2)$	Infinite Duration CTFT $$G(f) = \int_{-\infty}^{\infty} g(t)e^{-i2\pi ft}\, dt;$$ $f \in (-\infty, \infty)$

Table 2.1: Fourier Transform Types

$$f \in \{-\frac{N}{2N+1}, \cdots, -\frac{1}{2N+1}, 0, \frac{1}{2N+1}, \cdots, \frac{N}{2N+1}\},$$

and a two-sided finite duration CTFT

$$G(f) = \int_{\frac{-T}{2}}^{\frac{T}{2}} g(t)e^{-i2\pi ft}\, dt; \; f \in \{k/T; \; k \in \mathcal{Z}\}.$$

It is also common to replace the frequency domain for the infinite duration DTFT by $[0, 1)$. There is some arbitrariness in these choices, but as we shall see the key point is to use a frequency domain which suffices to invert the transform. The reader is likely to encounter an alternative notation for the DTFT. Many books that treat the DTFT as a variation on the z transform (which we will consider later) write $G(e^{i2\pi f})$ instead of the simpler $G(f)$.

A discrete time finite duration Fourier transform or finite duration DTFT defined for the frequencies $\{0, 1/N, \ldots, (N-1)/N\}$ is also called a *discrete Fourier transform* or *DFT* because of the discrete nature of both the time domain and frequency domain. It is common to express the transform as $G(k)$ instead of $G(k/N)$ in order to simplify the notation, but one should keep in mind that the frequency is the normalized k/N and not the integer k.

The DFT can be expressed in the form of vectors and matrices: Given a signal $g = \{g_n; \; n = 0, 1, \cdots, N-1\}$, suppose that we consider it as a column vector $\mathbf{g} = (g_0, g_1, \cdots, g_{N-1})^t$, where the superscript denotes the transpose of the vector (which makes the row vector written in line with the text a column vector). We will occasionally use boldface notation for vectors when we wish to emphasize that we are considering them as column vectors and we are doing elementary linear algebra using vectors and matrices. Similarly let $\mathbf{G}$ denote the DFT vector $(G(0), G(1/N), \cdots, G((N-1)/N))^t$. Lastly, define the $N \times N$ square matrix $\mathbf{W}$ by

$$\mathbf{W} = \{e^{-i\frac{2\pi}{N}kj}; k = 0, 1, \cdots, N-1; \; j = 0, 1, \cdots, N-1\}. \qquad (2.6)$$

Then the DFT can be written simply as

$$\mathbf{G} = \mathbf{Wg} =$$

$$\begin{bmatrix} 1 & 1 & 1 & \cdots & 1 \\ 1 & e^{-i\frac{2\pi}{N}} & e^{-i\frac{4\pi}{N}} & \cdots & e^{-i\frac{2\pi}{N}(N-1)} \\ 1 & e^{-i\frac{4\pi}{N}} & e^{-i\frac{8\pi}{N}} & \cdots & e^{-i\frac{4\pi}{N}(N-1)} \\ \vdots & \vdots & \vdots & \ddots & \vdots \\ 1 & e^{-i\frac{2\pi}{N}(N-1)} & e^{-i\frac{4\pi}{N}(N-1)} & \cdots & e^{-i\frac{2\pi}{N}(N-1)(N-1)} \end{bmatrix} \times$$

$$\begin{pmatrix} g_0 \\ g_1 \\ \vdots \\ g_{N-1} \end{pmatrix}, \qquad (2.7)$$

demonstrating the fact that the DFT is expressible as a matrix multiplication and hence is clearly a linear operation.

When considering infinite duration DTFTs with infinite sums and CTFTs with integrals, the Fourier transforms may not be well-defined if the limits defining the infinite sums or Riemann integrals do not exist. There is no such problem, however, with the DFT provided the signal can take on only finite values. Before considering some of the technicalities that can arise and providing some sufficient conditions for the existence of Fourier transforms, we consider examples of evaluation of the various types of Fourier transforms. In the examples we adopt a common viewpoint: First try to evaluate the sums or integrals. If the integrals or sums can be successfully evaluated, then the Fourier transform exists. If not, then it may be necessary to generalize the definition of the Fourier transform, e.g., by taking suitable limits or by generalizing the definition of an integral or sum. We begin with simple examples wherein the calculus of evaluating the sums or integrals is straightforward. We later consider numerical techniques for finding Fourier transforms and return to the more general issue of when the definitions make sense.

During much of the book we will attempt to avoid actually doing integration or summation to find transforms, especially when the calculus strongly resembles something already done. Instead the properties of transforms will be combined with an accumulated collection of simple transforms to obtain new, more complicated transforms. The simple examples to be treated can be considered as a "bootstrap" for this approach; a modicum of calculus now will enable us to take many shortcuts later.

2.2 Simple Examples

The simplest possible example is trivial: the all zero signal $g = \{g(t);\ t \in \mathcal{T}\}$ defined by $g(t) = 0$ for all $t \in \mathcal{T}$. In this case obviously $G(f) = 0$ for all real f, regardless of the choice of the frequency domain.

The simplest nontrivial discrete time signal is the Kronecker delta or unit pulse (or unit sample). If we consider the infinite duration signal $g = \{\delta_n;\ n \in \mathcal{Z}\}$, then it is trivial calculus to conclude that for any real f

$$G(f) = \sum_{n=-\infty}^{\infty} \delta_n e^{-i2\pi f n} = 1$$

and hence we have the Fourier transform

$$\{\delta_n;\ n \in \mathcal{Z}\} \supset \{1;\ f \in [-\frac{1}{2}, \frac{1}{2})\}, \tag{2.8}$$

where we have restricted the frequency domain to $\mathcal{S}_{DTID}^{(2)}$. If instead we consider the discrete time finite duration signal $\{\delta_n;\ n = 0, 1, \cdots, N - 1\}$, then the transform relation becomes

$$\{\delta_n;\ n = 0, 1, \cdots, N - 1\} \supset \{1;\ f = \frac{k}{N} \text{ for } k = 0, 1, \cdots, N - 1\}. \tag{2.9}$$

In both cases the conclusion is the same, the transform of a Kronecker delta is a constant for all suitable values of frequency. The difference in the Fourier transforms of the two signals representing two types is only in the choice of the frequency domain. As mentioned before, this difference in the definition of the frequency domain will be explained when the inversion of the Fourier transform is developed. Essentially, the frequency domain consists of only those frequencies necessary for the reconstruction of the original signal, and that will differ depending on the signal type.

A simple variation on a Kronecker delta at the origin is a Kronecker delta at time l, that is, the shifted Kronecker delta function

$$\delta_{n-l} = \begin{cases} 1 & n = l \\ 0 & \text{else.} \end{cases} \tag{2.10}$$

This example is only slightly less trivial and yields

$$G(f) = \sum_{n=-\infty}^{\infty} \delta_{n-k} e^{-i2\pi fn} = e^{-i2\pi fk},$$

a complex exponential. This is a special case of a general result that will be seen in Chapter 4: shifting a signal causes the Fourier transform to be multiplied by a complex exponential with frequency proportional to the time shift. We therefore have the new transform

$$\{\delta_{n-l}; \ n \in \mathcal{Z}\} \supset \{e^{-i2\pi fl}; \ f \in [-\frac{1}{2}, \frac{1}{2})\}. \tag{2.11}$$

In a similar manner we can show that for the DFT and the cyclic shift that for any $l \in \mathcal{Z}_N$

$$\{\delta_{n-l}; \ n = 0, 1, \cdots, N-1\} \supset \{e^{-i2\pi fl}; \ f = \frac{k}{N} \text{ for } k = 0, 1, \cdots, N-1\}, \tag{2.12}$$

where as usual for this domain the shift is the cyclic shift.

The delta function examples provide an interpretation of a Fourier transform for discrete time signals. A single shifted delta function has as a Fourier transform a single complex exponential with a frequency proportional to the shift. Any discrete time signal can be expressed as a linear combination of scaled and shifted delta functions since we can write

$$g_n = \sum_l a_l \delta_{n-l}, \tag{2.13}$$

where $a_l = g_l$. Since summations are linear, taking the Fourier transform of the signal g thus amounts to taking a Fourier transform of a sum of scaled and shifted delta functions, which yields a sum of scaled complex exponentials. Each sample of the input signal yields a single scaled complex exponential in the Fourier domain, so the entire signal yields a weighted sum of complex exponentials.

Consider next the infinite duration continuous time signal

$$g(t) = \begin{cases} e^{-\lambda t} & t \geq 0 \\ 0 & \text{otherwise} \end{cases}$$

which can be written more compactly as

$$g(t) = e^{-\lambda t} u_{-1}(t); \ t \in \mathcal{R},$$

where $\lambda > 0$. From elementary calculus we have that

$$G(f) = \int_0^{\infty} e^{-(\lambda + i2\pi f)t} \, dt = \frac{1}{\lambda + i2\pi f} \tag{2.14}$$

for $f \in \mathcal{R}$. Thus the transform relation is

$$\{e^{-\lambda t} u_{-1}(t); \ t \in \mathcal{R}\} \supset \{\frac{1}{\lambda + i2\pi f}; \ f \in \mathcal{R}\}. \tag{2.15}$$

The magnitude and phase of this transform are depicted in Figure 2.1. (The units of phase are radians here.) In this example the Fourier transform

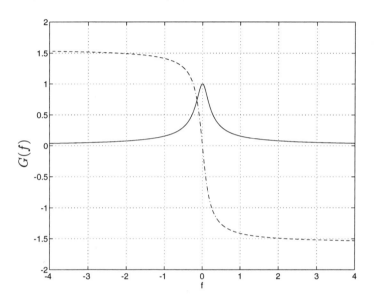

Figure 2.1: Fourier Transform of Exponential Signal: solid line = magnitude, dashed line = phase

exhibits strong symmetries:

• The magnitude of the transform is symmetric about the origin, i.e.,

$$|G(-f)| = |G(f)|; \quad \text{all } f \in \mathcal{S}. \tag{2.16}$$

• The phase of the transform is antisymmetric about the origin, i.e.,

$$\angle G(-f) = -\angle G(f); \quad \text{all } f \in \mathcal{S}. \tag{2.17}$$

In Chapter 4 symmetry properties such as these will be considered in detail.

Next consider an analogous discrete time signal, the finite duration geometric signal of (1.5). For any f we have, using the geometric progression summation formula, that the Fourier transform (here the finite duration

DTFT) is given by

$$G(f) = \sum_{n=0}^{N-1} r^n e^{-i2\pi fn} = \frac{1 - r^N e^{-i2\pi fN}}{1 - re^{-i2\pi f}}. \tag{2.18}$$

There is a potential problem with the above evaluation if the possibility $re^{-i2\pi f} = 1$ arises, which it can if $r = 1$ and f is an integer. In this case the solution is to evaluate the definition directly rather than attempt to apply the geometric progression formula:

$$G(f) = \sum_{n=0}^{N-1} 1^n = N \text{ if } re^{-i2\pi f} = 1. \tag{2.19}$$

Alternatively, L'Hôpital's rule of calculus can be used to find the result from the geometric progression. Applying the appropriate frequency domain of definition $\mathcal{S}_{DTFD}$ we have found the Fourier transform (here the DFT):

$$\{r^n; n = 0, 1, \ldots, N - 1\} = \{1, r, \ldots, r^{N-1}\} \supset \tag{2.20}$$

$$\{G_k \triangleq G(\frac{k}{N}) = \frac{1 - r^N}{1 - re^{-i2\pi k/N}}; \ k = 0, 1, \cdots, N - 1\},$$

where the answer holds provided we exclude the possibility of $r = 1$. The subscripted notation for the DFT is often convenient and in common use, it replaces the frequency by the index of the frequency component and saves a little writing. The magnitude and phase of the DFT are plotted as o's and *'s, respectively, in Figure 2.2. Observe the even symmetry of the magnitude and the odd symmetry of the phase about the central frequency .5.

It is common in the digital signal processing literature to interchange the left and right hand parts of the figure. This simply takes advantage of the periodicity of the DFT in f to replace the original frequencies $[.5, 1)$ by $[-.5, 0)$ and hence replace the frequency domain $[0, 1)$ by $[-.5, .5)$. Most software packages for computing Fourier transforms have a command to perform this frequency shift. In Matlab it is fftshift, in Image it is Blockshift. The exchange produces the alternative picture of Figure 2.3. The advantage of this form is that low frequencies are grouped together and the symmetry of the figure is more evident. In particular, all frequencies "near" 0 or dc are adjacent.

For this simple signal for which a closed form formula for the general definition of $G(k/N)$ could be found, the previous figure could have been generated simply by plugging in the 32 possible values of frequency into the formula and plotting the magnitude and phase. If we had not had such a

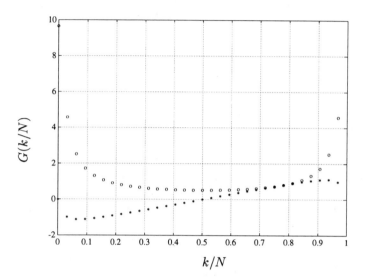

Figure 2.2: DFT of Finite Duration Geometric Signal: o=magnitude, *=phase

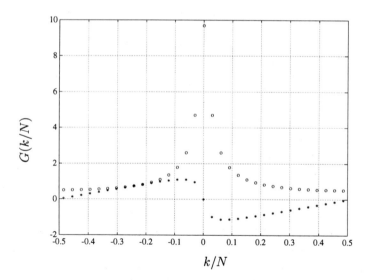

Figure 2.3: DFT of Finite Duration Geometric Signal with shifted frequency domain: o=magnitude, *=phase

useful formula, however, we would have in principle had to do a brute force calculation of the definition

$$G(\frac{k}{32}) = \sum_{n=0}^{31} g_n e^{-i2\pi \frac{k}{32} n}$$

(2.21)

for $k = 0, 1, \cdots, 31$. To produce the plot we would have had to perform roughly 32 multiply/adds for each frequency and hence $32^2 = 1024$ multiply adds to perform the entire transform. This would be the case, for example, if we wished to compute the Fourier transform of the randomly generated signal of Figure 1.18 or the sum of the sine and random noise of Figure 1.20. This would be too much computation if done with a pocket calculator, but it would be easy enough if programmed for a computer. It is easy to see, however, that even computers can get strained if we attempt such brute force computation for very large numbers of points, e.g., if N were several thousand (which is not unreasonable for many applications). In fact Figure 2.2 was produced using a program called a "fast Fourier transform" or FFT which uses the structure of the definition of the DFT to take many shortcuts in order to compute the values of (2.21) for all values of k of interest. We shall later describe in some detail how the FFT computes the DFT of an N-point signal in roughly $N \log N$ multiply/adds instead of the N^2 multiply/adds needed to calculate the DFT by brute force. As examples of the DFT for less structured signals, the FFT is used to compute the DFT of the signals of Figure 1.18 and Figure 1.20 and the results are displayed in Figures 2.4 and 2.5, respectively. Note how the DFT of the purely random signal itself appears to be quite random. In fact, a standard exercise of probability theory could be used to find the probability density functions of the magnitude and phase of the DFT of a sequence of independent Gaussian variables. The sine plus noise, however, exhibits important structure that is not evident in the signal itself: There are clearly two components in the magnitude of the DFT that stand out above the largely random behavior. To see what these components are, consider the DFT of the sinusoidal term alone, that is, the DFT of the signal depicted in Figure 1.3. This is shown in Figure 2.6. The DFT of the sinusoid alone shows the same two peaks in the magnitude at frequencies of 1/8 and 7/8 along with some smaller frequency components. The point is that these two peaks are still visible in the DFT of the signal plus noise, although the sinusoidal nature of the time signal is less visible. This example provides a simple demonstration of the ability of a Fourier transform to extract structure from a signal where such structure is not apparent in the time domain. The two peaks in the signal plus noise at 1/8 and 7/8 suggest that the noisy signal contains a discrete time sinusoid of period 8.

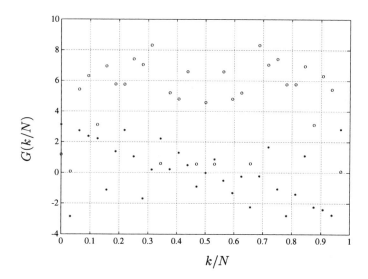

Figure 2.4: DFT of a Random Signal: o=magnitude, *=phase

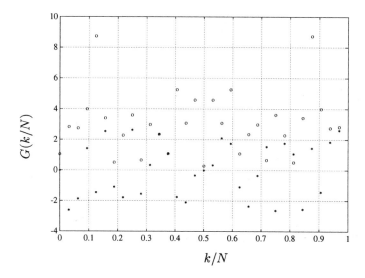

Figure 2.5: DFT of the Sum of a Random Signal and Sinusoid: o=magnitude, *=phase

George Tejera

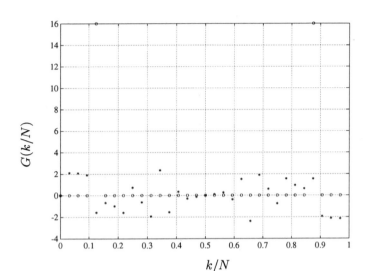

Figure 2.6: DFT of Finite Duration One-Sided Sinuosoid: o=magnitude, *=phase

Returning to the original example of a geometric signal, if $r = 1$ and hence the signal is a constant value of 1, then from (2.18) and (2.19) we have that

$$\{1; n = 0, 1, \ldots, N - 1\} \supset \{G_k = G(\frac{k}{N}) = N\delta_k;\ k = 0, 1, \cdots, N - 1\};$$
(2.22)

that is, the DFT of a unit constant is proportional to the Kronecker delta. In fact this is the *dual* of the previous result that the Kronecker delta time signal has a constant Fourier transform. We shall later see how all Fourier transform results provide dual results by reversing the roles of time and frequency.

If instead of the finite duration signal g, we wished to find the Fourier transform of the infinite duration signal $\bar{g}$ formed by zero filling, i.e., $\bar{g}_n = r^n$ for $n = 0, 1, \cdots, N - 1$ and $\bar{g}_n = 0$ otherwise, then, assuming $|r| \neq 1$, the Fourier transform relation would instead be

$$\{g_n;\ n \in \mathcal{Z}\} \supset \{\frac{1 - r^N e^{-i2\pi f N}}{1 - r e^{-i2\pi f}};\ f \in [-\frac{1}{2}, \frac{1}{2})\}$$
(2.23)

because of the different frequency domains (here $\mathcal{S}_{DTID}^{(2)}$).

Next consider the DTFT of the infinite duration signal

$$g_n = \begin{cases} r^n & n = 0, 1, \dots \\ 0 & n < 0, \end{cases} \tag{2.24}$$

where now we require that $|r| < 1$. We can write this signal more compactly using the unit step function as

$$g_n = r^n u_{-1}(n); \ n \in \mathcal{Z}. \tag{2.25}$$

Observe that the functional form of the time dependence is the same here as in the previous finite duration example; the difference is that now the functional form is valid for all integer times rather than just for a finite set. The DTFT is

$$G(f) = \sum_{n=0}^{\infty} r^n e^{-i2\pi fn},$$

which is given by the geometric progression formula as

$$G(f) = \frac{1}{1 - re^{-i2\pi f}}. \tag{2.26}$$

Note that the spectrum is again complex valued. Restricting frequencies to $\mathcal{S}_{DTID}$ we have the Fourier transform relation

$$\{r^n u_{-1}(n); \ n \in \mathcal{Z}\} \supset \left\{ \frac{1}{1 - re^{-i2\pi f}}; \ f \in [-\tfrac{1}{2}, \tfrac{1}{2}) \right\}. \tag{2.27}$$

This transform is the discrete time version of (2.15). Note the transforms do not clearly resemble each other as much as the original signals do.

As another example of a DTFT consider the two-sided discrete time box function

$$g_n = \square_N(n) \triangleq \begin{cases} 1 & \text{if } |n| \le N \\ 0 & n = \pm(N+1), \pm(N+2), \cdots \end{cases} \tag{2.28}$$

Application of the geometric progression formula and some algebra and trigonometric identities then yield

$$\begin{aligned} G(f) &= \sum_{n=-N}^{N} e^{-i2\pi fn} \\ &= \frac{\cos(2\pi fN) - \cos(2\pi f(N+1))}{1 - \cos(2\pi f)} \\ &= \frac{\sin(2\pi f(N + \tfrac{1}{2}))}{\sin(\pi f)}. \end{aligned} \tag{2.29}$$

Applying the appropriate frequency domain of definition yields the transform

$$\{\Box_N(n);\ n \in \mathcal{Z}\} \supset \{\frac{\sin(2\pi f(N + \frac{1}{2}))}{\sin(\pi f)};\ f \in [-\frac{1}{2}, \frac{1}{2})\}. \tag{2.30}$$

This spectrum for the case of Figure 1.6 is thus purely real and $(N = 5)$ is plotted in Figure 2.7. Note the resemblance to the sinc function. In

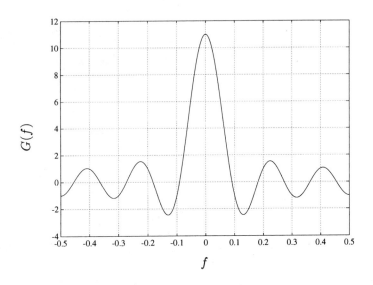

Figure 2.7: Transform of $\{\Box_5(n);\ n \in \mathcal{Z}\}$

fact, this function is often referred to as a "discrete time sinc" function because of the similarity of form (and because, as we shall see, the analogous continuous time Fourier transform of a box is indeed a sinc). The Fourier transform for this example is purely real.

One can also use the box signal to describe a finite duration constant signal $\{g_n = 1;\ n = -N, \ldots, N\}$. The functional form of (2.29) then simplifies for the frequencies $\{0, 1/(2N + 1), \cdots, 2N/(2N + 1)\}$ to $G(k/N) = (2N + 1)\delta(k)$. This provides a two-sided version of (2.22).

As an example of the finite duration CTFT, recall the continuous time ramp function of Figure 1.41. The solution to the integral defining the CTFT for all real frequencies is given by

$$\int_0^1 t e^{-i2\pi f t} dt = \begin{cases} \frac{1}{2} & \text{if } f = 0 \\ \frac{ie^{-i2\pi f}}{2\pi f} + \frac{e^{-i2\pi f} - 1}{(2\pi f)^2} & \text{if } f \neq 0 \end{cases}. \tag{2.31}$$

Note that the calculus would yield $G(f)$ if we considered g to be an infinite duration signal for which $g(t) = 0$ for t not in the interval $[0, 1)$. Restricting the result to the frequency domain of definition for finite duration signals yields the transform

$$\{g(t); t \in [0, 1)\} \supset \left\{ G(k) = \begin{cases} \frac{1}{2}, & k = 0 \\ \frac{i}{2\pi k} & k \in \mathcal{Z}, \, k \neq 0 \end{cases} \right\}. \qquad (2.32)$$

If we consider instead the zero filled extension $\bar{g}$ defined by $\bar{g}(t) = t$ for $t \in [0, 1)$ and $\bar{g}(t) = 0$ for $t \neq [0, 1)$, then the transform becomes

$$\{\bar{g}(t); t \in \mathcal{R}\} \supset \left\{ \begin{cases} \frac{1}{2} & \text{if } f = 0 \\ \frac{ie^{-i2\pi f}}{2\pi f} + \frac{e^{-i2\pi f} - 1}{(2\pi f)^2} & \text{if } f \in \mathcal{R}, \, f \neq 0. \end{cases} \right\}. \qquad (2.33)$$

As another simple example of a CTFT, consider the continuous time analog of the box of the previous DTFT example. This example can be considered finite or infinite duration.

Recall the definition of the rectangle function $\sqcap(t)$:

$$\sqcap(t) = \begin{cases} 1 & \text{if } |t| < \frac{1}{2} \\ \frac{1}{2} & \text{if } t = \pm\frac{1}{2} \\ 0 & \text{otherwise} \end{cases}. \qquad (2.34)$$

For the signal $g(t) = \sqcap(t)$ we have that

$$G(f) = \int_{-\infty}^{\infty} \sqcap(t) e^{-i2\pi ft} dt = \int_{-\frac{1}{2}}^{\frac{1}{2}} e^{-i2\pi ft} dt$$

$$= \left. \frac{e^{-i2\pi ft}}{-i2\pi f} \right|_{-\frac{1}{2}}^{\frac{1}{2}} = \frac{\sin(\pi f)}{\pi f} \overset{\Delta}{=} \text{sinc}(f), \qquad (2.35)$$

where we have recalled the definition of the sinc function. If we consider the infinite duration case, the Fourier transform is

$$\{\sqcap(t); \, t \in \mathcal{R}\} \supset \{\text{sinc}(f); \, f \in \mathcal{R}\}. \qquad (2.36)$$

The sinc function was shown in Figure 1.10. Although that was $\text{sinc}(t)$ and this is $\text{sinc}(f)$, obviously the name of the independent variable has nothing to do with the shape of the function.

Riemann integrals are not affected by changing the integrand by a finite amount at a finite number of values of its argument. As a result, the Fourier transforms of the two box functions $\square_{1/2}(t)$ and $\sqcap(t)$, which differ in value only at the two endpoints, are the same; hence, we cannot count on inverting the Fourier transform in an unambiguous fashion. This fact is important and merits emphasis:

> Different signals may have the same Fourier transform and hence the Fourier transform may not have a unique inverse. As in the box example, however, it turns out that two signals with the same Fourier transform must be the same for most values of their argument ("almost everywhere," to be precise).

More generally, given $T > 0$ the Fourier transform of $\Box_T(t)$ (and of $\sqcap(t/2T)$) is given by

$$
\begin{aligned}
G(f) &= \int_{-\infty}^{\infty} \Box_T(t) e^{-i2\pi ft}\, dt = \int_{-T}^{T} e^{-i2\pi ft}\, dt \\[2mm]
&= \left. \frac{e^{-i2\pi ft}}{-i2\pi f} \right|_{-T}^{T} = \frac{\sin(2\pi fT)}{\pi f} = 2T\operatorname{sinc}(2Tf), \qquad (2.37)
\end{aligned}
$$

and hence

$$
\{\Box_T(t);\ t \in \mathcal{R}\} \supset \{2T\operatorname{sinc}(2Tf);\ f \in \mathcal{R}\}. \qquad (2.38)
$$

This is another example of a real-valued spectrum. Note the different forms of the spectra of the discrete time box function of (2.29) and the continuous time transform of (2.37). We also remark in passing that the spectrum of (2.37) has the interesting property that its samples at frequencies of the form $k/2T$ are 0:

$$
G(\frac{k}{2T}) = 0;\ k = \pm 1, \pm 2, \cdots, \qquad (2.39)
$$

as can be seen in Figure 1.10. Thus the zeros of the sinc function are uniformly spaced.

We have completed the evaluation, both analytically and numerically, of a variety of Fourier transforms. Many of the transforms just developed will be seen repeatedly throughout the book. The reader is again referred to the Appendix where these and other Fourier transform relations are summarized.

2.3 Cousins of the Fourier Transform

The Cosine and Sine Transforms

Observe that using Euler's formulas we can rewrite the CTFT as

$$
\mathcal{F}_f(g) = G(f)
$$

$$= \int_{t \in T} g(t)[\cos(-2\pi ft) + i\sin(-2\pi ft)]dt$$

$$= \int_{t \in T} g(t)\cos(2\pi ft)dt - i\int_{t \in T} g(t)\sin(2\pi ft)dt$$

$$= C_f(g) - iS_f(g), \tag{2.40}$$

where $C_f(g)$ is called the *cosine transform* of $g(t)$ and $S_f(g)$ is called the *sine transform*. (There often may be a factor of 2 included in the definitions.) Observe that if the signal is real-valued, then the Fourier transform can be found by evaluating two real integrals. The cosine transform is particularly important in image processing where a variation of the two-dimensional discrete time cosine transform is called the discrete cosine transform or DCT in analogy to the DFT. Its properties and computation are studied in detail in Rao and Yip [27].

The Hartley Transform

If instead of combining the sine and cosine transforms as above to form the Fourier transform, one changes $-i$ to 1 and defines

$$\mathcal{H}_f(g) = \int_{t \in T} g(t)[\cos(2\pi ft) + \sin(2\pi ft)]\,dt$$

$$= C_f(g) + S_f(g) \tag{2.41}$$

then the resulting transform is called the *Hartley transform* and many of its properties and applications strongly resemble those of the Fourier transform since the Hartley transform can be considered as a simple variation on the Fourier transform.

An alternative way to express the Hartley transform is by defining the cas function

$$\text{cas}(x) \stackrel{\Delta}{=} \cos(x) + \sin(x) \tag{2.42}$$

and then write

$$\mathcal{H}_f(g) = \int_{t \in T} g(t)\,\text{cas}(2\pi ft)\,dt. \tag{2.43}$$

The Hartley transform has the advantage that it is real-valued for a real-valued signal, but it lacks some of the important theoretical properties that we shall find for Fourier transforms. In particular, it does not prove a decomposition of compex signals into a linear combination of eigenfunctions of linear time invariant systems. A thorough treatment of Hartley transforms can be found in R. Bracewell's *The Hartley Transform* [7].

z and Laplace Transforms

The Fourier transform is an exponential transform; that is, it is found by integrating or summing a signal multiplied by a complex exponential involving a frequency variable. So far only real values of frequency have been permitted, but an apparently more general class of transforms can be obtained if the frequency variable is allowed to be complex. The resulting generalizations provide transforms that will exist for some signals which do not have Fourier transforms and the generalized transforms are useful in some applications, especially those emphasizing one-sided signals and systems with initial conditions. The added generality comes at a cost, however; the generalized transforms do not exist for all values of their arguments (z for a z-transform and s for a Laplace transform) and as a result they can be much harder to invert than Fourier transforms and their properties can be more complicated to describe. Most problems solvable by the general transforms can also be solved using ordinary Fourier transforms. For completeness, however, we describe the two most important examples for engineering applications, the z transform for discrete time signals and the Laplace transform for continuous time signals, so that their connection to the Fourier transform can be seen.

Given a discrete time signal $g = \{g_n; n \in \mathcal{T}\}$, its z transform is defined by

$$G_Z(z) = \sum_{n \in \mathcal{T}} g_n z^{-n}, \qquad (2.44)$$

provided that the sum exists, i.e., that the sum converges to something finite. The region of z in the complex plane where the sum converges is called the *region of convergence* or *ROC*. When the sum is taken over a two-sided index set $\mathcal{T}$ such as the set $\mathcal{Z}$ of all integers or $\{-N, \ldots, -1, 0, 1, \ldots, N\}$, the transform is said to be *two-sided* or *bilateral*. If it is taken over the set of nonnegative integers or a set of the form $0, 1, \cdots, N$, it is said to be *one-sided* or *unilateral*. Both unilateral and bilateral transforms have their uses and their properties tend to be similar, but there are occasionally differences in details. We shall focus on the bilateral transform as it is usually the simplest. If $G(f)$ is the Fourier transform of g_n, then formally we have that

$$G(f) = G_Z(e^{i2\pi f});$$

that is, the Fourier transform is just the z transform evaluated at $e^{i2\pi f}$. For this reason texts treating the z-transform as the primary transform often use the z-transform notation $G(e^{i2\pi f})$ to denote the DTFT, but we prefer the simpler Fourier notation of $G(f)$. If we restrict f to be real, then the Fourier transform is just the z-transform evaluated on the unit circle in the

z plane. If, however, we permit f to be complex, then the two are equally general and are simply notational variants of one another.

Why let f be complex or, equivalently, let z be an arbitrary complex number? Provided $|z| \neq 0$, we can write z in magnitude-phase notation as $z = r^{-1}e^{i\theta}$. Then the z transform becomes

$$G_Z(z) = \sum_{n \in \mathcal{T}} (g_n r^n) e^{-in\theta}.$$

This can be interpreted as a Fourier transform of the new signal $g_n r^n$ and this transform might exist even if that of the original signal g_n does not since the r^n can serve as a damping factor. Put another way, a Fourier transform was said to exist if $G(f)$ made sense for *all* f, but the z transform does not need to exist for all z to be useful, only within its ROC. In fact, the existence of the Fourier transform of an infinite duration signal is equivalent to the ROC of its z-transform containing the unit circle $\{z : |z| = 1\}$, which is the region of all $z = e^{-i2\pi f}$ for real f.

As an example, consider the signal $g_n = u_{-1}(n)$ for $n \in \mathcal{Z}$. Then the ordinary Fourier transform of g does not exist for all f, e.g., it blows up for $f = 0$. Choosing $|r| < 1$, however, yields a modified signal $g_n r^n$ which, as we have seen, has a transform. In summary, the z transform will exist for some region of possible values of z even though the Fourier transform may not. The two theories, however, are obviously intimately related.

The Laplace transform plays the same role for continuous time waveforms. The Laplace transform of a continuous time signal g is defined by

$$G_L(s) = \int_{t \in \mathcal{T}} g(t) e^{-st} \, dt. \tag{2.45}$$

As for the z transform, for the case where $\mathcal{T}$ is the real line, one can define a bilateral and a unilateral transform, the latter being the bilateral transform of $g(t)u_{-1}(t)$. As before, we focus on the bilateral case.

If we replace $i2\pi f$ in the Fourier transform by s we get the Laplace transform, although the Laplace transform is more general since s can be complex instead of purely imaginary. Letting the f in a Fourier transform take on complex values is equivalent in generality. Once again, the Laplace transform can exist more generally since, with proper choice of s, the original signal is modified in an exponentially decreasing fashion before taking a Fourier transform.

The primary advantage of Laplace and z-transforms over Fourier transforms in engineering applications is that the one-sided transforms provide a natural means of incorporating initial conditions into linear systems analysis. Even in such applications, however, two-sided infinite duration Fourier transforms can be used if the initial conditions are incorporated using delta functions.

It is natural to inquire why these two variations of the Fourier transform with the same general goal are accomplished in somewhat different ways. In fact, one could define a Laplace transform for discrete time signals by replacing the integral by a sum and one could define a z transform for continuous time signals by replacing the sum by an integral. In fact, the latter transform is called a *Mellin transform* and it is used in the mathematical theory of Dirichlet series. The differences of notation and approach for what are clearly closely related transforms is attributable to history; they arose in different fields and were developed independently.

2.4   Multidimensional Transforms

The basic definition of a Fourier transform extends to two-dimensional and higher dimensional transforms. Once again the transform is different depending on whether the index set $\mathcal{T}$ is discrete or continuous. As usual a signal is defined as a function $g = \{g(t); t \in \mathcal{T}\}$. In the multidimensional case $\mathcal{T}$ has dimension K. The case $K = 2$ is the most common because of its importance in image processing, but three-dimensional transforms are useful in X-ray diffraction problems and higher dimensions are commonly used in probabilistic systems analysis when characterizing multidimensional probability distributions. To ease the distinction of the multidimensional case from the one-dimensional case, in this section we use boldface for the index to emphasize that it is a column vector, i.e., $\mathbf{t} = (t_1, t_2, \ldots, t_K)^t$ and the signal is now $g = \{g(\mathbf{t}); \mathbf{t} \in \mathcal{T}\}$. When transforming a signal with a K-dimensional parameter we also require that the frequency variable be a K-dimensional parameter, e.g., $\mathbf{f} = (f_1, f_2, \ldots, f_K)^t$.

The Fourier transforms of the signal g are defined exactly as in the one-dimensional case with two exceptions: the one-dimensional product ft in the exponential $e^{-i2\pi ft}$ is replaced by an *inner product* or *dot product* or *scalar product*

$$\mathbf{t} \cdot \mathbf{f} = \mathbf{t}^t \mathbf{f} = \sum_{k=1}^{K} t_k f_k$$

and the one dimensional integral or sum over t is replaced by the K-dimensional integral or sum over $\mathbf{t}$. The inner product is also denoted by $< \mathbf{t}, \mathbf{f} >$. Thus in the continuous parameter case we have that

$$G(\mathbf{f}) = \int_{\mathcal{T}} g(\mathbf{t}) e^{-i2\pi \mathbf{t} \cdot \mathbf{f}} \, d\mathbf{t} \qquad (2.46)$$

which in the two-dimensional case becomes

$$G(f_X, f_Y) = \int \int g(x, y) e^{-i2\pi(xf_X + yf_Y)} \, dx \, dy; \quad f_X, f_Y \in \mathcal{R}. \quad (2.47)$$

The limits of integration depend on the index set $\mathcal{T}$; they can be finite or infinite. Likewise the frequency domain is chosen according to the nature of $\mathcal{T}$.

In the discrete parameter case the transform is the same with integrals replaced by sums:

$$G(\mathbf{f}) = \sum_{\mathbf{t} \in \mathcal{T}} g(\mathbf{t}) e^{-i2\pi \mathbf{t} \cdot \mathbf{f}} \quad (2.48)$$

which in the two dimensional case becomes

$$G(f_X, f_Y) = \sum_x \sum_y g(x, y) e^{-i2\pi(xf_X + yf_Y)}. \quad (2.49)$$

Multidimensional Fourier transforms can be expressed as a sequence of ordinary Fourier transforms by capitalizing on the property that the exponential of a sum is a product of exponentials. This decomposition can be useful when evaluating Fourier transforms and when deriving their properties. For example, consider the discrete time 2D Fourier transform and observe that

$$
\begin{aligned}
G(f_X, f_Y) &= \sum_x \sum_y g(x, y) e^{-i2\pi(xf_X + yf_Y)} \\
&= \sum_x e^{-i2\pi xf_X} \sum_y g(x, y) e^{-i2\pi yf_Y} \\
&= \sum_x e^{-i2\pi xf_X} G_x(f_Y) \quad (2.50)
\end{aligned}
$$

where

$$G_x(f_Y) \stackrel{\Delta}{=} \sum_y g(x, y) e^{-i2\pi yf_Y} \quad (2.51)$$

is the ordinary one-dimensional Fourier transform with respect to y of the signal $\{g(x, y); \ y \in \mathcal{T}_Y\}$. Thus one can find the 2D transform of a signal by first performing a collection of 1D transforms G_x for every value of x and then transforming G_x with respect to x. This sequential computation has an interesting interpretation in the special case of a finite-duration 2D transform of a signal such as $\{g(k, n); \ k \in \mathcal{Z}_N, \ n \in \mathcal{Z}_N\}$. Here the signal can be thought of as a matrix describing a 2D sampled image raster. The

2D Fourier transform can then be written as

$$G(f_X, f_Y) = \sum_k e^{-i2\pi f_X k} \left(\sum_n e^{-i2\pi f_Y n} g(k, n) \right);$$

$$f_X, f_Y \in \{\frac{k}{N}; \ k \in \mathcal{Z}_N\},$$

where the term in parentheses can be thought of as the 1D transform of the kth column of the matrix g. Thus the 2D transform is found by first transforming the columns of the matrix g to form a new matrix. Taking the 1D transforms of the rows of this matrix then gives the final transform. Continuous parameter 2D Fourier transforms can be similarly found as a sequence of 1D transforms.

Separable 2D Signals

The evaluation of Fourier transforms of 2D signals is much simplified in the special case of 2D signals that are separable in rectangular coordinates; i.e., if

$$g(x, y) = g_X(x) g_Y(y),$$

then the computation of 2D Fourier transforms becomes particularly simple. For example, in the continuous time case

$$
\begin{aligned}
G(f_X, f_Y) &= \int_{-\infty}^{\infty} \int_{-\infty}^{\infty} g_X(x) g_Y(y) e^{-i2\pi(f_X x + f_Y y)} \, dx \, dy \\
&= \left(\int_{-\infty}^{\infty} g_X(x) e^{-i2\pi f_X x} \, dx \right) \left(\int_{-\infty}^{\infty} g_Y(y) e^{-i2\pi f_Y y} \, dy \right) \\
&= G_X(f_X) G_Y(f_Y),
\end{aligned}
$$

the product of two 1-D transforms. In effect, separability in the space domain implies a corresponding separability in the frequency domain. As a simple example of a 2D Fourier transform, consider the continuous parameter 2D box function of (1.17): $g(x, y) = \Box_T(x) \Box_T(y)$ for all real x and y. The separability of the signal makes its evaluation easy:

$$
\begin{aligned}
G(f_X, f_Y) &= \int \int g(x, y) e^{-i2\pi(x f_X + y f_Y)} \, dx \, dy \\
&= \left(\int \Box_T(x) e^{-i2\pi x f_X} \, dx \right) \left(\int \Box_T(y) e^{-i2\pi y f_Y} \, dy \right) \\
&= \mathcal{F}_{f_X}(\{\Box_T(x); \ x \in \mathcal{R}\}) \mathcal{F}_{f_Y}(\{\Box_T(x); \ x \in \mathcal{R}\}) \\
&= (2T \operatorname{sinc}(2T f_X))(2T \operatorname{sinc}(2T f_Y)), \qquad\qquad (2.52)
\end{aligned}
$$

where we have used (2.38) for the individual 1D transforms.

A similar simplification results for signals that are separable in polar coordinates. Consider the 2D disc of (1.22) with a radius of 1: $g(x, y) = \square_1(\sqrt{x^2 + y^2})$ for all real x, y. In other words, the signal is circularly symmetric and in polar coordinates we have that $g(r, \theta) = g_R(r)$. To form the Fourier transform we convert into polar coordinates

$$
\begin{aligned}
x &= r \cos \theta \\
y &= r \sin \theta \\
dx\, dy &= r\, dr\, d\theta.
\end{aligned}
$$

We also transform the frequency variables into polar coordinates as

$$
\begin{aligned}
f_X &= \rho \cos \phi \\
f_Y &= \rho \sin \phi
\end{aligned}
$$

Substituting into the Fourier transform:

$$
\begin{aligned}
G(\rho, \phi) &= \int_0^{2\pi} d\theta \int_0^{\infty} dr\, r g_R(r) e^{-i2\pi[r\rho \cos \theta \cos \phi + r\rho \sin \theta \sin \phi]} \\
&= \int_0^{\infty} dr\, r g_R(r) \int_0^{2\pi} d\theta e^{-i2\pi r\rho \cos(\theta - \phi)}.
\end{aligned}
$$

To simplify this integral we use an identity for the zero-order Bessel function of the first kind:

$$
J_0(a) = \frac{1}{2\pi} \int_0^{2\pi} e^{-ia \cos(\theta - \phi)}\, d\theta. \qquad (2.53)
$$

Thus

$$
G(\rho, \phi) = 2\pi \int_0^{\infty} r g_R(r) J_0(2\pi r \rho)\, dr = G(\rho).
$$

Thus the Fourier transform of a circularly symmetric function is itself circularly symmetric. The above formula is called the *zero-order Hankel transform* or the *Fourier-Bessel transform* of the signal.

For our special case of $g_R(r) = \square_1(r)$:

$$
\begin{aligned}
G(\rho) &= 2\pi \int_0^{\infty} r \square_1(r) J_0(2\pi r \rho)\, dr \\
&= 2\pi \int_0^{1} r J_0(2\pi r \rho)\, dr
\end{aligned}
$$

Make the change of variables $r' = 2\pi r\rho$ so that $r = r'/2\pi\rho$ and $dr = dr'/2\pi\rho$. Then

$$G(\rho) = \frac{1}{2\pi\rho^2} \int_0^{2\pi\rho} r' J_0(r') dr'.$$

It is a property of Bessel functions that

$$\int_0^x \zeta J_0(\zeta) d\zeta = x J_1(x),$$

where $J_1(x)$ is a first-order Bessel function. Thus

$$G(\rho) = \frac{J_1(2\pi\rho)}{\rho} = \pi \left[2\frac{J_1(2\pi\rho)}{2\pi\rho} \right].$$

The bracketed term is sometimes referred to as the Bessinc function and it is related to the jinc function by a change of variables as $G(\rho) = 4\text{jinc}2\rho$.

The Discrete Cosine Transform (DCT)

A variation on the 2D discrete Fourier transform is the discrete cosine transform or DCT that forms the heart of the international JPEG (Joint Photographic Experts Group) standard for still picture compression [26]. (The "Joint" emphasizes the fact that the standard was developed by cooperative effort of two international standards organizations, ISO and ITU-T, formerly CCITT.) Consider a finite-domain discrete parameter 2D signal $g = \{g(k, j); k = 0, 1, \ldots, N - 1, j = 0, 1, \ldots, N - 1\}$. Typically the $g(k, j)$ represent intensity values at picture elements (pixels) in a scanned and sampled image. Often these values are chosen from a set of $2^8 = 256$ possible levels of gray. The DCT $G = \{G(l, m); l = 0, 1, \ldots, N - 1, m = 0, 1, \ldots, N - 1\}$ is defined by

$$G(l, m) = \frac{1}{4} C(l)C(m) \sum_{k=0}^{N-1} \sum_{j=0}^{N-1} g(k, j) \cos \frac{(2k+1)l\pi}{2N} \cos \frac{(2j+1)m\pi}{2N},$$

$$(2.54)$$

where

$$C(n) = \begin{cases} \frac{1}{\sqrt{2}} & \text{if } n = 0 \\ 1 & \text{otherwise} \end{cases}$$

As with the Hartley transform, the prime advantage of the DCT over the ordinary DFT is that it is purely real. An apparent additional advantage is that an image of N^2 samples yields a DCT also having N^2 real values while an FFT yields twice that number of real values; i.e., N^2 magnitudes and N^2 phases. It will turn out, however, that not all $2N^2$ values of the FFT are necessary for reconstruction because of symmetry properties.

2.5 ⋆ The DFT Approximation to the CTFT

We have now surveyed a variety of basic signal examples with Fourier transforms that can be evaluated using ordinary calculus and algebra. Analytical tools are primarily useful when the signals are elementary functions or simple combinations of linear functions. Numerical techniques will be needed for random signals, physically measured signals, unstructured signals, or signals which are simply not amenable to closed form integration or summation.

The discrete Fourier transform (DFT) plays a role in Fourier analysis that is much broader than the class of discrete time, finite duration signals for which it is defined. When other signal types such as infinite duration continuous time signals are being considered, one may still wish to compute Fourier transforms numerically. If the Fourier transforms are to be computed or manipulated by a digital computer or digital signal processing (DSP) system, then the signal must be sampled to form a discrete time signal and only a finite number of samples can be used. Thus the original continuous time infinite duration signal is approximated by a finite duration discrete time signal and hence the numerical evaluation of a Fourier transform is in fact a DFT, regardless of the original signal type! For this reason it is of interest to explore the complexity of evaluating a DFT and to find low complexity algorithms for finding DFTs.

Suppose that $g = \{g(t); \ t \in \mathcal{R}\}$ is an infinite duration continuous time signal and we wish to compute approximately its Fourier transform. Since a digital computer can only use a finite number of samples and computes integrals by forming Riemann sums, a natural approximation can be found by taking a large number $N = 2M + 1$ of samples

$$\{g(-MT), \cdots, g(0), \cdots, g(MT)\}$$

with a small sampling period T so that the total time windowed NT is large. We then form a Riemann sum approximation to the integral as

$$
\begin{aligned}
G(f) &= \int_{-\infty}^{\infty} g(t)e^{-i2\pi ft}\, dt \\
&\approx \sum_{n=-M}^{M} g(nT)e^{-i2\pi fnT}T;
\end{aligned}
\tag{2.55}
$$

that is, we make the Riemann approximation that $dt \approx T$ for T small enough. This has the general form of a DFT. We can put it into the exact form of the one-sided DFT emphasized earlier if we define the discrete time signal

$$\hat{g}_n = g(nT - MT); \ n = 0, \cdots, N - 1.$$

(By construction $\hat{g}_0 = g(-MT)$ and $\hat{g}_{N-1} = g(MT)$.) Then the Riemann sum approximation yields

$$
\begin{aligned}
G(f) &\approx \sum_{n=-M}^{M} g(nT)e^{-i2\pi fnT}T \\
&= T\sum_{n=0}^{N-1} g(nT - MT)e^{-i2\pi f(n-M)T} \\
&= e^{i2\pi fMT}T\sum_{n=0}^{N-1} \hat{g}_n e^{-i2\pi fnT} \\
&= e^{i2\pi fMT}T\hat{G}(fT), \qquad\qquad (2.56)
\end{aligned}
$$

where $\hat{G}$ is the DFT of $\hat{g}$. The usual frequency domain of definition for the DFT is the set $\mathcal{S}_{DTFD}^{(1)} = \{0, 1/N, \cdots, (N-1)/N\}$ which means that (2.56) provides an approximation to $G(f)$ for values of f for which $fT \in \{0, 1/N, \cdots, M/N, (M+1)/N, \cdots, (N-1)/N = 2M/N\}$, that is, for $f \in \{0, 1/NT, \cdots, (N-1)/NT\}$. It is useful, however, to approximate $G(f)$ not just for a collection of positive frequencies, but for a collection of positive and negative frequencies that increasingly fills the real line as T becomes small and NT becomes large. Recalling that the DFT is periodic in f with period 1, we can add or subtract 1 to any frequency without affecting the value of the DFT. Hence we can also take the frequency domain to be

$$
\mathcal{S}_{DTFD}^{(2)} = \{\frac{M+1}{N} - 1 = -\frac{M}{N}, \cdots, (\frac{N-1}{N}) - 1 = -\frac{1}{N}, 0, \frac{1}{N}, \cdots, \frac{M}{N}\}.
$$
$$(2.57)$$

This can be viewed as performing the frequency shift operation on the DFT.

Combining these facts we have the following approximation for the infinite duration CTFT based on the DFT:

$$
G(\frac{k}{NT}) \approx e^{i2\pi kM/N}T\hat{G}(\frac{k}{N}) = e^{i2\pi kM/N}T\hat{G}_k; \; k = -M, \cdots, 0, \cdots, M,
$$
$$(2.58)$$

which provides an approximation for a large discrete set of frequencies that becomes increasingly dense in the real line as T shrinks and NT grows. The approximation is given in terms of the DFT scaled by a complex exponential; i.e., a phase term with unit magnitude, and by the sampling period.

The above argument makes the point that the DFT is useful more generally than in its obvious environment of discrete time finite duration signals. It can be used to numerically evaluate the Fourier transform of continuous time signals by approximating the integrals by Riemann sums.

2.6 The Fast Fourier Transform

We consider first the problem of computing the DFT in the most straightforward and least sophisticated manner, paying particular attention to the number of computational operations required. We then take advantage of the structure of a DFT to significantly reduce the number of computations required.

Recall that given a discrete time finite duration signal $g = \{g_n; n = 0, 1, \ldots, N - 1\}$, the DFT is defined by

$$G(\frac{m}{N}) = \sum_{n=0}^{N-1} g_n e^{-i2\frac{\pi}{N}nm}; \quad m = 0, 1, \ldots, N - 1. \tag{2.59}$$

Since g is in general complex-valued, computation of a single spectral coefficient $G(m/N)$ requires N complex multiplications and $N - 1$ complex additions. Note that a complex multiply has the form

$$(a + ib)(c + id) = (ac - bd) + i(bc + ad),$$

which consists of four real multiplies and two real adds, and a complex addition has the form

$$(a + ib) + (c + id) = (a + c) + i(b + d),$$

which consists of two real adds. As an aside, we mention that it is also possible to multiply two complex numbers with three real multiplies and five real additions. To see this, form $A = (a + b)(c - d)$, which takes two additions and one multiply to give $A = ac + bc - ad - bd$, $B = bc$ (one multiply), and $C = bd$ (one multiply). Then the real part of the product is found as $A - B + C = ac - bd$ (two additions), and the imaginary part is given by $B + C = bc + ad$ (one addition). The total number of operations is then three multiplies and five additions, as claimed.

This method may be useful when the time required for a real multiplication is greater than three times the time required for a real addition.

Complexity can be measured by counting the number of operations, but the reader should keep in mind that the true cost or complexity of an algorithm will depend on the hardware used to implement the arithmetic operations. Depending on the hardware, a multiplication may be vastly more complicated than an addition or it may be comparable in complexity. As a compromise, we will measure complexity by the number of complex-multiply-and-adds, where each such operation requires four real multiplies and four real adds. The complexity of straightforward computation of a single DFT coefficient is then approximately N complex-multiply-and-adds.

Computation of all N DFT coefficients $G(0), \ldots, G((N-1)/N)$ then requires a total of N^2 complex-multiply-and-adds, or $4N^2$ real multiplies and $4N^2$ real adds.

If k represents the time required for one complex-multiply-and-add, then the computation time T_d required for this "direct" method of computing a DFT is $T_d = kN^2$. An approach requiring computation proportional to $N \log_2 N$ instead of N^2 was popularized by Cooley and Tukey in 1965 and dubbed the *fast Fourier transform* or *FFT* [13]. The basic idea of the algorithm had in fact been developed by Gauss and considered earlier by other authors, but Cooley and Tukey are responsible for introducing the algorithm into common use. The reduction from N^2 to $N \log_2 N$ is significant if N is large. These numbers do not quite translate into proportional computation time because they do not include the non-arithmetic operations of shuffling data to and from memory.

The Basic Principle of the FFT Algorithm

There are several forms of the FFT algorithm. We shall focus on one, known as "decimation in time," a name which we now explain. A discrete time signal $g = \{g_n; \ n \in \mathcal{T}\}$ can be used to form a new discrete time signal by *downsampling* or *subsampling* in much the same way that a continuous time signal could produce a discrete time signal by sampling. Downsampling means that we form a new process by taking a collection of regularly spaced samples from the original process. For example, if M is an integer, then downsampling g by M produces a signal $g^{(M)} = \{g_n^{(M)}; \ n \in \mathcal{T}^{(M)}\}$ defined by

$$g_n^{(M)} = g_{nM} \qquad\qquad (2.60)$$

for all n such that $nM \in \mathcal{T}$. Thus $g_n^{(M)}$ is formed by taking every Mth sample of g_n. This new signal is called a *downsampled* version of the original signal. Downsampling is also called *decimation* after the Roman army practice of decimating legions with poor performance (by killing every tenth soldier). We will try to avoid this latter nomenclature as it leads to silly statements like "decimating a signal by a factor of 3" which is about as sensible as saying "halving a loaf into three parts." Furthermore, it is an incorrect use of the term since the decimated legion referred to the survivors and hence to the 90% of the soldiers who remained, not to the every tenth soldier who was killed. Unfortunately, however, the use of the term is so common that we will need to use it on occasion to relate our discussion to the existing literature.

We change notation somewhat in this section in order to facilitate the introduction of several new sequences that arise and in order to avoid repeating redundant parameters. Specifically, let $g(n)$; $n = 0, 1, \ldots, N - 1$, denote the signal and $G(m)$; $m = 0, 1, \ldots, N - 1$ denote the DFT coefficients. Thus we have replaced the subscripts by the parenthetical index. Note that $G(m)$ is not the Fourier transform of the sequence in the strict sense because m is not a frequency variable, $G(m)$ is the value of the spectrum at the frequency m/N. In the notation used through most of the book, $G(m)$ would be G_m or $G(m/N)$, but for the current manipulations it is not worth the bother of constantly writing m/N.

The problem now is to compute

$$G(m) = \sum_{n=0}^{N-1} g(n) e^{-i\frac{2\pi}{N}nm} = \sum_{n=0}^{N-1} g(n) W^{nm}, \qquad (2.61)$$

where $W = e^{-i\frac{2\pi}{N}}$, and $m = 0, 1, \ldots, N - 1$. We assume that N is a power of two — the algorithm is most efficient in this case. Our current example is for $N = 8$.

Step 1: Downsample $g(n)$ using a sampling period of 2; that is, construct two new signals $g_0(n)$ and $g_1(n)$ by

$$g_0(n) \stackrel{\Delta}{=} g(2n); \; n = 0, 1, \ldots, \frac{N}{2} - 1 \qquad (2.62)$$

$$g_1(n) \stackrel{\Delta}{=} g(2n + 1); \; n = 0, 1, \ldots, \frac{N}{2} - 1. \qquad (2.63)$$

For an 8-point $g(n)$, the downsampled signals $g_0(n)$ and $g_1(n)$ have four points as in Figure 2.8.

The direct Fourier transforms of the downsampled signals g_0 and g_1, say G_0 and G_1, can be computed as

$$G_0(m) = \sum_{n=0}^{\frac{N}{2}-1} g_0(n) e^{-i\frac{2\pi}{N/2}mn} = \sum_{n=0}^{\frac{N}{2}-1} g(2n) W^{2mn} \qquad (2.64)$$

$$G_1(m) = \sum_{n=0}^{\frac{N}{2}-1} g_1(n) e^{-i\frac{2\pi}{N/2}mn} = \sum_{n=0}^{\frac{N}{2}-1} g(2n + 1) W^{2mn}, \qquad (2.65)$$

where $m = 0, 1, \ldots, \frac{N}{2} - 1$. As we have often done before, we now observe that the Fourier sums above can be evaluated for all integers m and that the sums are periodic in m, the period here being $N/2$. Rather than formally define the periodic extensions of G_0 and G_1 and cluttering the notation further (in the past we put tildes over the function being extended), here we just consider the above sums to define G_0 and G_1 for *all* m and keep in mind that the functions are periodic.

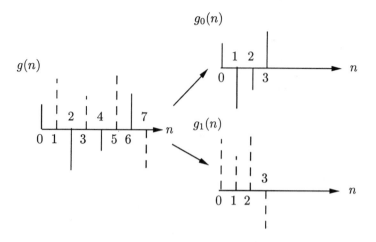

Figure 2.8: Downsampled Signals

Now relate G_0 and G_1 to G: For $m = 0, 1, \ldots, N-1$

$$
\begin{aligned}
G(m) &= \sum_{n=0}^{N-1} g(n) e^{-i\frac{2\pi}{N} mn} \\
&= \sum_{n=0}^{\frac{N}{2}-1} g(2n) e^{-i\frac{2\pi}{N} m(2n)} + \sum_{n=0}^{\frac{N}{2}-1} g(2n+1) e^{-i\frac{2\pi}{N} m(2n+1)} \\
&= G_0(m) + G_1(m) e^{-i\frac{2\pi}{N} m} \\
&= G_0(m) + G_1(m) W^m.
\end{aligned}
\tag{2.66}
$$

Note that this equation makes sense because we extended the definition of $G_0(m)$ and $G_1(m)$ from $Z_{\frac{N}{2}-1}$ to Z_N.

We have thus developed the scheme for computing $G(m)$ given the smaller sample DFTs G_0 and G_1 as depicted in Figure 2.9 for the case of $N = 8$. In the figure, arrows are labeled by their gain, with unlabeled arrows having a gain of 1. When two arrow heads merge, the signal at that point is the sum of the signals entering through the arrows. The connection pattern in the figure is called a *butterfly pattern*.

The total number of complex-multiply-and-adds in this case is now

$$
\underbrace{2(\frac{N}{2})^2}_{\text{Two } N' = 4 \text{ DFTs}} + \underbrace{N}_{\text{Combining Step}} .
$$

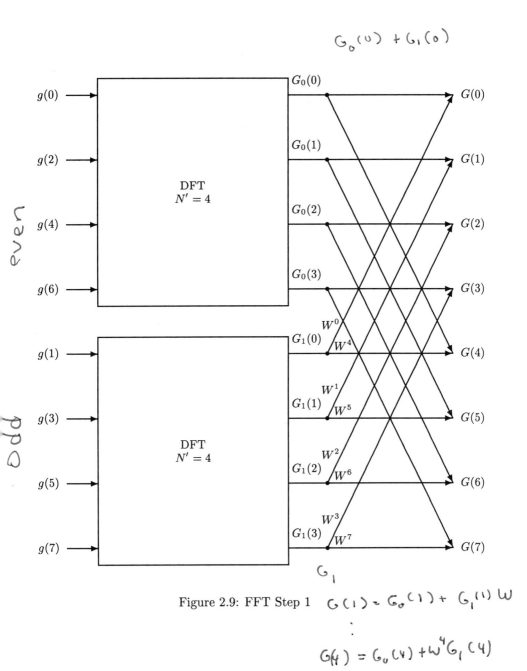

Figure 2.9: FFT Step 1

$$G_0(0) + G_1(0)$$

$$G(1) = G_0(1) + G_1(1) W$$

$$\vdots$$

$$G(4) = G_0(4) + W^4 G_1(4)$$
$$= G_0(0) - G_7(0)$$

If N is large, then this is approximately $N^2/2$ and the decomposition of the DFT into two smaller DFTs plus a combination step has roughly halved the computation.

Step 2: We reduce the computation even further by using the same trick to compute the DFTs G_0 and G_1, that is, by further downsampling the downsampled signals and repeating the procedure described so far. This can be continued until we are transforming individual input signal symbols. We now define

$$
\begin{aligned}
g_{00}(n) &= g_0(2n) = g(4n) \\
g_{01}(n) &= g_0(2n+1) = g(4n+2) \\
g_{10}(n) &= g_1(2n) = g(4n+1) \\
g_{11}(n) &= g_1(2n+1) = g(4n+3)
\end{aligned}
$$

for $n = 0, 1, \ldots, \frac{N}{4} - 1$. In the $N = 8$ example, each of these signals has only two samples. It can be shown in a straightforward manner that

$$
G_0(m) = G_{00}(m) + G_{01}(m)W^{2m}; \ m = 0, 1, \ldots, \frac{N}{2} - 1 \quad (2.67)
$$

$$
G_1(m) = G_{10}(m) + G_{11}(m)W^{2m}; \ m = 0, 1, \ldots, \frac{N}{2} - 1. \quad (2.68)
$$

This implies that we can expand on the left side of the previous flow graph to get the flow graph shown in Figure 2.10. Recall that any unlabeled branch has a gain of 1. We preserve the branch labels of W^0 to highlight the structure of the algorithm.

The number of computations now required is

$$
\underbrace{4(\frac{N}{4})^2}_{4\ N'' = 2\ \text{DFTs}} + \underbrace{N}_{\text{This Step Combination}} + \underbrace{N}_{\text{Previous Step Combination}}
$$

If N is large, the required computation is on the order of $N^2/4$, again cutting the computation by a factor of two.

Step 3: We continue in this manner, further downsampling the sequence in order to compute DFTs. In our example with $N = 8$, this third step is the final step since downsampling a signal with two samples yields a signal with only a single sample:

$$
\begin{aligned}
g_{000}(0) &= g(0) & g_{100}(0) &= g(1) \\
g_{001}(0) &= g(4) & g_{101}(0) &= g(5) \\
g_{010}(0) &= g(2) & g_{110}(0) &= g(3) \\
g_{011}(0) &= g(6) & g_{111}(0) &= g(7).
\end{aligned} \quad (2.69)
$$

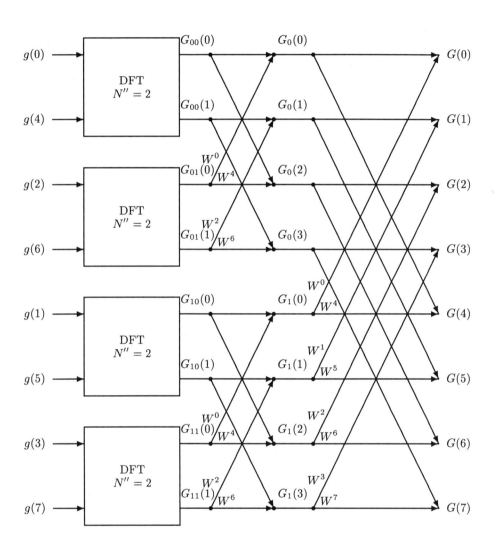

Figure 2.10: FFT Step 2

The DFT of a single sample signal is trivial to find, i.e.,

$$G_{000}(0) = g_{000}(0)W^0 = g(0) \qquad G_{100}(0) = g_{100}(0)W^0 = g(1)$$
$$G_{001}(0) = g_{001}(0)W^0 = g(4) \qquad G_{101}(0) = g_{101}(0)W^0 = g(5)$$
$$G_{010}(0) = g_{010}(0)W^0 = g(2) \qquad G_{110}(0) = g_{110}(0)W^0 = g(3)$$
$$G_{011}(0) = g_{011}(0)W^0 = g(6) \qquad G_{111}(0) = g_{111}(0)W^0 = g(7).$$

$$(2.70)$$

In other words, the DFTs of the one-point signals are given by the signals themselves. We can now work backwards to find G_{00}, G_{01}, G_{10}, and G_{11} as follows:

$$G_{00}(m) = g(0) + g(4)W^{4m}; \, m = 0, 1$$
$$G_{01}(m) = g(2) + g(6)W^{4m}; \, m = 0, 1$$
$$G_{10}(m) = g(1) + g(5)W^{4m}; \, m = 0, 1$$
$$G_{11}(m) = g(3) + g(7)W^{4m}; \, m = 0, 1.$$

The complete flow chart is shown in Fig. 2.11.

Either by counting arrows and junctions or studying the operations required in each step, it is seen that each stage involves N multiplications (not counting N of the trivial multiplications by 1) and N additions to combine the signal. We assumed that $N = 2^R$ is a power of two and there are $R = \log_2 N$ stages. Thus the overall complexity (as measured by multiply/adds) is

$$\boxed{N \log_2 N \text{ complex-multiply-and-adds.}}$$

This is the generally accepted computational complexity of the FFT. There are, however, further tricks and variations that can yield lower complexity. For example, of the $N \log_2 N$ multiplies, $\frac{N}{2} \log_2 N$ can be eliminated by noting that $W^4 = -W^0$, $W^5 = -W^1$, $W^6 = -W^2$, and $W^7 = -W^3$, allowing half the multiplies to be replaced by inverters as shown in Fig. 2.12.

The astute reader will have observed a connection between the binary vector subscripts of the final single sample signals (and the corresponding trivial DFTs) with the corresponding sample of the original signal in Eq. 2.69. If the index of the original signal is represented in binary and then reversed, one gets the subscript of the single sample final downsampled sequence. For example, writing the argument 3 of $g(3)$ in binary yields 011 which yields 110 when the bits are reversed so that $g(3) = g_{110}(0)$.

The reduction of computation from roughly N^2 to $N \log_2 N$ may not seem all that significant at first glance. A simple example shows that it is

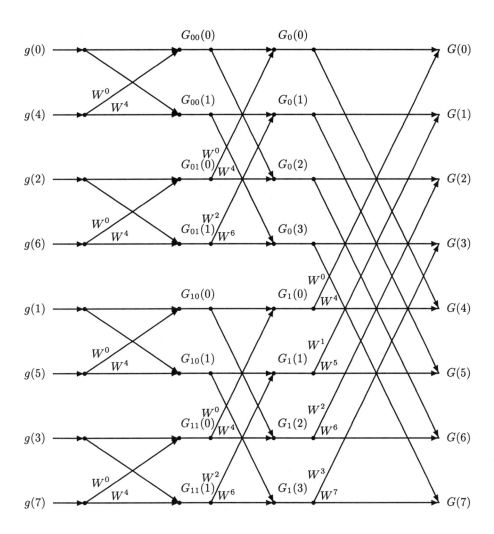

Figure 2.11: FFT Algorithm Flow Chart: N=8

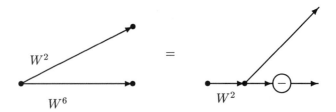

Figure 2.12: Inversion

indeed an enormous help in ordinary tasks. Consider the examples of the
Mona Lisa and MR images of Figures 1.31 and 1.32. These images consist
of 256×256 square arrays of pixels, for a total of $N = 2^{16} = 65536$ pixels.
A brute force evaluation of the DFT by evaluating the double sums of
exponentials would take on the order of N^2 operations and the FFT would
take on the order of $N \log_2 N$ operations. Hence the brute force method
requires

$$\frac{N^2}{N \log_2 N} = \frac{N}{\log_2 N}$$

times as many computations. In our example with $N = 2^{16}$, this means the
brute force approach will take roughly $2^{16}/2^4 = 2^{12} = 4096$ times as long.
On a Macintosh IIci the Matlab FFT of one of these images took about 39
seconds. A brute force evaluation would take more than 45 hours! (In fact
it can take much longer because of the optimized code of an FFT and the
brute force evaluation of the powers of the complex exponentials required
before the multiply and adds.)

As a more extreme example, a typical digitized x-ray image has $2048 \times
2048$ pixels, yielding a computational complexity of 22×2^{22} with the FFT
in comparison to 2^{44} for the brute force method!

FFT Examples

We have already considered 1D examples of the FFT when we computed
the DFT of the random signal and the sinusoid plus the random signal
in Figures 2.4–2.5. The advantages of the FFT become more clear when
computing 2D DFTs, e.g., of image signals. Before doing so, however, we
point out an immediate problem. Since images are usually represented as
a nonnegative signal (the intensity at each pixel is a nonnegative number),
they tend to have an enormous DC value. In other words, the value of
the Fourier transform for $(f_X, f_Y) = (0,0)$ is just the average of the entire
image, which is often a large positive number. This large DC value dwarfs

the values at other frequencies and can make the resulting DFT look like a spike at the origin with 0 everywhere else. For this reason it is common to weight plots of the spectrum so as to deemphasize the low frequencies and enhance the higher frequencies. The most common such weighting is logarithmic: instead of plotting the actual magnitude spectrum $|G(f_X, f_Y)|$, it is common to instead plot

$$G_{\log}(f_X, f_Y) = \log(1 + |G(f_X, f_Y)|). \qquad (2.71)$$

The term 1 in this expression is added to assure that when $|G|$ has value zero, so will $G_{\log}(f_X, f_Y)$. Although other weightings are possible (the 1 and the magnitude spectrum can be multiplied by constants or one can use a power of the magnitude spectrum rather than the log), this form seems the most popular. Figures 2.13–2.24 show the Fourier transforms of several of the 2D signal examples. Both mesh and image plots are shown for the log weighted and unweighted versions of the simple box and disk functions. For the Mona Lisa and MR images the mesh figures are not shown as the number of pixels is so high as to render the mesh figures too black.

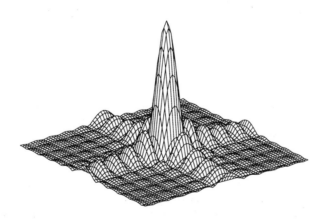

Figure 2.13: FFT of 2D Box of Figure 1.25: mesh

An optical illusion obscures the circular shape of the inner contour curves. This example provides a warning that visual interpretation can be misleading.

Figure 2.14: FFT of 2D Box of Figure 1.25: image

In these artificial images, the dc component of the DFT is not so strong as to completely wipe out the effects of the nonzero frequencies. One can see in both the mesh and image plots the falling off of the magnitude spectrum. The log weighting, however, clearly enhances the relative values of the nonzero frequencies.

The unweighted FFT of natural images typically shows only a single bright spot at the origin. In order to display the full scale, all nonzero frequency components become so small as to be invisible. With the logarithmic weighting the nonzero frequencies become visible and the structure of the spectrum becomes more apparent.

2.7   ⋆ Existence Conditions

Recall that the Fourier transform of a finite duration discrete time signal always exists provided only that the values of the signal are finite. In this case the Fourier transform is simply a weighted sum of a finite number of terms and there are no limits to concern us. In all of the other cases, the

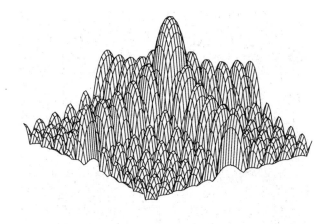

Figure 2.15: Log Weighted FFT of 2D Box of Figure 1.25: mesh

existence of the Fourier transform is not as obvious. The Fourier transform in those cases is defined either as a sum over an infinite set of indices, which is a limit of simple sums, or as an integral, which is itself a limit of sums. Whenever a quantity is defined as a limit, it is possible that the limit might not exist and hence the corresponding Fourier transform might not be well-defined. Existence conditions are conditions on a signal which force it to be "nice" enough for the Fourier transform to exist for all (or perhaps at least some) relevant frequency values. There is no single theorem providing easy-to-use necessary and sufficient conditions for a transform to exist. Instead we must be content with a few of the most important results providing sufficient conditions for a Fourier transform to exist.

To make matters even worse, there are many ways to define the existence of a limit and these different notions of limits can give different definitions of limiting sums and integrals and hence of the Fourier transform. Delving into these mathematical details is beyond the intended scope of this text, but we can at least provide some simple conditions under which we are guaranteed to have no problems and the Fourier transform exists in the usual sense. Unfortunately these conditions exclude some interesting classes of signals, so we also briefly describe a more general class for which the Fourier transforms exist if we use a more general notion of integration

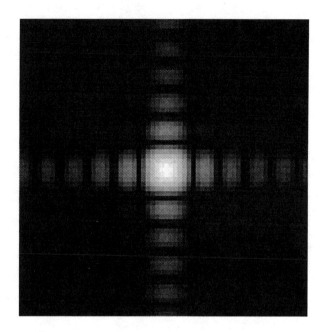

Figure 2.16: Log Weighted FFT of 2D Box of Figure 1.25: image

and infinite summation. Some of the details are described in the starred subsections. For those who do not wish to (or are not asked to) read these sections, the key points are summarized below.

- Discrete Time Signals.

  - A sufficient condition for a discrete time infinite duration signal $\{g(n);\ n \in \mathcal{T}\}$ to have a Fourier transform is that it be *absolutely summable* in the sense

$$\sum_{n \in \mathcal{T}} |g(n)| < \infty.$$

If this sum is finite and equals, say, M, then the spectrum $G(f)$ exists for all $f \in \mathcal{S}_{DTID}$ and

$$|G(f)| < M, \quad \text{all } f. \tag{2.72}$$

  - If a discrete time infinite duration signal $\{g(n);\ n \in \mathcal{T}\}$ has

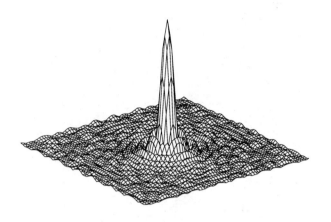

Figure 2.17: FFT of 2D Disk of Figure 1.27: mesh

finite energy in the sense that

$$\mathcal{E}_g = \sum_{n=-\infty}^{\infty} |g(n)|^2 < \infty, \tag{2.73}$$

then the Fourier transform need not exist for all frequencies. For example, the signal $g(n) = 1/n$ for $n > 0$ does not have a Fourier transform at 0. Finite energy is, however, a sufficient condition for the existence of a Fourier transform in a *mean square* sense or *limit in the mean* sense or L_2 sense: There exists a function $G(f)$ with the property that

$$\lim_{N \to \infty} \int_{-\frac{1}{2}}^{\frac{1}{2}} |G(f) - \sum_{n=-N}^{N} g(n)e^{-i2\pi fn}|^2 \, df = 0. \tag{2.74}$$

(The function is unique in that any two functions satisfying the formula must be the same except for a set of f of zero measure, e.g., they can differ on a finite collection of points of f.) In other words, the truncated Fourier transforms

$$G_N(f) = \sum_{n=-N}^{N} g(n)e^{-i2\pi fn}$$

Figure 2.18: FFT of 2D Disk of Figure 1.27: image

might not converge in the ordinary sense as $N \to \infty$, but they do converge in the sense that there is a frequency function $G(f)$ for which the error energy

$$\epsilon^2 = \int_{-\frac{1}{2}}^{\frac{1}{2}} |G(f) - G_N(f)|^2 \, df$$

between $G(f)$ and the approximations $G_N(f)$ converges to 0 as $N \to \infty$.

- Continuous Time Signals.

    - A sufficient condition for a continuous time signal $\{g(t); \ t \in \mathcal{T}\}$ (finite or infinite duration) to have a Fourier transform is that it be *absolutely integrable* in the sense

$$\int_{t \in \mathcal{T}} |g(t)| \, dt < \infty.$$

If this integral is finite and equals, say, M, then the spectrum $G(f)$ exists for all $f \in \mathcal{S}_{CTFD}$ for finite duration or $f \in \mathcal{S}_{CTID}$

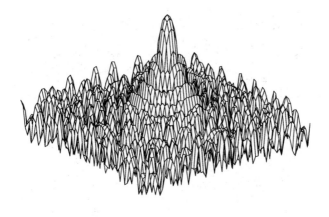

Figure 2.19: Log Weighted FFT of 2D Disk of Figure 1.27: mesh

for infinite duration and

$$|G(f)| < M, \quad \text{all } f. \tag{2.75}$$

– If a continuous time signal $\{g(t); \ t \in \mathcal{T}\}$ has *finite energy* in the sense that

$$\mathcal{E}_g = \int_{\mathcal{T}} |g(t)|^2 \, dt < \infty, \tag{2.76}$$

then the Fourier transform need not exist for all frequencies. For example, the signal $g(t) = 1/t$ for $0 < t < 1$ and 0 otherwise does not have a Fourier transform at $f = 0$. Finite energy is, however, a sufficient condition for the existence of a Fourier transform in a *mean square* sense or *limit in the mean* or L_2 sense analogous to the discrete time case. We will not treat such transforms in much detail because of the complicated analysis required. We simply point out that the mathematical machinery exists to generalize most results of this book from the absolutely integrable case to the finite energy case.

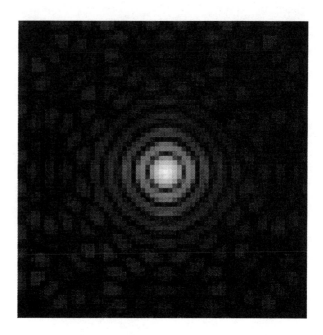

Figure 2.20: Log Weighted FFT of 2D Disk of Figure 1.27: image

⋆ Discrete Time

The most common infinite duration DTFT has $\mathcal{T}$ equal to the set of all integers and hence

$$G(f) = \sum_{n=-\infty}^{\infty} g(n)e^{-i2\pi fn}. \tag{2.77}$$

This infinite sum is in fact a limit of finite sums and the limit may blow up or not converge for some values of f. Thus the DTFT may or may not exist, depending on the signal. To be precise, an infinite sum is defined as

$$\sum_{n=-\infty}^{\infty} a_n = \lim_{N\to\infty, K\to\infty} \sum_{n=-K}^{N} a_n$$

if the double limit exists, i.e., converges. Mathematically, the double limit exists and equals, say, a if for any $\epsilon > 0$ there exist numbers M and L such

Figure 2.21: FFT of Mona Lisa of Figure 1.31

that if $N \geq M$ and $K \geq L$, then

$$\left| \sum_{n=-N}^{K} a_n - a \right| \leq \epsilon.$$

Note that this means that if we fix either N or K large enough (bigger than M or L above, respectively) and let the other go to ∞, then the sum cannot be more than ϵ from its limit. For example, the sum $\sum_{n=-N}^{K} 1 = K + N + 1$ does not have a limit, since if we fix N the sum blows up as $K \to \infty$.

A Fourier transform is often said to exist if the limiting sum exists in the more general Cauchy sense or Cauchy principal value sense, that is, if the limit

$$\sum_{n=-\infty}^{\infty} a_n = \lim_{N \to \infty} \sum_{n=-N}^{N} a_n$$

exists. We will not dwell on the differences between these limits; we simply point out that care must be taken in interpreting and evaluating infinite sums. The infinite sum $\sum_{n=-\infty}^{\infty} n$ does exist in the Cauchy principal value

Figure 2.22: Log Weighted FFT of Mona Lisa of Figure 1.31

sense (it is 0). Similarly, the sum

$$\sum_{k=-\infty, k\neq 0}^{\infty} \frac{1}{k}$$

does not exist in the strict sense, but it does exist in the Cauchy sense. Several of the generalizations of Fourier transforms encountered here and in the literature are obtained by using a weaker or more general definition for an infinite sum.

A sufficient condition for the existence of the sum (and hence of the transform) can be shown (using real analysis) to be

$$\sum_{n=-\infty}^{\infty} |g(n)| < \infty; \tag{2.78}$$

that is, if the signal is *absolutely summable* then the Fourier transform

Figure 2.23: FFT of Magnetic Resonance Image of Figure 1.32

exists in the usual sense of convergence of sums. For example, if

$$\sum_{n=-\infty}^{\infty} |g(n)| = M < \infty,$$

then application of the inequality

$$|\sum_{k} a_k| \le \sum_{k} |a_k| \tag{2.79}$$

implies that

$$\begin{aligned}
|G(f)| &= |\sum_{n=-\infty}^{\infty} g_n e^{-i2\pi fn}| \\
&\le \sum_{n=-\infty}^{\infty} |g_n e^{-i2\pi fn}| \\
&= \sum_{n=-\infty}^{\infty} |g_n||e^{-i2\pi fn}|
\end{aligned}$$

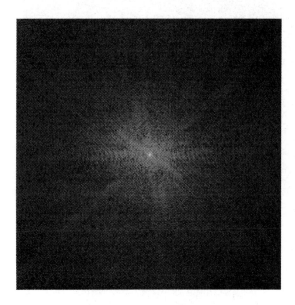

Figure 2.24: Log Weighted FFT of Magnetic Resonance Image of Figure 1.32

$$= \sum_{n=-\infty}^{\infty} |g_n| = M.$$

It should be kept in mind that absolute summability is a sufficient but not necessary condition. We will encounter sequences which violate absolute summability yet have an ordinary Fourier transform. We will also encounter sequences which violate the condition and do not have an ordinary Fourier transform, but which do have a Fourier transform in a suitably generalized sense. As examples of the absolute summability condition, note that the signal $\{r^k; \ k = 0, 1, \ldots\}$ is absolutely summable if $|r| < 1$. It is not absolutely summable if $|r| \geq 1$. The sequence $\{k^{-1}; \ k = 1, 2, \ldots\}$ is not absolutely summable, but the sequence $\{k^{-2}; \ k = 1, 2, \ldots\}$ is.

We shall encounter several means of generalizing the Fourier transform so that a suitably defined transform holds for a larger class of signals. This is accomplished by generalizing the idea of an infinite sum (or integral). An important example of such a generalization is the following. Recall that a discrete time signal is said to be square summable or to have finite energy if it satisfies (2.73). For example, the sequence $\{g(n) = 1/n;\ n = 1, 2, \ldots\}$ has finite energy but is not absolutely summable. If a signal has finite energy, then it can be proved that the Fourier transform $G(f)$ exists in the following sense:

$$\lim_{N \to \infty} \int_{-\frac{1}{2}}^{\frac{1}{2}} |G(f) - \sum_{n=-N}^{N} g(n)e^{-i2\pi fn}|^2\, df = 0. \qquad (2.80)$$

When this limit exists we say that the Fourier transform $G(f)$ exists in the sense of the *limit in the mean*. This is sometimes expressed as

$$G(f) = \underset{N \to \infty}{\text{l.i.m.}} \sum_{n=-N}^{N} g(n)e^{-i2\pi fn}$$

where "l.i.m." stands for "limit in the mean." Even when this sense is intended, we often write the familiar and simpler form

$$G(f) = \sum_{n=-\infty}^{\infty} g(n)e^{-i2\pi fn},$$

but if finite energy signals are being considered, the infinite sum should be interpreted as an abbreviation of (2.80). Note in particular that when a Fourier transform of this type is being considered, we cannot say anything about the ordinary convergence of the sum $\sum_{n=-K}^{N} g(n)e^{-i2\pi fn}$ for any *particular* frequency f as K and N go to ∞, we can only know that an integral of the form (2.80) converges.

It is not expected that this definition will be natural at first glance, but the key point is that one can extend Fourier analysis to finite energy infinite duration signals, but that the definitions are somewhat different. We also note that discrete time finite energy signals are sometimes called l^2 sequences in the mathematical literature.

In the discrete time case, the property of finite energy is indeed more general than that of absolutely summable; that is, absolute summability implies finite energy but not vice versa. For example, if

$$\sum_{n} |g_n| \leq M < \infty,$$

then

$$M^2 \geq \left(\sum_{n} |g_n|\right)^2 = \sum_{n} \sum_{k} |g_n||g_k|$$

$$= \sum_{n} |g_n|^2 + \sum_{n \neq k} |g_n||g_k| \geq \sum_{n} |g_n|^2.$$

⋆ Continuous Time

The most common finite duration continuous time Fourier transform considers a signal of the form $g = \{g(t); \ t \in [0, T]\}$ and has the form

$$G(f) = \int_0^T g(t)e^{-i2\pi ft} dt. \tag{2.81}$$

Unfortunately, unlike the finite duration DTFT case we cannot guarantee that this integral always exists. It can be shown that the integral exists for all f if the signal itself is *absolutely integrable* in the sense that

$$\int_0^T |g(t)| dt < \infty. \tag{2.82}$$

As in the DTFT case, this condition is sufficient but not necessary. An example of a signal violating absolute integrability is $\{t^{-1}; t \in (0, 1)\}$. As in the discrete time case, we can extend the definition of a Fourier transform to include signals with finite energy. We postpone that discussion until the infinite duration case.

The most common infinite duration CTFT considers a signal of the form $g = \{g(t); \ t \in \mathcal{R}\}$ and the transform is given by

$$G(f) = \int_{-\infty}^{\infty} g(t)e^{-i2\pi ft} dt. \tag{2.83}$$

The usual definition of an improper (infinite limit) Riemann integral is

$$\int_{-\infty}^{\infty} g(t)e^{-i2\pi ft} dt = \lim_{S \to \infty} \lim_{T \to \infty} \int_{-S}^{T} g(t)e^{-i2\pi ft} dt, \tag{2.84}$$

if the limits exist. For such a double limit to exist, one must get the same answer when taking S and T to their limits separately in any manner whatsoever.

We formalize the statement of the basic existence theorem so as to ease comparison with later existence theorems. No proof is given (it is standard integration theory).

Theorem 2.1 *A sufficient condition for the existence of the transform in the strict sense (that is, an ordinary improper Riemann integral) is that the signal be absolutely integrable:*

$$\int_{-\infty}^{\infty} |g(t)| dt < \infty. \tag{2.85}$$

As in the finite duration case, this is sufficient but not necessary for the existence of the CTFT. Signals violating the condition include $\{t(1 + t^2)^{-1}; \ t \in \mathcal{R}\}$ and $\{\text{sinc } t; \ t \in \mathcal{R}\}$. Observe that if a signal is absolutely integrable, then its transform is bounded in magnitude:

$$
\begin{aligned}
|G(f)| &= \left| \int_{-\infty}^{\infty} g(t) e^{-i2\pi ft} \, dt \right| \le \int_{-\infty}^{\infty} |g(t) e^{-i2\pi ft}| \, dt \\
&= \int_{-\infty}^{\infty} |g(t)| |e^{-i2\pi ft}| \, dt = \int_{-\infty}^{\infty} |g(t)| \, dt < \infty.
\end{aligned}
$$

Note the use of the basic integration inequality

$$
\left| \int g(t) \, dt \right| \le \int |g(t)| \, dt
$$

in the previous inequality chain.

As a first generalization consider the case where the improper integral exists in the Cauchy principal value sense; that is,

$$
\int_{-\infty}^{\infty} g(t) e^{-i2\pi ft} \, dt = \lim_{T \to \infty} \int_{-T}^{T} g(t) e^{-i2\pi ft} \, dt. \tag{2.86}
$$

This is more general than the usual notion. (That is, if the integral exists as an improper Riemann integral, then it also exists in the Cauchy principal value sense. The converse, however, is not always true.) The following theorem gives sufficient conditions for the Fourier transform to exist in this sense. A proof may be found in Papoulis [24].

Theorem 2.2 *If $g(t)$ has the form*

$$
g(t) = f(t) \sin(\omega_0 t + \phi_0) \tag{2.87}
$$

for some constants ω_0 and ϕ_0, where $|f(t)|$ is monotonically decreasing as $|t| \to \infty$, and where $g(t)/t$ is absolutely integrable for $|t| > A > 0$, i.e.,

$$
\int_{A}^{\infty} \left| \frac{g(t)}{t} \right| \, dt < \infty \tag{2.88}
$$

then the Fourier transform $G(f)$ of $g(t)$ exists in the Cauchy principal value sense.

This general condition need not hold for all interesting signals. For example, if $g(t)$ is equal to 1 for all t, the condition of (2.88) is violated. The most important example of a signal meeting the conditions of this theorem is $g(t) = \text{sinc}(t)$. This signal is not absolutely integrable, but it meets the conditions of the theorem with $\phi_0 = 0$, $\omega_0 = \pi$, and $f(t) = 1/(\pi t)$.

To prepare for a final existence theorem (for the present, at least), we say that the Fourier transform of a signal $g = \{g(t); \ t \in \mathcal{T}\}$ exists in a limit-in-the-mean sense if the following conditions are satisfied:

- The signal can be approximated arbitrarily closely by a sequence of signals $g_N = \{g_N(t); \ t \in \mathcal{T}\}$ in the sense that the error energy goes to zero as $N \to \infty$:

$$\lim_{N\to\infty} \int_{\mathcal{T}} |g(t) - g_N(t)|^2 \, dt = 0, \qquad (2.89)$$

  i.e., g is the limit in the mean of g_N;

- all of the signals g_N have Fourier transforms in the usual sense, i.e.,

$$G_N(f) = \int_{\mathcal{T}} g_N(t) e^{-i2\pi f t} \, dt \qquad (2.90)$$

  is well-defined for all f in the frequency domain of definition; and

- G_N has a limit in the mean, i.e., if $\mathcal{T} = [0, T)$ and $\mathcal{S} = \{k/T; k \in \mathcal{Z}\}$, then there is a function $G(f)$ for which

$$\lim_{N\to\infty} \sum_{f \in \mathcal{S}} |G(f) - G_N(f)|^2 = 0; \qquad (2.91)$$

  if $\mathcal{T} = \mathcal{R}$ and $\mathcal{S} = \mathcal{R}$, then there is a function $G(f)$ for which

$$\lim_{N\to\infty} \int_{\mathcal{S}} |G(f) - G_N(f)|^2 \, df = 0 \qquad (2.92)$$

  (The convergence is called l_2 for the sum in the first case and L_2 for the integral in the second case.)

If these conditions are met then G is a Fourier transform of g. (We say "a Fourier transform" not "the Fourier transform" since it need not be unique. For example, changing a G in a finite way at a finite collection of points yields another G with the desired properties since Riemann integrals are not affected by changing the integrand at a finite number of points.) This is usually expressed formally by writing

$$G(f) = \int_{\mathcal{T}} g(t) e^{-i2\pi f t} \, dt,$$

but in the current situation this formula is an abbreviation for the more exact definition above.

Theorem 2.3 *A sufficient condition for a continuous time signal g to possess a Fourier transform in the sense of limit-in-the-mean is that it have finite energy (or be square integrable):*

$$\mathcal{E}_g = \int_{\mathcal{T}} |g(t)|^2 dt < \infty. \tag{2.93}$$

Finite energy neither implies nor is implied by absolutely integrability. For example, the signal $\frac{\sin(\pi t)}{\pi t}$ is square integrable but not absolutely integrable, while the signal $\{[\sqrt{t}(1+t)]^{-1}; \; t \in (0, \infty)\}$ is absolutely integrable but not square integrable. Square integrable functions are often called L^2 functions in the mathematical literature (like the notation l^2 for square summable functions). Once again, finite energy signals have a Fourier theory, it is just different (and somewhat more complicated) than that used for absolutely integrable functions.

2.8 Problems

2.1. Let g be the signal in problem 1.11 Find the Fourier transforms of g and its zero-filled extension $\bar{g}$.

2.2. Find the DTFT of the following infinite duration signals ($\mathcal{T} = \mathcal{Z}$) signals:

   (a) $g_n = r^{|n|}$, where $|r| < 1$. What happens if $r = 1$?

   (b) $g_n = \delta_{n-k}$, the shifted Kronecker delta function, where k is a fixed integer.

   (c) $g_n = \sum_{k=-N}^{N} \delta_{n-k}$.

   (d) $g_n = a$ for $|n| \leq N$ and $g_n = 0$ otherwise.

   (e) $g_n = r^{n-a}$ for $n \geq a$ and $g_n = 0$ otherwise. Assume that $|r| < 1$.

2.3. Find the DFT of the following sequences:

   (a) $\{g_n\} = \{1, 1, 1, 1, 1, 1, 1, 1\}$

   (b) $\{g_n\} = \{1, 0, 0, 0, 0, 0, 0, 0\}$

   (c) $\{g_n\} = \{0, -1, 0, 0, 0, 0, 0, 1\}$

   (d) $\{g_n\} = \{1, -1, 1, -1, 1, -1, 1, -1\}$

2.4. Find the DFT of the following 8-point sequences:

   (a) $\{g_n\} = \{e^{i\pi \frac{n}{4}}\}$

(b) $\{g_n\} = \{\cos(\frac{\pi n}{4})\}$

(c) $\{g_n\} = \{\sin(\frac{\pi n}{4})\}$

2.5. Find the DFTs of the following signals:

   (a) $g_1 = \{2, 2, 2, 2, 2, 2, 2, 2\}$

   (b) $g_2 = \{e^{i\pi n/2}; n = 0, 1, 2, 3, 4, 5, 6, 7\}$

   (c) $g_3 = \{e^{i\pi(n-2)/2}; n = 0, 1, 2, 3, 4, 5, 6, 7\}$

   (d) $g_4 = \{0, 0, 0, 0, 0, 0, 1, 0\}$

2.6. Find the DFTs of the following sequences:

   (a) $\{g_n\} = \{0, 0, 0, 1, 0, 0, 0, 0\}$

   (b) $\{g_n\} = \{\cos(\pi n); n = 0, 1, \ldots, 7\}$

   (c) $\{g_n\} = \{\cos(\frac{\pi n}{2}); n = 0, 1, \ldots, 7\}$

   (d)
   $$g_n = \begin{cases} 1 - \frac{n}{4}, & n = 0, 1, 2, 3 \\ \frac{n}{4} - 1, & n = 4, 5, 6, 7. \end{cases}$$

2.7. Suppose that g is an infinite duration continuous time signal that is nonzero only for the interval $[0, T)$. Find an approximation to the Fourier transform $G(f)$ of g in terms of the DFT $\hat{G}$ of the sampled signal $\{\hat{g}_n; n \in \mathcal{Z}_N\}$. For what values of frequency f does the approximation hold?

2.8. Find the CTFT of the following signals using the following special signals ($\mathcal{T}$ is the real line in all cases): The rectangle function

$$\sqcap(t) = \begin{cases} 1 & \text{if } -\frac{1}{2} < t < \frac{1}{2} \\ \frac{1}{2} & \text{if } |t| = \frac{1}{2} \\ 0 & \text{otherwise} \end{cases}$$

and the unit step function

$$H(t) = \begin{cases} 1 & t > 0 \\ \frac{1}{2} & t = 0 \\ 0 & \text{otherwise} \end{cases}$$

   (a) $g(t) = e^{-\lambda t} H(t)$

   (b) $g(t) = \sqcap(\frac{t}{a} - b)$ where $a > 0$.

(c)

$$g(t) = \begin{cases} c(1 - \frac{|t|}{a}) & -a \leq t \leq a \\ 0 & \text{otherwise} \end{cases},$$

where c is a fixed constant.

(d) The raised cosine pulse $g(t) = \sqcap(t)(1 + \cos(2\pi t)); \ t \in \mathcal{R}$.

(e) The Gaussian pulse $g(t) = e^{-\pi t^2}; \ t \in \mathcal{R}$. (*Hint:* Use the technique of "completing the square.")

2.9. Find the CTFT of the following signals.

(a) $g(t) = r^{|t|}; \ t \in \mathcal{R}$, where $0 < r < 1$.

(b) $g(t) = \text{sgn}(t); \ t \in [-1, 1)$.

(c) $g(t) = \sin(2\pi f_0 t); \ t \in [-\pi, \pi]$.

(d) The pulse defined by $p(t) = A$ if $t \in [0, \tau]$ and 0 otherwise, $t \in \mathcal{R}$.

2.10. By direct integration, find the Fourier transform of the infinite duration continuous time signal $g(t) = t \wedge (t)$.

2.11. Find the Fourier transform of the signals $\{e^{-\lambda|t|}; \ t \in \mathcal{R}\}$ and $\{\text{sgn}(t)e^{-\lambda|t|}; \ t \in \mathcal{R}\}$.

2.12. Find the CTFT of the following signals:

(a) $g(t) = At$ for $t \in [-T/2, T/2]$ and 0 for $t \in \mathcal{R}$ but $t \notin [-T/2, T/2]$.

(b) $\{|\sin t|; |t| < \pi\}$.

(c)

$$g(t) = \begin{cases} A & \text{for } 0 < t < T/2 \\ 0 & \text{for } t = 0 \\ -A & \text{for } -T/2 < t < 0 \\ 0 & \text{for } |t| \geq T/2 \end{cases}.$$

2.13. An important property of the Fourier transform is *linearity*. If you have two signals g and h with Fourier transforms G and H, respectively, and if a and b are two constants (i.e., complex numbers), then the Fourier transform of $ag + bh$ is $aG + bH$.

(a) Prove this result for the special cases of the DFT (finite duration discrete time Fourier transform) and the infinite duration CT Fourier transform.

(b) Suppose that g is the continuous time one-sided exponential $g(t) = e^{-\lambda t}u_{-1}(t)$ for all real t and that h is the signal defined by $h(t) = g(-t)$. Sketch the signals $g + h$ and $g - h$ and find the Fourier transforms of each.

2.14. An important property of the Fourier transform is *linearity*. If you have two signals g and h with Fourier transforms G and H, respectively, and if a and b are two constants (i.e., complex numbers), then the Fourier transform of $ag + bh$ is $aG + bH$.

(a) Prove this result for the special cases of the infinite duration DTFT.

(b) Use this result to evaluate the Fourier transform of the signal $g = \{g_n;\ n \in \mathcal{Z}\}$ defined by

$$g_n = \begin{cases} (3^{-n} + (-3)^{-n})\,u_{-1}(n) & n = 0, 1, \ldots \\ 0 & \text{otherwise} \end{cases}$$

and verify your answer by finding the Fourier transform directly.

2.15. Suppose that g is the finite duration discrete time signal $\{\delta_n;\ n = 0, \cdots, N-1\}$. Find $\mathcal{F}(\mathcal{F}(g))$, that is, the Fourier transform of the Fourier transform of g. Repeat for the signal h defined by $h_n = 1$ for $n = k$ (k a fixed integer in $\{0, \cdots, N-1\}$) and $h_n = 0$ otherwise.

2.16. Suppose you know the Fourier transform of a real-valued signal. How can you find the Hartley transform? (*Hint:* Combine the transform and its complex conjugate to find the sine and cosine transforms.) Can you go the other way, that is, construct the Fourier transform from the Hartley transform?

2.17. Suppose that $g = \{g(h, v);\ h \in [0, H], v \in [0, V]\}$ (h represents horizontal and v represents vertical) represents the intensity of a single frame of a video signal. Suppose further that g is entirely black ($g(h, v) = 0$) except for a centered white rectangle ($g(h, v) = 1$) of width αH and height αV ($\alpha < 1$). Find the two-dimensional Fourier transform $\mathcal{F}(g)$.

2.18. Suppose that $g = \{g(h, v);\ h \in [0, H), v \in [0, V)\}$ is a two dimensional signal (a continuous parameter image raster). The independent variables h and v stand for "horizontal" and "vertical", respectively. The signal $g(h, v)$ can take on three values: 0 for black, 1/2 for grey, and 1 for white. Consider the specific signal of Figure 2.25.

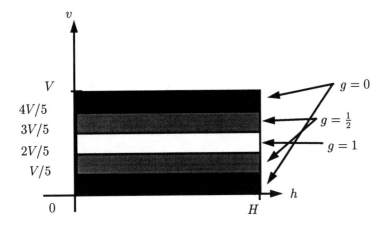

Figure 2.25: 2D "Stripes"

(a) Write a simple expression for $g(h, v)$ in terms of the box function

$$\Box_T(x) = \begin{cases} 1 & |x| \leq T \\ 0 & \text{otherwise} \end{cases}.$$

(You can choose g to have any convenient values on the bound-aries as this makes no difference to the Fourier transform.)

(b) Find the 2-D Fourier transform $G(f_h, f_v)$ of g.

2.19. Find the Fourier transform of the binary image below.

```
0000
1001
0000
1111
```

2.20. Suppose that $g = \{g(k, n); \ k, n \in \mathcal{Z}_N\}$ is an image and $G = \{G(\frac{k}{N}, \frac{n}{N}); \ k, n \in \mathcal{Z}_N\}$ is its Fourier transform. Let $\mathbf{W}$ denote the complex exponential matrix defined in (2.6). Is it true that

$$G = \mathbf{W}g^t\mathbf{W}^t? \tag{2.94}$$

Note that $g^t\mathbf{W}^t = (\mathbf{W}g)^t$.

2.21. Consider an $N = 4$ FFT.

(a) Draw a flow graph for the FFT.

(b) Draw a modified flow graph using inverters to eliminate half the multiplies.

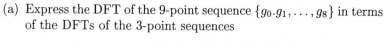

2.22. (a) Express the DFT of the 9-point sequence $\{g_0.g_1, \ldots, g_8\}$ in terms of the DFTs of the 3-point sequences

$$
\begin{aligned}
g_a(n) &= \{g_0, g_3, g_6\} \\
g_b(n) &= \{g_1, g_4, g_7\} \\
g_c(n) &= \{g_2, g_5, g_8\}.
\end{aligned}
$$

(b) Draw a legible flow graph for the "base 3" method for computing the FFT, as suggested above, for $N = 9$.

2.23. Suppose you have an infinite duration discrete time signal $g = \{g_n; \ n \in \mathcal{Z}\}$ and that its Fourier transform is $G = \{G(f); \ f \in [-1/2, 1/2)\}$. Consider the new signals

$$
\begin{aligned}
h_n &= g_{2n}; \ n \in \mathcal{Z} \\
w_n &= g_{2n+1}; \ n \in \mathcal{Z} \\
v_n &= \begin{cases} g_{n/2} & \text{if } n \text{ is an even number} \\ 0 & \text{otherwise} \end{cases}
\end{aligned}
$$

with Fourier transforms H, W, and V, respectively. h and w are examples of *downsampling* or *subsampling*. v is called *upsampling*. Note that downsampling and upsampling are not generally inverse operations, i.e., downsampling followed by upsampling need not recover the original signal.

(a) Find an expression for V in terms of G.

(b) Find an expression for G in terms of W and H. (This is a variation on the fundamental property underlying the FFT.)

(c) Suppose now that $r = \{r_n; \ n \in \mathcal{Z}\}$ is another signal and that ρ is the upsampled version of r, i.e., $\rho_n = r_{n/2}$ for even n and 0 otherwise. We now form a signal x defined by

$$
x_n = v_n + \rho_{n+1}; \ n \in \mathcal{Z}.
$$

This signal can be considered as a *time-division multiplexed* (TDM) version of g and r since alternate samples come from one or the other. Find the Fourier transform X in terms of G and R.

(d) Evaluate the above transforms and verify the formulas for the case $g_n = 2^{-n}u_{-1}(n)$, $r_n = 4^{-n}u_{-1}(n)$.

2.24. Consider the signal $g = \{g_n = nr^n u_{-1}(n); \ n \in \mathcal{Z}\}$, where $|r| < 1$.

(a) Is this signal absolutely summable?

(b) Find a simple upper bound to $|G(f)|$ that holds for all f.

(c) Find the Fourier transform of the signal g.

(d) Consider the signal $h = \{h_n; \ n \in \mathcal{Z}\}$ defined by $h_n = g_{2n}$. (h is a downsampled version of g.) Find the DTFT of h.

(e) Find a simple upper bound to $|H(f)|$ that holds for all f.

(f) Consider the signal $w = \{w_n; \ n \in \mathcal{Z}\}$ defined by $w_{2n} = h_n$ and $w_{2n+1} = 0$ for all integer n. w is called an upsampled version of h. Find the DTFT of w.

(g) Find the DTFT of the signal $g - w$.

Chapter 3

Fourier Inversion

Having defined the Fourier transform and examined several examples, the next issue is that of invertibility: if $G = \mathcal{F}(g)$, can g be recovered from the spectrum G? More specifically, is there an *inverse Fourier transform* $\mathcal{F}^{-1}$ with the property that

$$\mathcal{F}^{-1}(\mathcal{F}(g)) = g? \tag{3.1}$$

When this is the case, we shall call g and G a *Fourier transform pair* and write

$$g \leftrightarrow G, \tag{3.2}$$

where the double arrow notation emphasizes that the signal and its Fourier transform together form a Fourier transform pair. We have already seen that Fourier transforms are not always invertible in the strict sense, since changing a continuous signal at a finite number of points does not change the value of the Riemann integral giving the Fourier transform. For example, $\{\square_{1/2}(t);\ t \in \mathcal{R}\}$ and $\{\sqcap(t);\ t \in \mathcal{R}\}$ have the same transform. In this chapter we shall see that except for annoying details like this, the Fourier transform can usually be inverted.

The first case considered is the DFT, the finite duration discrete time Fourier transform. This case is considered first since an affirmative answer is easily proved by a constructive demonstration. The remaining cases are handled with decreasing rigor, but the key ideas are accurate.

3.1 Inverting the DFT

Given a finite duration discrete time signal $g = \{g_n;\ n \in \{0, 1, \ldots, N-1\}\}$ the DFT is defined for all real f by $G(f) = \sum_{n=0}^{N-1} g_n e^{-i2\pi fn}$. Although this formula is well defined for all real f, we earlier restricted the frequency domain of definition of $G(f)$ for a discrete time finite duration signal to the subset of real frequencies $\mathcal{S}_{DTFD}^{(1)} = \{0, 1/N, 2/N, \ldots, (N-1)/N\}$. We now proceed to justify this selection by showing that this set of frequencies suffices to recover the original signal. We point out that this set is not unique; there are other choices that work equally well. It is, however, simple and the most common.

The present goal is this: suppose that we know a signal g has a spectrum $G = \{G(k/N); k = 0, 1, \ldots, N-1\}$ or, equivalently, $G = \{G(f); f = 0, 1/N, \ldots, (N-1)/N\}$, but we do not know the signal itself. How do we reconstruct g from G?

Consider for $n = 0, 1, \ldots, N-1$ the sequence

$$y_n = \frac{1}{N} \sum_{k=0}^{N-1} G(\frac{k}{N}) e^{i2\pi \frac{k}{N} n}. \tag{3.3}$$

Here y_n can be considered as N^{-1} times the DFT of the signal G with $-i$ replacing i in the exponential. Since

$$G(\frac{k}{N}) = \sum_{l=0}^{N-1} g_l e^{-i2\pi \frac{k}{N} l} \tag{3.4}$$

we have that

$$y_n = \frac{1}{N} \sum_{k=0}^{N-1} \sum_{l=0}^{N-1} g_l e^{-i2\pi \frac{k}{N} l} e^{i2\pi \frac{k}{N} n} \tag{3.5}$$

and hence exchanging the order of summation, which is always valid for finite sums, yields

$$y_n = \sum_{l=0}^{N-1} g_l \frac{1}{N} \sum_{k=0}^{N-1} e^{i2\pi \frac{k}{N}(n-l)}. \tag{3.6}$$

To evaluate this sum, consider

$$\frac{1}{N} \sum_{k=0}^{N-1} e^{i2\pi \frac{k}{N} m}.$$

If $m = 0$ (or is any multiple of N) the sum is 1. If m is not a multiple of N, then the geometric progression formula implies that

$$\frac{1}{N} \sum_{k=0}^{N-1} e^{i2\pi \frac{k}{N} m} = \frac{1}{N} \frac{1 - e^{i2\pi \frac{m}{N} N}}{1 - e^{i2\pi \frac{m}{N}}} = 0.$$

Readers may recognize this as a variation on the fact that the sum of all roots of unity of a particular order is 0. Alternatively, adding up N equally spaced points on a circle gives their center of gravity, which is just the origin of the circle. Observe that the $m = 0$ result is consistent with the $m \neq 0$ result if we apply L'Hôpital's rule to the latter.

Recalling the definition of the Kronecker delta function δ_m:

$$\delta_m = \left\{ \begin{array}{ll} 1 & \text{if } m = 0 \\ 0 & \text{otherwise} \end{array} \right. , \tag{3.7}$$

then

$$\frac{1}{N} \sum_{k=0}^{N-1} e^{i2\pi \frac{k}{N} m} = \delta_{m \bmod N}; \text{ for any integer } m. \tag{3.8}$$

Since both n and l in (3.6) are integers between 0 and $N-1$, the sum inside (3.6) becomes

$$\frac{1}{N} \sum_{k=0}^{N-1} e^{i2\pi \frac{k}{N}(n-l)} = \delta_{n-l}; \ n, l \in \mathbb{Z}_N \tag{3.9}$$

and hence

$$y_n = \sum_{l=0}^{N-1} g_l \delta_{n-l} = g_n; \tag{3.10}$$

that is, the original sequence has been perfectly recovered.

Thus given a spectrum $G = \{G_k = G(k/N); \ k = 0, 1, \ldots, N-1\}$ it is reasonable to define the *inverse discrete Fourier transform* (or *IDFT*) by

$$g_n = \frac{1}{N} \sum_{k=0}^{N-1} G(\frac{k}{N}) e^{i2\pi \frac{k}{N} n}; \ n = 0, 1, \cdots, N-1, \tag{3.11}$$

and we have shown that

$$g = \mathcal{F}^{-1}(G) = \mathcal{F}^{-1}(\mathcal{F}(g)). \tag{3.12}$$

In summary we have shown that the following are a Fourier transform pair:

$$G(\frac{k}{N}) = \sum_{n=0}^{N-1} g_n e^{-i2\pi \frac{k}{N} n}; \ k \in \mathbb{Z}_N \tag{3.13}$$

$$g_n = \frac{1}{N} \sum_{k=0}^{N-1} G(\frac{k}{N}) e^{i2\pi \frac{k}{N} n}; \ n \in \mathbb{Z}_N. \tag{3.14}$$

Several observations on this result are worth making.

- Recall from the matrix form of the DFT of (2.7) that $\mathbf{G} = \mathbf{Wg}$, where $\mathbf{W} = \{e^{-i\frac{2\pi}{N}kj}; \ k = 0, 1, \cdots, N-1; \ j = 0, 1, \cdots, N-1\}$. From elementary linear algebra this implies that

$$\mathbf{g} = \mathbf{W}^{-1}\mathbf{G}.$$

This is readily identified as the vector/matrix form of (3.14) if we make the identification

$$\begin{aligned} \mathbf{W}^{-1} &= N^{-1}\{e^{+i\frac{2\pi}{N}kj}; \ k = 0, 1, \cdots, N-1; \ j = 0, 1, \cdots, N-1\} \\ &= N^{-1}\mathbf{W}^*, \end{aligned} \tag{3.15}$$

where the $*$ attached to a matrix or vector indicates the conjugate transpose, that is, the matrix formed by taking the complex conjugate and then transposing. The matrix inversion of (3.15) is just the matrix formulation of (3.9).

- Instead of the frequency domain of definition $\{0, 1/N, \cdots, (N-1)/N\}$, we could use any shift of this domain by an integer K, that is, any frequency domain of the form $\{K, K+1/N, \cdots, K+(N-1)/N\}$. With this choice the Fourier transform of $\{g_n; \ n = 0, \cdots, N-1\}$ becomes $\{G(k/N); \ k = KN, KN+1, \cdots, KN+N-1\}$ and the IDFT becomes

$$g_n = \frac{1}{N}\sum_{k=KN}^{KN+N-1} G(\frac{k}{N})e^{i2\pi\frac{k}{N}n} = \frac{1}{N}\sum_{k=0}^{N-1} G(\frac{k}{N})e^{i2\pi\frac{k}{N}n}$$

because of the periodicity of complex exponentials; that is, because for each l

$$e^{i2\pi\frac{KN+l}{N}n} = e^{i2\pi\frac{l}{N}n}.$$

- Different choices for the time domain of definition lead to similar alterations in the inversion formula. For example, if the signal is two-sided: $\{g_n; \ n = -N, \cdots, -1, 0, 1, \cdots, N\}$, then the DFT is commonly taken as

$$\{G(\frac{k}{2N+1}); \ k = -N, -N+1, \cdots, -1, 0, 1, N-1, N\}; \tag{3.16}$$

that is, the frequency domain of definition is

$$S_{DTFD}^{(2)} = \{-\frac{N}{2N+1}, \cdots, -\frac{1}{2N+1}, 0, \frac{1}{2N+1}, \cdots, \frac{N}{2N+1}\}.$$

As an exercise you should prove the corresponding inversion formula:

$$G(f) = \sum_{n=-N}^{N} g_n e^{-i2\pi fn}; \; f \in \mathcal{S}_{DTFD}^{(2)} \qquad (3.17)$$

$$g_n = \frac{1}{2N+1} \sum_{k=-N}^{N} G(\frac{k}{2N+1}) e^{i2\pi \frac{k}{2N+1} n}; \qquad (3.18)$$

$$n \in \{-N, \cdots, 0, \cdots, N\}.$$

- The key property yielding the inversion is the *orthogonality* of the discrete time exponentials. A collection of finite duration discrete time complex valued signals $\{\phi_n^{(m)}; \; n \in \mathcal{Z}_N\}$ for some collection of integers m is said to be *orthogonal* if

$$\sum_{k=0}^{N-1} \phi_k^{(n)} \phi_k^{(l)*} = C_n \delta_{n-l} \qquad (3.19)$$

for $C_n \neq 0$; i.e., two signals are orthogonal if the sum of the coordinate products of one signal with the complex conjugate of the other is 0 for different signals, and nonzero for two equal signals. If $C_n = 1$ for all appropriate n, the signals are said to be *orthonormal*. Eq. (3.9) implies that the exponential family $\{e^{i2\pi \frac{k}{N}m}; \; k = 0, 1, \ldots, N-1\}$ for $m = 0, 1, \ldots, N-1$ are orthogonal and the scaled signals

$$\frac{1}{\sqrt{N}} \{e^{i2\pi \frac{k}{N}m}; \; k = 0, 1, \ldots, N-1\}$$

are orthonormal since for $l, n \in \mathcal{Z}_N$

$$\frac{1}{N} \sum_{k=0}^{N-1} e^{-i2\pi \frac{k}{N}l} e^{i2\pi \frac{k}{N}n} = \delta_{l-n}. \qquad (3.20)$$

The matrix form of this relation also crops up. Let $\mathbf{W}$ be the exponential matrix defined in (2.6) and define

$$\mathbf{U} = \frac{1}{\sqrt{N}} \mathbf{W}. \qquad (3.21)$$

Then (3.20) becomes in matrix form

$$\mathbf{U}^{-1} = \mathbf{U}^*; \qquad (3.22)$$

that is, the inverse of the matrix $\mathbf{U}$ is its complex conjugate transpose. A matrix with this property is said to be *unitary*. Many of the properties of Fourier transforms for finite duration sequences (or, equivalently, finite dimensional vectors) generalize to unitary transformations of the form $\mathbf{G} = \mathbf{Hg}$ for a unitary matrix $\mathbf{H}$.

- The factor of $1/N$ in the IDFT could have easily been placed in the definition of the Fourier transform instead of in the inverse. Alternatively we could have divided it up into symmetric factors of $1/\sqrt{N}$ in each transform. These variations are used in the literature and also referred to as the discrete Fourier transform.

- The Fourier inversion formula provides a representation or a decomposition into complex exponentials for an arbitrary discrete time, finite duration signal. Eq. (3.14) describes a signal as a weighted sum of exponentials, where the weighting is given by the spectrum. As an example, consider the signal $g = \{g_n = r^n; \ n \in \mathcal{Z}_N\}$. The DFT is given by (2.20) and hence (3.14) becomes

$$r^n = \frac{1 - r^N}{N} \sum_{k=0}^{N-1} \frac{e^{i2\pi \frac{k}{N} n}}{1 - re^{-i2\pi \frac{k}{N}}}; \ n \in \mathcal{Z}_N, \qquad (3.23)$$

a formula that might appear somewhat mysterious out of context. The formula decomposes r^n into a weighted sum of complex exponentials of the form

$$g_n = \sum_{k=0}^{N-1} c_k e^{i2\pi \frac{k}{N} n}; \ n \in \mathcal{Z}_N. \qquad (3.24)$$

This can be viewed as one way to represent the signal as a simple linear combination of primitive components. It is an example of a *Fourier series* representation of a signal, here a discrete time finite duration signal. This representation will be extremely important for finite duration signals and for periodic signals and hence we formalize the idea in the next section.

As another example of a Fourier series, (3.9) immediately provides a Fourier series of the Kronecker delta defined on a finite domain; that is, the signal $\{\delta_n; \ n = 0, 1, \cdots, N - 1\}$ has a Fourier series

$$\delta_n = \frac{1}{N} \sum_{k=0}^{N-1} e^{i2\pi \frac{k}{N} n}; \ n \in \mathcal{Z}_N. \qquad (3.25)$$

This again has the general form of (3.24), this time with $c_k = 1/N$ for all $k \in \mathcal{Z}_N$.

3.2 Discrete Time Fourier Series

We can rewrite (3.14) as

$$g_n = \sum_{k=0}^{N-1} c_k e^{i2\pi \frac{k}{N} n}; \ n \in \mathcal{Z}_N \tag{3.26}$$

where

$$c_k = \frac{1}{N} G(\frac{k}{N}) = \frac{1}{N} \sum_{l=0}^{N-1} g_l e^{-i2\pi l \frac{k}{N}}; \ k \in \mathcal{Z}_N. \tag{3.27}$$

This decomposition of a signal into a linear combination of complex exponentials will be important in the study of linear systems where complex exponentials play a fundamental role. An exponential sum of this form is called a *Fourier series* representation of the discrete time finite duration signal g. This representation has an interesting and useful extension. Although the left-hand side of (3.26) is defined only for $n \in \mathcal{Z}_N$, the right-hand side gives a valid function of n for all integers n. Furthermore, the right-hand side is clearly periodic in n with period N and hence consists of periodic replications of the finite duration signal g. In other words, we have a convenient representation for the periodic extension $\tilde{g}$ of g:

$$\tilde{g}_n = \sum_{k=0}^{N-1} \frac{1}{N} G(\frac{k}{N}) e^{i2\pi \frac{k}{N} n}; \ n \in \mathcal{Z}. \tag{3.28}$$

Conversely, if we are given a discrete time periodic signal $\tilde{g}_n$, then we can find a Fourier series in the above form by using the Fourier transform of the truncated signal consisting of only a single period of the periodic signal; that is,

$$\tilde{g}_n = \sum_{k=0}^{N-1} c_k e^{i2\pi \frac{k}{N} n}; \ n \in \mathcal{Z} \tag{3.29}$$

$$c_k = \frac{1}{N} \sum_{l=0}^{N-1} \tilde{g}_l e^{-i2\pi l \frac{k}{N}}; \ k \in \mathcal{Z}_N. \tag{3.30}$$

This provides a Fourier representation for discrete time infinite duration periodic signals. This is a fact of some note because the signal, an infinite duration discrete time signal, violates the existence conditions for ordinary Fourier transforms (unless it is trivial, i.e., $g_n = 0$ for all n). If g_n is ever nonzero and it is periodic, then it will not be absolutely summable nor will it have finite energy. Hence such periodic signals do not have Fourier transforms in the strict sense. We will later see in Chapter 5 that they have a generalized Fourier transform using Dirac delta functions (generalized functions), but for the moment we have a perfectly satisfactory Fourier representation without resort to generalizations.

3.3 Inverting the Infinite Duration DTFT

We next turn to the general DTFT with infinite duration. The general idea will be the same as for the DFT, but the details will be different. Suppose now that we have a signal $g = \{g_n; \; n \in \mathcal{Z}\}$, where as usual $\mathcal{Z}$ is the set of all integers. Assume that the Fourier transform

$$G(f) = \sum_{k=-\infty}^{\infty} g_k e^{-i2\pi f k} \qquad (3.31)$$

exists. Since $G(f)$ is a periodic function of f with period 1 (because $e^{-i2\pi f k}$ is periodic in f with period 1), we can consider $\mathcal{S} = [-1/2, 1/2)$ (or any other interval of unit length such as $[0, 1)$) to be the domain of definition of the spectrum and we can consider the spectrum to be $G = \{G(f); \; f \in [-1/2, 1/2)\}$. We have that

$$
\begin{aligned}
y_n &= \int_{-\frac{1}{2}}^{\frac{1}{2}} G(f) e^{i2\pi f n} \, df \\
&= \int_{-\frac{1}{2}}^{\frac{1}{2}} \left(\sum_{k=-\infty}^{\infty} g_k e^{-i2\pi k f} \right) e^{i2\pi f n} \, df \\
&= \sum_{k=-\infty}^{\infty} g_k \int_{-\frac{1}{2}}^{\frac{1}{2}} e^{i2\pi f(n-k)} \, df. \qquad (3.32)
\end{aligned}
$$

To complete the evaluation observe for $m = 0$ that

$$\int_{-\frac{1}{2}}^{\frac{1}{2}} e^{i2\pi m f} \, df = \int_{-\frac{1}{2}}^{\frac{1}{2}} df = 1$$

and for any nonzero integer m that

$$\int_{-\frac{1}{2}}^{\frac{1}{2}} e^{i2\pi m f} \, df = 0$$

and hence

$$\int_{-\frac{1}{2}}^{\frac{1}{2}} e^{i2\pi m f} \, df = \delta_m; \; m \in \mathcal{Z}. \qquad (3.33)$$

Thus inserting (3.33) into (3.32) we have that

$$
\begin{aligned}
y_n &= \sum_{k=-\infty}^{\infty} g_k \int_{-\frac{1}{2}}^{\frac{1}{2}} e^{i2\pi f(n-k)} \, df \\
&= \sum_{k=-\infty}^{\infty} g_k \delta_{n-k} \\
&= g_n,
\end{aligned}
\tag{3.34}
$$

recovering the signal from its transform..

Observe the close resemblance of (3.33) to (3.8). As in the case of the DFT, the key property of exponentials leading to the inversion formula is the *orthogonality* of the signals $\{e^{i2\pi kf}; \ f \in [-1/2, 1/2)\}$ for $k \in \mathcal{Z}$ in the sense that

$$
\int_{-\frac{1}{2}}^{\frac{1}{2}} e^{i2\pi kf} e^{-i2\pi lf} \, df = \delta_{k-l}; \ k, l \in \mathcal{Z},
\tag{3.35}
$$

the continuous time finite duration analog of (3.20). It must be admitted that we have been somewhat cavalier in assuming that the sum and the integral could be exchanged above, but under certain conditions the exchange can be mathematically justified, e.g., if the signal is absolutely summable.

We have shown that

$$
g_n = \int_{-\frac{1}{2}}^{\frac{1}{2}} G(f) e^{i2\pi fn} \, df; \ n \in \mathcal{Z}
\tag{3.36}
$$

and hence that the right hand side above indeed gives the inverse Fourier transform. For example, application of the discrete time infinite duration inversion formula to the signal $r^n u_{-1}(n); \ n \in \mathcal{Z}$ for $|r| < 1$ yields

$$
r^n u_{-1}(n) = \int_{-\frac{1}{2}}^{\frac{1}{2}} (1 - re^{-i2\pi f})^{-1} e^{i2\pi fn} \, df; \ n \in \mathcal{Z}.
\tag{3.37}
$$

Unlike (3.14) where a discrete time finite duration signal was represented by a weighted sum of complex exponentials (a Fourier series), here a discrete time infinite duration signal is represented as a weighted *integral* of complex exponentials, where the weighting is the spectrum. Instead of a Fourier series, in this case we have a *Fourier integral* representation of a signal. Intuitively, a finite duration signal can be perfectly represented by only a finite combination of sinuosoids. An infinite duration signal, however, requires a continuum of frequencies in general. As an example, (3.33) provides the infinite duration analog to the Fourier series representation (3.25) of a

finite duration discrete time Kronecker delta: the infinite duration discrete
time signal $\{\delta_n;\ n \in \mathcal{Z}\}$ has the Fourier integral representation

$$\delta_n = \int_{-\frac{1}{2}}^{\frac{1}{2}} e^{i2\pi nf}\, df;\ n \in \mathcal{Z}. \tag{3.38}$$

To summarize the infinite duration discrete time Fourier transform and
inversion formula, the following form a Fourier transform pair:

$$G(f) = \sum_{n=-\infty}^{\infty} g_n e^{-i2\pi fn};\ f \in [-\frac{1}{2}, \frac{1}{2}) \tag{3.39}$$

$$g_n = \int_{-\frac{1}{2}}^{\frac{1}{2}} G(f) e^{i2\pi fn}\, df;\ n \in \mathcal{Z}. \tag{3.40}$$

With this definition the Fourier inversion of (3.12) extends to the more
general case of the two-sided DTFT. Note, however, the key difference between these two cases: in the case of the DFT the spectrum was *discrete* in
that only a finite number of frequencies were needed and the inverse transform, like the transform itself, was a sum. This resulted in a Fourier series
representation for the original signal. In the infinite duration DTFT case,
however, only time is discrete, the frequencies take values in a continuous
interval and the inverse transform is an integral, resulting in a Fourier integral representation of the signal. We can still interpret the representation
of the original signal as a weighted average of exponentials, but the average
is now an integral instead of a sum.

We have not really answered the question of how generally the result of
(3.39)–(3.40) is valid. We have given without proof a sufficient condition
under which it holds: if the original sequence is absolutely summable then
(3.39)–(3.40) are valid.

One might ask at this point what happens if one begins with a spectrum
$\{G(f);\ f \in \mathcal{S}\}$ and defines the sequence via the inverse Fourier transform.
Under what conditions on $G(f)$ will the formulas of (3.39)–(3.40) still hold;
that is, when will one be able to recover g_n from $G(f)$ using the given
formulas? This question may now appear academic, but it will shortly gain
in importance. Unfortunately we cannot give an easy answer, but we will
describe some fairly general analogous conditions when we treat the infinite
duration CTFT.

Analogous to the remarks following the DFT inversion formula, we could
also consider different frequency domains of definition, in particular any
unit length interval such as $[0, 1)$ would work, and we could consider different time domains of definition, such as the nonnegative integers which

yield one-sided signals of the form $\{g_n; \ n = 0, 1, \cdots\}$. These alternatives will not be considered in detail. Instead we consider another variation on the choice of a frequency domain of definition and the resulting inversion formula. This alternative will prove very useful in our subsequent consideration of continuous time Fourier transforms.

Frequency Scaling

Instead of having a frequency with values in an interval of unit length, we can scale the frequency by an arbitrary positive constant and adjust the formulas accordingly. For example, we could fix $f_0 > 0$ and define a Fourier transform as

$$G_{f_0}(f) = G(\frac{f}{f_0}) = \sum_{n=-\infty}^{\infty} g_n e^{-i2\pi \frac{f}{f_0} n}; \ f \in [0, f_0),$$

and hence, changing variables of integration,

$$g_n = \frac{1}{f_0} \int_0^{f_0} G_{f_0}(f) e^{i2\pi \frac{f}{f_0} n} \, df; \ n \in \mathcal{Z}.$$

We now restate the Fourier transform pair relation with the scaled frequency value and change the name of the signal and the spectrum in an attempt to minimize confusion. Replace the sequence g_n by h_n and the transform $G_{f_0}(f)$ by $H(f)$. We have now proved that for a given sequence h, the following are a Fourier transform pair (provided the technical assumptions used in the proof hold):

$$H(f) = \sum_{n=-\infty}^{\infty} h_n e^{-i2\pi \frac{f}{f_0} n}; \ f \in [0, f_0) \qquad (3.41)$$

$$h_n = \frac{1}{f_0} \int_0^{f_0} H(f) e^{i2\pi \frac{f}{f_0} n} \, df; \ n \in \mathcal{Z}. \qquad (3.42)$$

The idea is that we can scale the frequency parameter and change its range and the Fourier transform pair relation still holds with minor modifications. The most direct application of this form of transform is in sampled data systems when one begins with a continuous time signal $\{g(t); \ t \in \mathcal{R}\}$ and forms a discrete time signal $\{g(nT); \ n \in \mathcal{Z}\}$. In this situation the scaled frequency form of the Fourier transform is often used with a scaling $f_0 = 1/T$. More immediately, however, the scaled representation will prove useful in the next case treated.

3.4   Inverting the CTFT

The inversion of discrete time Fourier transforms is both straightforward
and easy. In both cases the inversion formula follows from the orthogonality
of complex exponentials. For the finite duration case it is their orthogonality using a sum over a set of integers. For the infinite duration case
it is their orthogonality using an integral over a unit interval. Although
the basic orthogonality idea extends to continuous time, the proofs become
significantly harder and, in fact, the details are beyond the prerequisites
assumed for this course. They are typically treated in a course on integration or real analysis. Hence we shall first summarize the inversion formulas
and conditions for their validity without proof, but we emphasize that they
are essentially just an extension of the discrete time formulas to continuous
time, that the orthogonality of exponentials is still the key idea, and that
the formulas are clear analogs to those for discrete time. In optional subsections we present some "almost proofs" that, it is hoped, provide some
additional insight into the formulas.

Just stating the inversion theorems requires some additional ideas and
notation, which we now develop.

Piecewise Smooth Signals

Let $g = \{g(t); \ t \in \mathcal{T}\}$ be a continuous time signal. g is said to have a *jump
discontinuity* at a point t if the following two limits, called the *upper limit*
and *lower limit*, respectively, exist:

$$g(t^+) \overset{\Delta}{=} \lim_{\epsilon \to 0} g(t + |\epsilon|) \tag{3.43}$$

$$g(t^-) \overset{\Delta}{=} \lim_{\epsilon \to 0} g(t - |\epsilon|), \tag{3.44}$$

but are not equal. If the limits exist and equal g at t, $g(t) = g(t^+) = g(t^-)$,
then $g(t)$ is continuous at t.

A real valued signal $\{g(t); \ t \in \mathcal{T}\}$ is said to be *piecewise continuous* on
an interval $(a, b) \subset \mathcal{T}$ if it has only a finite number of jump discontinuities
in (a, b) and if the lower limit exists at b and the upper limit exists at a.

A real valued signal $\{g(t); \ t \in \mathcal{T}\}$ is said to be *piecewise smooth* on an
interval (a, b) if its derivative $dg(t)/dt$ is piecewise continuous on (a, b).

A real valued signal $\{g(t); \ t \in \mathcal{T}\}$ is said to be *piecewise continuous*
(*piecewise smooth*) if it is piecewise continuous (piecewise smooth) for all
intervals $(a, b) \subset \mathcal{T}$. Piecewise smooth signals are a class of "nice" signals
for which an extension of Fourier inversion works.

One further detail is required before we can formally treat the inversion
of finite duration continuous time signals. Suppose that $\mathcal{T} = [0, T)$. What

if the discontinuity occurs at the origin where only the upper limit $g(t^+)$ makes sense? (We do not need to worry about the point T because we have purposefully excluded it from the time domain of definition $[0, T)$.) We somewhat arbitrarily redefine the lower limit of a signal defined on $[0, T)$ at 0 by

$$g(0^-) = \lim_{\epsilon \to 0} g(T - |\epsilon|); \qquad (3.45)$$

that is, the limit of $g(t)$ as t approaches T. This definition can be interpreted as providing the ordinary lower limit for the periodic extension $\bar{g}(t) = g(t \bmod T)$ of the finite duration signal. Alternatively, it can be considered as satisfying the ordinary definition (3.44) if we interpret addition and subtraction of time modulo T; that is, $t - \tau$ means $(t - \tau) \bmod T$. This will later be seen to be a reasonable interpretation when we consider shifts for finite duration signals.

We can now state the inversion theorems for continuous time signals. The reader should concentrate on their similarities to the discrete time analogs for the moment, the chief difference being the special treatment given to jump discontinuities in the signal.

Finite Duration Signals

Theorem 3.1 *The Finite Duration Fourier Integral Theorem*
Suppose that $g = \{g(t); \ t \in [0, T)\}$ is a finite duration continuous time signal such that

- *g is absolutely integrable; that is,*

$$\int_0^T |g(t)| \, dt < \infty, \qquad (3.46)$$

- *g is piecewise smooth.*

Define the Fourier transform by

$$G(f) = \int_0^T g(t) e^{-i 2\pi f t} \, dt, \ f \in \{k/T; \ k \in \mathcal{Z}\} = \mathcal{S}_{CTFD}^{(2)}. \qquad (3.47)$$

Then

$$\frac{g(t^+) + g(t^-)}{2} = \sum_{n=-\infty}^{\infty} \frac{G(\frac{n}{T})}{T} e^{i 2\pi \frac{n}{T} t}. \qquad (3.48)$$

If g is continuous at t, then

$$g(t) = \sum_{n=-\infty}^{\infty} \frac{G(\frac{n}{T})}{T} e^{i 2\pi \frac{n}{T} t}. \qquad (3.49)$$

As it is somewhat awkward to always write the midpoints for possible jump discontinuities, (3.48) is often abbreviated to

$$g(t) \sim \sum_{n=-\infty}^{\infty} \frac{G(\frac{n}{T})}{T} e^{i2\pi \frac{n}{T} t}. \qquad (3.50)$$

This formula means that the two sides are equal at points t of continuity of $g(t)$, but that the more complicated formula (3.48) holds if $g(t)$ has a jump discontinuity at t. With this notation we can easily summarize the theorem as stating that under suitable conditions, the following is a Fourier transform pair:

$$G(f) = \int_0^T g(t) e^{-i2\pi ft} \, dt, \ f \in \{k/T; \ k \in \mathcal{Z}\}, \qquad (3.51)$$

$$g(t) \sim \sum_{n=-\infty}^{\infty} \frac{G(\frac{n}{T})}{T} e^{i2\pi \frac{n}{T} t}; \ t \in [0, T). \qquad (3.52)$$

As an example of the inversion formula for a finite duration continuous time signal, consider the signal $\{t; \ t \in [0, 1)\}$. From the inversion formula and (2.32) we have that

$$t = \frac{1}{2} + \sum_{k \neq 0, k \in \mathcal{Z}} \frac{i}{2\pi k} e^{i2\pi kt}; \ t \in (0, 1). \qquad (3.53)$$

Note that the spectrum $\{G(k); \ k \in \mathcal{Z}\}$ is not absolutely summable, yet it has an inverse transform!

The theorem remains true if the time domain of definition is replaced by a two-sided index set of the same length, i.e., $[-T/2, T/2)$. Thus if $g = \{g(t); \ t \in [-T/2, T/2)\}$ is absolutely integrable and piecewise smooth, then we have the Fourier transform pair

$$G(f) = \int_{\frac{-T}{2}}^{\frac{T}{2}} g(t) e^{-i2\pi ft} \, dt, \ f \in \{k/T; \ k \in \mathcal{Z}\}, \qquad (3.54)$$

$$g(t) \sim \sum_{n=-\infty}^{\infty} \frac{G(\frac{n}{T})}{T} e^{i2\pi \frac{n}{T} t}; \ t \in [-\frac{T}{2}, \frac{T}{2}). \qquad (3.55)$$

As an example of inversion of a two-sided signal, consider the two-sided continuous time finite duration rectangle function $\{\sqcap(t); \ t \in [-T/2, T/2)\}$,

$T > 1$, which has discontinuities at $\pm 1/2$. From (2.35) the Fourier transform for this signal is found by restricting the frequency domain to the integer multiples of k/T, that is,

$$\{\sqcap(t);\ t \in [-T/2, T/2)\} \supset \{\text{sinc}(\frac{k}{T});\ k \in \mathcal{Z}\}.$$

Using the continuous time finite duration inversion formula then yields

$$\sqcap(t) = \sum_{k=-\infty}^{\infty} \frac{1}{T} \text{sinc}(\frac{k}{T}) e^{i2\pi t \frac{k}{T}};\ t \in [-\frac{T}{2}, \frac{T}{2}). \tag{3.56}$$

Since the rectangle function has values at discontinuities equal to the midpoints of the upper and lower limits, the Fourier inversion works. Although $\square_{1/2}(t)$ shares the same Fourier transform, (3.56) does not hold with $\square_{1/2}$ replacing $\sqcap(t)$ because the right-hand side does not agree with the $\square_{1/2}$ signal at the discontinuities. There are two common ways to handle this difficulty. The first is to modify all interesting signals with discontinuities so that their values at the discontinuities are the midpoints. The second is to simply realize that the Fourier inversion formula for continuous time signals can only be trusted to hold at times where the signal is continuous. This latter approach is accomplished by recalling the notation

$$\square_{1/2}(t) \sim \sum_{k=-\infty}^{\infty} \frac{1}{T} \text{sinc}(\frac{k}{T}) e^{i2\pi t \frac{k}{T}};\ t \in [-\frac{T}{2}, \frac{T}{2}), \tag{3.57}$$

to denote the fact that the right hand side gives the left hand side only at points of continuity. The right hand side gives the midpoints of the left hand side's upper and lower limits at points of discontinuity.

Infinite Duration Signals

Next suppose that $g = \{g(t);\ t \in \mathcal{T}\}$ is an infinite duration continuous time signal. We will focus on the most important special case of an infinite duration signal with $\mathcal{T} = \mathcal{R}$.

Theorem 3.2 *The Fourier Integral Theorem*
Given a continuous time infinite duration signal $g = \{g(t);\ t \in \mathcal{R}\}$, suppose that g is absolutely integrable and piecewise smooth. Define the Fourier transform

$$G(f) = \int_{-\infty}^{\infty} g(t) e^{-i2\pi tf}\ dt;\ f \in \mathcal{R}. \tag{3.58}$$

Then

$$\int_{-\infty}^{\infty} G(f)e^{i2\pi ft}\,df = \begin{cases} g(t) & \text{if } t \text{ is a point of continuity} \\ \frac{g(t^+)+g(t^-)}{2} & \text{otherwise.} \end{cases} \quad (3.59)$$

Note the strong resemblance of the inverse transform to that of the discrete time infinite duration case. In both cases the inversion formula is an integral of the Fourier transform times an exponential having the opposite sign than that used to define the transform. The only difference is that here the integral is over all $f \in (-\infty, \infty)$ instead of over $[-1/2, 1/2)$. Note also the resemblance of the inverse CTFT to the CTFT itself—only the sign of the exponential has been changed.

Again we summarize the theorem by saying simply that subject to suitable conditions, the following is a Fourier transform pair:

$$G(f) = \int_{-\infty}^{\infty} g(t)e^{-i2\pi ft}\,dt; \ f \in \mathcal{R}, \quad (3.60)$$

$$g(t) \sim \int_{-\infty}^{\infty} G(f)e^{i2\pi ft}; \ t \in \mathcal{R}\,df. \quad (3.61)$$

As an example of the inversion formula for an infinite duration continuous time signal, we have using Eq. 2.15 that

$$e^{-t}H(t) = \int_{-\infty}^{\infty} \frac{e^{i2\pi ft}}{1 + 2\pi i f}\,df; \ t \in \mathcal{R}. \quad (3.62)$$

This is a Fourier integral representation of the given signal.

The following subsections discuss these inversion formulas in some depth.

⋆ Inverting the Finite Duration CTFT

We begin with an educated guess on how to recover a finite duration continuous time signal $g = \{g(t); \ t \in [0, T)\}$ from its Fourier transform. Observe that we now have a finite duration continuous time signal and take an integral transform thereof. This is analogous to the earlier case where we had a finite duration continuous frequency spectrum and took an inverse Fourier transform that was an integral transform. Since there is nothing magic about the interpretation of time and frequency (mathematically they are just independent variables), we can view this situation as a dual of the previous case and guess that our transform will have discrete parameters (previously time, now frequency) and that the Fourier transform pair will look like (3.41)–(3.42) except that the roles of time and frequency will be reversed. One slight additional change is that the signs of the exponentials have also been reversed since we are switching the Fourier transform

with its inverse. Putting this together yields the guess that the appropriate
Fourier transform pair is given by

$$G(\frac{k}{T}) = \int_0^T g(t) e^{-i2\pi \frac{k}{T} t} dt; \; k \in \mathcal{Z} \tag{3.63}$$

$$g(t) = \sum_{n=-\infty}^{\infty} \frac{G(\frac{n}{T})}{T} e^{i2\pi \frac{n}{T} t}; \; t \in [0, T). \tag{3.64}$$

Note that as in the DFT inversion formula, we again have the original
signal represented as a Fourier series. It is not expected that the above pair
should be obvious given the corresponding result for the DTFT, only that
it should be a plausible guess. We now prove that it is in fact equivalent
to the DTFT result. To do this we make the substitutions of Table 3.1 in
the scaled-frequency DTFT Fourier transform pair.

Eq. 3.42	Eq. 3.47
f (frequency)	t (time)
$H(f)$ (spectrum)	$g(t)$ (signal)
f_0	T
h_n (signal)	$\frac{1}{T} G(-\frac{n}{T})$ (spectrum)

Table 3.1: Duality of the Infinite Duration DTFT and the Finite Duration
CTFT

With these definitions (3.41) becomes

$$g(t) = \sum_{n=-\infty}^{\infty} \frac{G(-\frac{n}{T})}{T} e^{-i2\pi \frac{n}{T} t}.$$

Changing the summation dummy variable sign yields

$$g(t) = \sum_{n=-\infty}^{\infty} \frac{G(\frac{n}{T})}{T} e^{i2\pi \frac{n}{T} t}.$$

Similarly (3.42) becomes

$$\frac{G(-\frac{n}{T})}{T} = \int_0^T \frac{g(t)}{T} e^{i2\pi \frac{t}{T} n} dt.$$

Thus evaluating the formula at $-n$ yields

$$G(\frac{n}{T}) = \int_0^T g(t)e^{-i2\pi\frac{n}{T}t}\,dt,$$

which shows that (3.63)–(3.64) follow from Eqs. (3.41)–(3.42).

As with the DTFT we have swept some critical details under the rug. The Fourier transform pair formula does not hold for all possible finite duration continuous time signals $g(t)$; they must be "nice" enough for the arguments used in the proof (the exchange of sums and integrals, for example) to hold. Sufficient conditions for (3.63)–(3.64) to hold were first developed by Dirichlet and were named after him. They strongly resemble those that will be encountered in the next section on the infinite duration CTFT. While the proof of these conditions is beyond the assumed prerequisites of this course, we can describe sufficient conditions for the theorem to hold and sketch some of the ideas in the proof.

Roughly speaking, the sufficient conditions for the inversion formula (3.64) to hold are threefold:

1. $g(t)$ is absolutely integrable,

2. $g(t)$ cannot "wiggle" too much in a way made precise below, and

3. $g(t)$ is a continuous function of t.

The continuity condition can be dropped at the expense of complicating the inversion formula.

The condition that we have adopted to constrain the wiggliness of g is that it be piecewise smooth. Other, similar, conditions are also seen in the literature. Perhaps the most common is that of bounded variation, that g can have at most a finite number of maxima and minima in any finite length interval.

In order to sketch a partial proof of the theorem that will strongly resemble the continuous time infinite duration, we consider the behavior of the partial sums

$$g_N(t) = \sum_{n=-N}^{N} \frac{G(\frac{n}{T})}{T}e^{i2\pi\frac{n}{T}t}$$

as $N \to \infty$. We have that

$$g_N(t) = \sum_{n=-N}^{N} \frac{1}{T}(\int_0^T g(\tau)e^{-i2\pi\tau\frac{n}{T}}\,d\tau)e^{i2\pi\frac{n}{T}t}$$

$$= \frac{1}{T}\int_0^T g(\tau)(\sum_{n=-N}^{N} e^{i2\pi(t-\tau)\frac{n}{T}})\,d\tau. \qquad (3.65)$$

The internal sum can be evaluated using (2.29) as

$$\sum_{n=-N}^{N} e^{i2\pi(t-\tau)\frac{n}{T}} = \frac{\sin(2\pi(\frac{t-\tau}{T})(N+\frac{1}{2}))}{\sin(\pi(\frac{t-\tau}{T}))} \tag{3.66}$$

so that

$$g_N(t) = \int_0^T g(\tau) \frac{\sin(2\pi(\frac{t-\tau}{T})(N+\frac{1}{2}))}{T\sin(\pi(\frac{t-\tau}{T}))} d\tau. \tag{3.67}$$

What happens as $N \to \infty$? The term multiplying the signal inside the integral can be expressed in terms of the *Dirichlet kernel* defined by

$$D_N(t) = \frac{\sin(\pi t(N+\frac{1}{2}))}{\sin(\pi t)}, \quad t \in [-\frac{1}{2}, \frac{1}{2}), \tag{3.68}$$

which should be familiar if we change the t to f as the DTFT of the box function $\{\Box_N(n);\ n \in \{-N, \ldots, N\}\}$. The Dirichlet kernel is symmetric about the origin and has unit integral. To see this latter fact, observe that the Fourier inversion for the DTFT of the box function evaluated at $n = 0$ yields

$$\int_{-\frac{1}{2}}^{\frac{1}{2}} \frac{\sin(\pi f(N+\frac{1}{2}))}{\sin(\pi f)} df = \Box_N(0) = 1. \tag{3.69}$$

(This is an example of a *moment property* as will be considered in Chapter 4.) We can now express (3.67) as

$$g_N(t) = \int_0^T g(\tau) \frac{1}{T} D_N(\frac{t-\tau}{T}) d\tau. \tag{3.70}$$

We point out in passing that the right hand integral is an example of a *convolution integral,* an operation that we will consider in some detail in Chapter 6. $D_N(t)$ is plotted for three values of N in Figure 3.1. As N increases, the signal becomes more and more concentrated around the origin. Its height at the origin grows without bound and its total area of 1 is in an increasingly narrow region around the origin. The signal $T^{-1}D_N(\frac{t-\tau}{T})$; $t \in [-\frac{T}{2}, \frac{T}{2})$ will behave similarly, except that its peak will be at $t = \tau$. Multiplying $T^{-1}D_N(\frac{t-\tau}{T})$ by $g(\tau)$ and integrating with N large will result in very little contribution being made to the integral except for $\tau \approx t$. If $g(t)$ is continuous at t, the fact that the integral of $T^{-1}D_N(\frac{t-\tau}{T})$ equals 1 implies that the integral of (3.70) will be approximately $g(t)$. If $g(t)$ has a jump discontinuity, then the approximation will be replaced by $1/2$ the upper limit plus $1/2$ the lower limit, yielding the midpoint of the upper and lower limits.

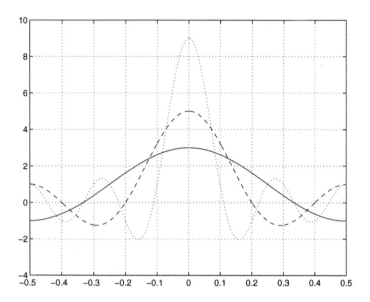

Figure 3.1: The Dirichlet Kernel $D_N(t)$: $N{=}1$ (solid line), 2 (dashed line), 4 (dash-dot line)

Making this line of argument precise, that is, actually proving that $g_N(t) \to g(t)$ as $N \to \infty$ would require the Riemann-Lebesgue Lemma, one of the fundamental results of Fourier series. Alternatively, it is proving that the limit of the Dirichlet kernels behave like a generalized function – the Dirac delta to be considered in Chapter 5. We here will content ourselves with the above intuitive argument, which is reinforced by the similarity of the result to the discrete time inversion formulas, which were proved in some detail.

The inversion formula can be extended from absolutely integrable signals to finite energy signals by considering the infinite sum to be a limit in the mean. In particular, it can be shown that if the signal g has finite energy, then it is true that

$$\lim_{N \to \infty} \int_0^T |g(t) - \sum_{n=-N}^{N} \frac{G(\frac{n}{T})}{T} e^{i2\pi \frac{n}{T} t}|^2 \, dt = 0. \qquad (3.71)$$

Note that this form of inversion says nothing about pointwise convergence of the sum to $g(t)$ for a particular value of t and is not affected by the values of $g(t)$ at simple isolated discontinuities.

⋆ Inverting the Infinite Duration CTFT

Next consider a continuous time signal $g = \{g(t); \ t \in \mathcal{R}\}$ with Fourier transform $G = \mathcal{F}(g)$ defined by

$$G(f) = \int_{-\infty}^{\infty} g(t)e^{-i2\pi tf} \, dt \tag{3.72}$$

for all real f.

We now sketch a proof of the Fourier integral theorem in a way resembling that for the finite duration continuous time inversion formula. Consider the function

$$g_a(t) = \int_{-a}^{a} G(f)e^{i2\pi ft} \, df \tag{3.73}$$

formed by truncating the region of integration in the inverse Fourier transform in a symmetric fashion. We will argue that under the assumed conditions,

$$\lim_{a \to \infty} g_a(t) = \frac{g(t^+) + g(t^-)}{2};$$

that is, that the integral

$$\int_{-\infty}^{\infty} G(f)e^{i2\pi ft} \, df$$

exists (in a Cauchy principal value sense) and has the desired value. As in the DFT and DTFT case, we begin by inserting the formula for $G(f)$ into the (truncated) inversion formula:

$$g_a(t) = \int_{-a}^{a} \left(\int_{-\infty}^{\infty} g(x)e^{-i2\pi fx} \, dx \right) e^{i2\pi ft} \, df. \tag{3.74}$$

If the function is suitably well behaved, we can change the order of integration:

$$g_a(t) = \int_{-\infty}^{\infty} dx g(x) \left(\int_{-a}^{a} e^{i2\pi f(t-x)} \, df \right) \tag{3.75}$$

The integral in parentheses is easily evaluated as

$$
\begin{aligned}
\frac{e^{i2\pi f(t-x)}}{i2\pi(t-x)}\Big|_{-a}^{a} &= \frac{\sin(2\pi a(t-x))}{\pi(t-x)} \\
&= 2a \operatorname{sinc} 2a(t-x) \\
&= \frac{\sin 2a(t-x)\pi}{(t-x)\pi} \tag{3.76} \\
&= F_a(t-x), \tag{3.77}
\end{aligned}
$$

where

$$F_a(t) = 2a \operatorname{sinc} 2at = \frac{\sin 2at\pi}{t\pi} \qquad (3.78)$$

is called the *Fourier integral kernel* and is simply the CTFT of the box function $\{\Box_T(t);\ t \in \mathcal{R}\}$ with t substituted for f. Like the Dirichlet kernel it is symmetric about the origin and has unit integral. Thus

$$g_a(t) = \int_{-\infty}^{\infty} g(x) F_a(t - x)\, dx. \qquad (3.79)$$

Once again we observe that the right hand integral is an example of a *convolution integral*. Figure 3.2 shows the Fourier integral kernel for several values of a. As with the Dirichlet kernel encountered in the finite duration continuous time case, the sinc functions become increasingly concentrated around their center as a becomes large. Thus essentially the same argument

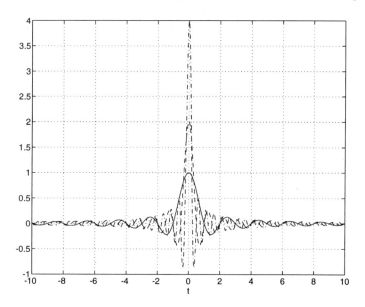

Figure 3.2: The Fourier Kernel $F_a(t)$: $a=1$ (solid line), 2 (dashed line), 4 (dash-dot line)

yields the inversion formula as in the finite duration case. The Riemann-Lebesgue Lemma implies that the limit of this function as $a \to \infty$ is as stated in the theorem under the given conditions.

We also consider an alternative intuitive "proof" of the Fourier integral theorem for the special case where the signal is continuous. The proof is a

simple limiting argument based on the finite duration inversion of (3.54)–
(3.55). Unfortunately the proof cannot easily be made rigorous and hence
is presented only as an interpretation.

Suppose that in (3.54)–(3.55) we let $T \to \infty$ in order to extend the
Fourier transform pair to infinite duration continuous time signals. In order
to do this define a frequency variable $f = n/T$. With these definitions (3.54)
becomes in the limit

$$G(f) = \lim_{T\to\infty} \int_{-\frac{T}{2}}^{\frac{T}{2}} g(t)e^{-i2\pi ft}\,dt = \int_{-\infty}^{\infty} g(t)e^{-i2\pi ft}\,dt,$$

the usual CTFT, and (3.55) can be considered as a Riemann sum ($1/T$
becomes df) which in the limit as $T \to \infty$ approximates the integral

$$
\begin{aligned}
g(t) &= \lim_{T\to\infty} g_T(t) \\
&= \lim_{T\to\infty} \sum_{k=-\infty}^{\infty} \frac{G(\frac{k}{T})}{T} e^{i2\pi \frac{k}{T} t} \\
&= \int_{-\infty}^{\infty} G(f)e^{i2\pi ft}\,df,
\end{aligned}
\tag{3.80}
$$

which is the claimed inverse transform. This "proof" at least points out
that the form of the infinite duration CTFT is consistent with that for the
finite duration CTFT. Thus the Fourier integral transform relations can be
viewed as a limiting form of the Fourier series relations for continuous time
signals. This is not, however, the way that such results are properly proved.
This is not a rigorous proof because the Riemann integrals are improper
(have infinite limits) and there is no guarantee that the Riemann sums will
converge to the integrals. Furthermore, the improper Riemann integrals in
the argument have only been considered in a Cauchy principal value sense.

3.5 Continuous Time Fourier Series

As in the discrete time finite duration case, the Fourier inversion formula
provides a representation for continuous time finite duration signals as a
weighted sum of complex exponentials, a Fourier series. For simplicity we
now assume that $g(t)$ is absolutely integrable and piecewise smooth. The
Fourier transform pair of (3.63)–(3.64) can also be written in the form

$$g(t) \sim \sum_{k=-\infty}^{\infty} c_k e^{i2\pi \frac{t}{T}k}; \quad t \in [0, T) \tag{3.81}$$

$$c_k = \frac{1}{T} \int_0^T g(t)e^{-i2\pi \frac{t}{T}k}\,dt. \tag{3.82}$$

As usual, the $\sim$ notation means equality if $g(t)$ is continuous at t and the midpoint of the upper and lower limits otherwise. This is a very common form for a Fourier series of a finite duration continuous time signal analogous to the discrete time Fourier series of (3.26)–(3.27). Note that all we have done is replace $G(\frac{k}{T})/T$ by c_k (which are also called Fourier coefficients). In some applications the exponential is expanded using Euler's formulas to form the *trigonometric series*

$$g(t) \sim \frac{a_0}{2} + \sum_{k=1}^{\infty} (a_k \cos(2\pi k \frac{t}{T}) + b_k \sin(2\pi k \frac{t}{T})). \qquad (3.83)$$

The a_k and b_k can be determined from the c_k and vice-versa. The details are left as an exercise.

As yet another form of this relation, consider the case where $\mathcal{T} = [-T/2, T/2)$ and replace c_k by d_k/T:

$$g(t) \quad \sim \quad \sum_{k=-\infty}^{\infty} \frac{d_k}{T} e^{i2\pi \frac{t}{T}k}; \; t \in [-T/2, T/2) \qquad (3.84)$$

$$d_k \quad = \quad \int_{-\frac{T}{2}}^{\frac{T}{2}} g(t) e^{-i2\pi \frac{t}{T}k} \, dt; \; k \in \mathcal{Z}. \qquad (3.85)$$

As a final observation, as with the inversion of the finite duration discrete time signal, the inversion formula for a finite duration continuous time signal is perfectly well defined for all real t and not just for the original $t \in [0, T)$. Also, as in the DFT case, the formula is periodic, now having period T. Thus we automatically have a Fourier series representation for the periodic extension $\tilde{g}$ of g:

$$\tilde{g}(t) \sim \sum_{k=-\infty}^{\infty} c_k e^{i2\pi \frac{t}{T}k}; \; t \in \mathcal{R}, \qquad (3.86)$$

$$c_k = \frac{1}{T} G(\frac{k}{T}) = \frac{1}{T} \int_{-\frac{T}{2}}^{\frac{T}{2}} g(t) e^{-i2\pi k \frac{t}{T}} \, dt; \; k \in \mathcal{Z}. \qquad (3.87)$$

Alternatively, given a continuous time periodic signal $\tilde{g}$ we can find a Fourier series as above using the Fourier transform of the truncated finite duration signal consisting of only a single period. As an example, consider the periodic signal $\{g(t) = t \bmod 1; \; t \in \mathcal{R}\}$. This has the Fourier series representation

$$g(t) \sim \sum_{k=-\infty}^{\infty} c_k e^{i2\pi \frac{t}{T}k}$$

with $T = 1$ and $c_0 = 1/2$ and $c_k = i/2\pi k$ for $k \neq 0$. Thus at all points of continuity of $t \bmod 1$ (all points except those where t is an integer) we have

$$t \bmod 1 = \frac{1}{2} + \sum_{k \neq 0, k \in \mathcal{Z}} \frac{i}{2\pi k} e^{i2\pi tk}; \quad t \in \mathcal{R}, t \notin \mathcal{Z}. \tag{3.88}$$

From the continuous time finite duration inversion formula this series will have a value of $1/2$ at the discontinuities which occur at integer t.

This series can be put in a somewhat simpler form by observing that the sum can be rewritten as

$$\sum_{k=1}^{\infty} \frac{i}{2\pi k} (e^{i2\pi tk} - e^{-i2\pi tk}) = -\sum_{k=1}^{\infty} \frac{\sin(2\pi kt)}{\pi k}$$

yielding the Fourier series

$$t \bmod 1 \sim \frac{1}{2} - \sum_{k=1}^{\infty} \frac{\sin(2\pi kt)}{\pi k}. \tag{3.89}$$

A simple application of Fourier series yields a famous equality involving Bessel functions that crops up regularly in the study of nonlinear systems. Consider the signal

$$g(\psi) = e^{iz \sin \psi}$$

for real ψ. The quantity z can be considered as a parameter. This signal is periodic in ψ with period 2π and hence can be expanded into a Fourier series

$$e^{iz \sin \psi} = \sum_{k=-\infty}^{\infty} c_k e^{i2\pi \frac{\psi}{2\pi} k} = \sum_{k=-\infty}^{\infty} c_k e^{i\psi k}$$

where

$$c_k = \frac{1}{2\pi} \int_0^{2\pi} e^{iz \sin \psi} e^{-i\psi k} \, d\psi = J_k(z), \tag{3.90}$$

the kth order Bessel function of (1.12). Thus we have the expansion

$$e^{iz \sin \psi} = \sum_{k=-\infty}^{\infty} J_k(z) e^{i\psi k}, \tag{3.91}$$

a result called the Jacobi-Anger expansion.

⋆ Continuous time Fourier Series and the DFT

We previously saw that the DFT could be used to approximate the Fourier transform of a continuous time signal. We here carry this discussion a bit further to point out that the DFT can be used to approximate the Fourier series for a continuous time function.

Suppose that $\{g(t); \ t \in [0, T)\}$ is a continuous time finite duration signal with Fourier transform G and hence that we have a Fourier series representation

$$g(t) = \frac{1}{T} \sum_{k=-\infty}^{\infty} G(\frac{k}{T}) e^{i2\pi \frac{k}{T} t}.$$

We can approximate the spectrum by the DFT of a sampled waveform $\{\hat{g}_l = g(lT/N); \ l = 0, \cdots, N-1\}$ using Riemann sum approximations as

$$
\begin{aligned}
G(\frac{k}{T}) &= \int_0^T g(t) e^{-i2\pi \frac{k}{T} t} \, dt \\
&\approx \frac{T}{N} \sum_{l=0}^{N-1} g(\frac{lT}{N}) e^{-i2\pi l \frac{k}{N}} \\
&= \frac{T}{N} \hat{G}(\frac{k}{N}), \ k \in \mathcal{Z}_N
\end{aligned}
$$

where $\hat{G}$ is the DFT of the sequence $\{\hat{g}_l; \ l \in \mathcal{Z}_N\}$. Approximating the remaining spectral terms by 0, we have the approximation

$$g(t) \approx \frac{1}{N} \sum_{k=0}^{N-1} \hat{G}(\frac{k}{N}) e^{i2\pi \frac{k}{T} t}, \tag{3.92}$$

which should be increasingly accurate as N becomes large. This is the way in which a digital computer approximately computes the Fourier series of a continuous time signal. (It must also digitize the samples, but that consideration is beyond the scope of this discussion.)

3.6 Duality

When inferring the inversion formula for the finite duration continuous time Fourier transform from that for the infinite duration discrete time Fourier transform, we often mentioned the *duality* of the two cases: interchanging the role of time and frequency turned one Fourier transform pair into another. We now consider this idea more carefully.

Knowing one transform often easily gives the result of a seemingly different transform. In fact, we have already taken advantage of this duality in proving the finite duration continuous time Fourier inversion formula by viewing it as a dual result to the infinite duration discrete time inversion result.

Suppose, for example, you know that the Fourier transform of an infinite duration continuous time signal $g = \{g(t); t \in \mathcal{R}\}$ is $G(f)$, e.g., we found that the transform of $g(t) = e^{-t}H(t)$ is $G(f) = (1 + i2\pi f)^{-1}$. Now suppose that you are asked to find the Fourier transform of the continuous time signal $r(t) = (1 + i2\pi t)^{-1}$. This is easily done by noting that $r(t) = G(t)$; that is, r has the same functional dependence on its argument that G has. We know, however, that the inverse Fourier transform of $G(f)$ is $g(t)$ and that the inverse Fourier transform in the infinite duration continuous time case is identical to a Fourier transform except for the sign of the exponent. Putting these facts together we know that the Fourier transform of $r(t) = G(t)$ will be $g(-f)$. The details might add some insight: Given $g(t)$ and $G(f)$, then the inversion formula says that

$$
g(t) \;=\; \int_{-\infty}^{\infty} G(f) e^{i2\pi ft}\, df
$$

$$
\;=\; \int_{-\infty}^{\infty} G(\alpha) e^{i2\pi \alpha t}\, d\alpha,
$$

where the name of the dummy variable is changed to help minimize confusion when interchanging the roles of f and t. Now if $r(t) = G(t)$, its Fourier transform is found using the previous formula to be

$$
R(f) \;=\; \int_{-\infty}^{\infty} r(t) e^{-i2\pi tf}\, dt = \int_{-\infty}^{\infty} G(t) e^{-i2\pi tf}\, dt
$$

$$
\;=\; \int_{-\infty}^{\infty} G(\alpha) e^{i2\pi \alpha (-f)}\, d\alpha = g(-f).
$$

To summarize:

If $\{g(t); \; t \in \mathcal{R}\}$ is an infinite duration continuous time signal with Fourier transform

$$
\mathcal{F}_f(\{g(t); \; t \in \mathcal{R}\}) = G(f); \; f \in \mathcal{R},
$$

then the Fourier transform of the infinite duration continuous time signal $\{G(t); \ t \in \mathcal{R}\}$ is

$$\mathcal{F}_f(\{G(t); \ t \in \mathcal{R}\}) = g(-f); \ f \in \mathcal{R}.$$

Thus for example, since

$$\mathcal{F}_f(\{e^{-t}H(t); \ t \in \mathcal{R}\}) = \frac{1}{1 + 2\pi i f},$$

then also

$$\mathcal{F}_f(\{\frac{1}{1 + 2\pi i t}; \ t \in \mathcal{R}\}) = e^f H(-f); \ f \in \mathcal{R}. \tag{3.93}$$

Similarly,

$$\mathcal{F}_f(\{\text{sinc}(t); \ t \in \mathcal{R}\}) = \sqcap(f); \ f \in \mathcal{R}. \tag{3.94}$$

Almost the same thing happens in the DFT case because both transforms have the same form; that is, they are both sums. Thus with suitable scaling, every DFT transform pair has its dual pair formed by properly reversing the role of time and frequency as above.

The finite duration continuous time and infinite duration discrete time are a little more complicated because the transforms and inverses are not of the same form: one is a sum and the other an integral. Note, however, that every transform pair for a finite duration continuous time signal has as its dual a transform pair for an infinite duration discrete time signal given appropriate scaling. Suppose, for example, that we have an infinite duration discrete time signal $\{g_n; \ n \in \mathcal{Z}\}$ with Fourier transform $\{G(f); \ f \in [0, 1)\}$ and hence

$$g_n = \int_0^1 G(f) e^{i2\pi f n} \ df. \tag{3.95}$$

As a specific case, suppose that $g_n = r^n u_{-1}(n); \ n \in \mathcal{Z}$ and hence $G(f) = (1 - r^{-i2\pi f})^{-1}; \ f \in [0, 1)$. Suppose that $\{x(t); \ t \in [0, T)\}$ is a finite duration continuous time signal defined by

$$x(t) = G(\frac{t}{T}); t \in [0, T)$$

(where now the scaling is needed to permit a time domain more general than $[0, 1)$). In the specific example the waveform becomes

$$x(t) = \frac{1}{1 - r^{-i2\pi \frac{t}{T}}}; \ t \in [0, T).$$

What is the Fourier transform $X(f)$ of x? Since the transform of a finite duration continuous time signal has a discrete frequency domain, we need to find

$$
\begin{aligned}
X(\frac{k}{T}) &= \int_0^T x(t)e^{-i2\pi \frac{k}{T}t}\, dt \\
&= \int_0^T x(\alpha)e^{-i2\pi \frac{k}{T}\alpha}\, d\alpha \\
&= \int_0^T G(\frac{\alpha}{T})e^{-i2\pi \frac{k}{T}\alpha}\, d\alpha \\
&= T\int_0^1 G(\beta)e^{-i2\pi k\beta}\, d\beta,
\end{aligned}
$$

where we have made the variable substitution $\beta = \alpha/T$. Combining this equation with (3.95) we have the duality result

$$
X(\frac{k}{T}) = Tg_{-k},
$$

which also has the flavor of the infinite duration continuous time result except that the additional scaling is required. We can summarize this duality result as follows.

If $\{g_n;\ n \in \mathcal{Z}\}$ is an infinite duration discrete time signal with Fourier transform

$$
\mathcal{F}_f(\{g_n;\ n \in \mathcal{Z}\}) = G(f);\ f \in [0,1),
$$

then the Fourier transform of the finite duration continuous time signal $\{G(t/T);\ t \in [0,T)\}$ is

$$
\mathcal{F}_{k/T}(\{G(t/T);\ t \in [0,T)\}) = Tg_{-k};\ k \in \mathcal{Z}.
$$

In our example this becomes

$$
X(\frac{k}{T}) = Tr^{-k}u_{-1}(-k) = \begin{cases} Tr^{|k|}; & k = 0, -1, -2, \cdots \\ 0 & \text{otherwise.} \end{cases}
$$

These simple tricks allow every transform pair we develop to play a double role.

3.7   Summary

The Fourier transform pairs for the four cases considered are summarized in Table 3.2. The $\sim$ notation in the continuous time inversion formulas can be changed to $=$ if the signals are continuous at t. Note in particular the ranges of the time and frequency variables for the four cases: both are discrete in the finite duration discrete time case (the DFT) and both are continuous in the infinite duration continuous time case. In the remaining two cases one parameter is continuous and the other discrete and these two cases can be viewed as duals in the sense that the roles of time and frequency have been reversed. These results assume that the transforms and inverses exist and that the functions of continuous parameters are piecewise continuous. Note also that the finite duration inversion formulas both involve a normalization by the length of the duration, a normalization not required in the infinite duration formulas. In many treatments, this normalization constant is divided between the Fourier and inverse Fourier transforms to make them more symmetric, e.g., both transform and inverse transform incorporate a normalization factor of $1/\sqrt{N}$ (discrete time) or $1/\sqrt{T}$ (continuous time). Such changes of the definitions by a constant do not affect any of the theory, but one should be consistent. If one uses a stretched frequency scale (instead of $[0, 1)$ or $[-1/2, 1/2)$) for the infinite duration discrete time case, then one needs a scaling of $1/S$ in the inversion formula, where S is the length of the frequency domain.

It is also informative to consider a similar table describing the nature of the frequency domains for the various signal types. This is done in Table 3.3. The focus is on the two attributes of the frequency domain $\mathcal{S}$. As with the time domain $\mathcal{T}$, we have seen that the frequency can be discrete or continuous. The table points out that also like the time domain, the frequency domain can be "finite duration" or "infinite duration" in the sense of being defined for a time interval of finite or infinite length. We dub these cases *finite bandwidth* and *infinite bandwidth* and observe that continuous time signals yield infinite bandwidth frequency domains and discrete time signals yield finite bandwidth frequency domains. These observations add to the duality of the time and frequency domains and between signals and spectra: as there are four basic signal types, there are also four basic spectra types. Lastly observe that discrete time finite duration signals yield discrete frequency finite bandwidth spectra and continuous time infinite duration signals yield continuous frequency infinite bandwidth spectra. Thus in these cases the behavior of the time and frequency domains are the same. On the other hand, continuous time finite duration signals yield discrete frequency infinite bandwidth spectra and discrete time infinite duration signals yield continuous frequency finite bandwidth spectra. In these cases the time and frequency domain behaviors are reversed.

Duration	Time	
	Discrete	Continuous
Finite	$G(\frac{k}{N}) = \sum\limits_{n=0}^{N-1} g_n e^{-i2\pi \frac{k}{N}n}; \ k \in \mathcal{Z}_N$ $g_n = \frac{1}{N} \sum\limits_{k=0}^{N-1} G(\frac{k}{N}) e^{i2\pi \frac{k}{N}n}; \ n \in \mathcal{Z}_N$	$G(\frac{k}{T}) = \int\limits_{0}^{T} g(t) e^{-i2\pi \frac{k}{T}t} \, dt; \ k \in \mathcal{Z}$ $g(t) \sim \sum\limits_{k=-\infty}^{\infty} \frac{G(\frac{k}{T})}{T} e^{i2\pi \frac{k}{T}t}; \ t \in [0,T)$
Infinite	$G(f) = \sum\limits_{n=-\infty}^{\infty} g_n e^{-i2\pi \frac{f}{S}n}; \ f \in [0, S]$ $g_n = \int\limits_{0}^{S} \frac{1}{S} G(f) e^{i2\pi \frac{f}{S}n} \, df; \ n \in \mathcal{Z}$	$G(f) = \int\limits_{-\infty}^{\infty} g(t) e^{-i2\pi tf} \, dt; \ f \in \mathcal{R}$ $g(t) \sim \int\limits_{-\infty}^{\infty} G(f) e^{i2\pi ft} \, df; \ t \in \mathcal{R}$

Table 3.2: Fourier Transform Pairs

As with the terminology "finite duration," there is potential confusion over the name "finite bandwidth" which could describe either a finite length domain of definition for the spectrum (as above) or a spectrum with infinite length domain of definition which just happens to be zero outside a finite length interval, e.g., a spectrum of an infinite duration signal which is zero outside of $[-W, W]$. We will refer to the latter case as a "band-limited" spectrum and the corresponding signal as a band-limited signal. This is analogous to our use of the term "time-limited" for infinite duration signals which are zero outside of some finite interval.

Duration	Time	
	Discrete	Continuous
Finite	Discrete Frequency Finite Bandwidth	Discrete Frequency Infinite Bandwidth
Infinite	Continuous Frequency Finite Bandwidth	Continuous Frequency Infinite Bandwidth

Table 3.3: Frequency Domain

3.8 ⋆ Orthonormal Bases

Many basic properties of the Fourier transform can be generalized by replacing complex exponentials by more general functions with similar properties.

We demonstrate this for the special case of discrete time finite duration signals. The case for continuous time finite duration signals is considered in the exercises and similar ideas extend to infinite duration signals. In the next section we consider an important special case, the discrete wavelet transform.

We here confine interest to signals of the form $g = \{g(n); \ n = 0, 1, \ldots, N-1\}$, where $N = 2^L$ for some integer L. Let $\mathcal{G}_N$ denote the collection of all such signals, that is, the space of all real valued discrete time signals of duration N.

A collection of signals $\psi_k = \{\psi_k(n); n = 0, 1, \ldots, N-1\}; k = 0, 1, \ldots, K-1$, is said to form an *orthonormal basis* for the space $\mathcal{G}_N$ if (1) the signals are *orthonormal* in the sense that

$$\sum_{n-0}^{N-1} \psi_k(n)\psi_l^*(n) = \delta_{k-l}, \qquad (3.96)$$

and (2), the set of signals is *complete* in the sense that any signal $g \in \mathcal{G}$ can be expressed in the form

$$g(n) = \sum_{k=0}^{K-1} a_k\psi_k(n). \qquad (3.97)$$

It follows from (3.9) that the discrete time exponentials

$$\psi_k = \{\frac{e^{i2\pi \frac{k}{N} n}}{\sqrt{N}}; \ n \in \mathcal{Z}_N\}, \ k = 0, \ldots, N-1, \qquad (3.98)$$

are orthonormal and the DFT inversion formula (3.14) guarantees that any signal $g \in \mathcal{G}_N$ can be written as a weighted sum of the ψ_k and hence the complex exponentials indeed form an orthonormal basis. Thus the idea of an orthonormal basis can be viewed as a generalization of the Fourier transform and its inverse.

The smallest integer K for which there exists an orthonormal basis for a space is called the *dimension* of the space. In the case of $\mathcal{G}_N$ the dimension is N. While we will not actually prove this, it should be believable since the Fourier example proves the dimension is not more than N.

The general case mimics the Fourier example in another way. If we wish to compute the linear weights a_k in the expansion, observe that

$$\sum_{n=0}^{N-1} \psi_l^*(n)g(n) \quad = \quad \sum_{n=0}^{N-1} \psi_l^*(n) \sum_{k=0}^{N-1} a_k\psi_k(n)$$

$$= \sum_{k=0}^{N-1} a_k \sum_{n=0}^{N-1} \psi_l^*(n)\psi_k(n)$$

$$= \sum_{k=0}^{N-1} a_k \delta_{l-k} = a_l$$

using the orthonormality. Thus the "coefficients" a_k are calculated in general in the same way as in the Fourier special case: multiply the signal by ψ_l^* and sum over time to get a_l. This is sometimes abbreviated using the *inner product* or *scalar product* notation

$$< g, \psi_k > \overset{\Delta}{=} \sum_{n=0}^{N-1} g(n)\psi_k^*(n) \qquad (3.99)$$

and we can write the general expansion as

$$g(n) = \sum_{k=0}^{K-1} < g, \psi_k > \psi_k(n). \qquad (3.100)$$

The inner product is often denoted by (g, ψ_k) as well as $< g, \psi_k >$.

The complex exponentials are not the only possible basis. A simpler basis is given by the shifted Kronecker delta functions $\psi_k(n) = \delta_{n-k}$.

Whenever we have an orthonormal basis we can think of the resulting coefficients $\{a_k; k = 0, \ldots, N - 1\}$ as forming the transform of the signal with respect to the basis, and (3.100) provides the inverse transform. Eq. (3.100) is sometimes called a *generalized Fourier series*.

3.9 ★ Discrete Time Wavelet Transforms

In this section we consider an alternative transform that has gained wide popularity in a variety of fields during recent years — the wavelet transform. For simplicity we consider only the special case of discrete time finite duration signals and we only provide a superficial treatment showing some of the similarities and differences with Fourier transforms. The interested reader can pursue the subject in detail in Daubechies [14], Rioul and Vetterli [29], Strang [31], or the references cited in these surveys.

A wavelet transform is a form of orthonormal transform with specific properties. In fact, it need not even be orthonormal, but we will only consider this important special case. A key idea is to construct all of the basis functions from a single continuous time signal called a *wavelet*. Let

$\psi(t)$; $t \in \mathcal{R}$ be a continuous time signal and consider the collection of $N = 2^L$ discrete time signals defined in terms of ψ as follows:

$$
\begin{aligned}
\psi_{0,0}(n) &= 2^{-\frac{L}{2}};\ n \in \mathcal{Z}_N \\
\psi_{m,k}(n) &= 2^{-\frac{m}{2}}\psi(2^{-m}n - k);\ n \in \mathcal{Z}_N; \\
&\quad m = 1, \ldots, L;\ k = 0, 1, \ldots, 2^{L-m} - 1.
\end{aligned}
\tag{3.101}
$$

Let $\mathcal{K}$ denote the set of all possible indices (m, k) specified in the above collection.

With the exception of the $(0,0)$ signal, these signals are all formed by *dilations* and *shifts* of the basic function ψ. m is called the dilation parameter and k is called the shift parameter. We say that $\psi(t)$ is a *wavelet* or *mother wavelet* if the collection $\{\psi_{m,k};\ (m, k) \in \mathcal{K}\}$ form an orthonormal basis for $\mathcal{G}_N$. (This is not the usual definition, but it will suit our purposes.) If this is the case, then we can expand any signal $g \in \mathcal{G}_N$ in a series of the form

$$
g(n) = \sum_{(m,k)\in\mathcal{K}} a_{m,k}\psi_{m,k}(n),
\tag{3.102}
$$

where $a_{m,k} =< g, \psi_{m,k} >$.

The first question is whether or not the definition makes sense, i.e., if functions ψ having this property exist. This is easily demonstrated by a basic example. Consider the function

$$
\psi(t) = \begin{cases} 1 & 0 \le t < 1/2 \\ -1 & 1/2 \le t < 1 \\ 0 & \text{otherwise} \end{cases}
\tag{3.103}
$$

depicted in Figure 3.3

In this case ψ is called the *Haar wavelet*, but the function and its basic properties go back to the early part of this century, long before it and its properties were unified with many previously disparate techniques and applications under the general topic of wavelets. For the special case where $L = 3$ and $N = 2^L = 8$, we have the following functions:

$$
\begin{aligned}
\psi_{1,0} &= \frac{1}{\sqrt{2}}(1, -1, 0, 0, 0, 0, 0, 0) \\
\psi_{1,1} &= \frac{1}{\sqrt{2}}(0, 0, 1, -1, 0, 0, 0, 0) \\
\psi_{1,2} &= \frac{1}{\sqrt{2}}(0, 0, 0, 0, 1, -1, 0, 0) \\
\psi_{1,3} &= \frac{1}{\sqrt{2}}(0, 0, 0, 0, 0, 0, 1, -1)
\end{aligned}
$$

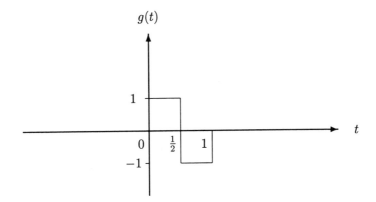

Figure 3.3: The Haar Wavelet

$$\psi_{2,0} = \frac{1}{\sqrt{4}}(1, 1, -1, -1, 0, 0, 0, 0)$$

$$\psi_{2,1} = \frac{1}{\sqrt{4}}(0, 0, 0, 0, 1, 1, -1, -1)$$

$$\psi_{3,0} = \frac{1}{\sqrt{8}}(1, 1, 1, 1, -1, -1, -1, -1)$$

$$\psi_{0,0} = \frac{1}{\sqrt{8}}(1, 1, 1, 1, 1, 1, 1, 1).$$

These signals are easily seen to be orthogonal. The fact that they form a basis is less obvious, but it follows from the fact that the space is known to have dimension N and hence N orthonormal signals must form a basis. It also follows from the fact that we can write every shifted delta function $\{\delta_{n-k}; n = 0, 1, \ldots, N-1\}$, $k = 0, 1, \ldots, N-1$, as a linear combination of the $\psi_{m,k}$ and hence since the shifted deltas form a basis, so do the $\psi_{m,k}$. For example,

$$
\begin{aligned}
\delta_n &= \frac{\sqrt{2}}{2}\psi_{1,0}(n) + \frac{\sqrt{4}}{4}\psi_{2,0}(n) + \frac{\sqrt{8}}{8}\psi_{3,0}(n) + \frac{\sqrt{8}}{8}\psi_{0,0}(n) \\
&= \frac{1}{\sqrt{2}}\psi_{1,0}(n) + \frac{1}{2}\psi_{2,0}(n) + \frac{1}{2\sqrt{2}}\psi_{3,0}(n) + \frac{1}{2\sqrt{2}}\psi_{0,0}(n).(3.104)
\end{aligned}
$$

To find relations for the shifted delta functions $\delta^j = \{\delta_{n-j}; n \in \mathcal{Z}_N\}$ for

$j \in \mathcal{Z}_N$, where as usual the shift is modulo N, verify the following formula:

$$\delta_{n-j} = \sum_{(m,k) \in \mathcal{K}} \left(\sum_{l=0}^{N-1} \psi_{m,k}(l) \delta_{l-j} \right) \psi_{m,k}(n)$$

$$= \sum_{(m,k) \in \mathcal{K}} \psi_{m,k}(j) \psi_{m,k}(n); \ j, n \in \mathcal{Z}_N. \qquad (3.105)$$

Why is the wavelet transform special? There are many reasons, among which is the fact that it provides a convenient *multiresolution* reconstruction of the original signal. A Fourier series representation gives an increasingly accurate approximation to the original signal as one adds up the terms in the sum. A wavelet reconstruction does this in a particularly useful way, providing progressively better approximations. The lowest resolution is achieved by including only the $\psi_{m,k}$ with large m in the reconstruction. As the lower m are added, the reconstructions becomes increasingly good. The wavelet transform is also of interest because many of the signals used in the decomposition have short duration in our example and in many other examples. This is in obvious contrast to the complex exponentials, which are non zero over almost the entire time domain. This means that errors in wavelet coefficients will usually have effects that are highly localized and not spread out over the full domain.

Wavelets also have the big advantage of simple and fast algorithms for their computation. In the simple example considered, the coefficients for a signal $g = (g_0, g_1, \ldots, g_9)$ are found as follows. First compute the coefficients $a_{1,k} = <g, \psi_{1,k}>$ as

$$a_{1,0} = <g, \psi_{1,0}> = g_0 - g_1$$
$$a_{1,1} = <g, \psi_{1,1}> = g_2 - g_3$$
$$a_{1,2} = <g, \psi_{1,2}> = g_4 - g_5$$
$$a_{1,3} = <g, \psi_{1,3}> = g_6 - g_7.$$

These coefficients can be found by taking successive differences $g_n - g_{n-1}$ in the signal and then looking at every other difference, that is, subsampling the differences by 2. Also compute another set of auxiliary coefficients which will be used to find $a_{0,0}$, but in the mean time will have other uses. Define

$$b_{1,0} = g_0 + g_1$$
$$b_{1,1} = g_2 + g_3$$
$$b_{1,2} = g_4 + g_5$$

$$b_{1,3} = g_6 + g_7.$$

Thus the $b_{1,k}$ sequence replaces the differences used for $a_{1,k}$ by sums. Note for later that summing up the $b_{1,k}$ will give $a_{0,0}$.

Next we wish to compute the coefficients $a_{2,k} = <g, \psi_{2,k}>$ as

$$a_{2,0} = <g, \psi_{2,0}> = g_0 + g_1 - g_2 - g_3$$
$$a_{2,1} = <g, \psi_{2,1}> = g_4 + g_5 - g_6 - g_7,$$

but we can use the computations already done to assist this. These coefficients can be found by forming differences (as we used to find the $a_{1,k}$ on the auxiliary $b_{1,k}$; i.e., form

$$a_{2,0} = b_{1,0} - b_{1,1}$$
$$a_{2,1} = b_{1,2} - b_{1,3}.$$

We can also form the auxiliary sequence as before by replacing these differences by sums.

$$b_{2,0} = b_{1,0} + b_{1,1}$$
$$b_{2,1} = b_{1,2} + b_{1,3}.$$

To finish, we now note that the coefficient $a_{3,0}$ can be evaluated as

$$a_{3,0} = <g, \psi_{3,0}>$$
$$= g_0 + g_1 + g_2 + g_3 - g_4 - g_5 - g_6 - g_7$$
$$= b_{2,0} - b_{2,1}$$

and $a_{0,0}$ as

$$a_{0,0} = <g, \psi_{0,0}>$$
$$= g_0 + g_1 + g_2 + g_3 + g_4 + g_5 + g_6 + g_7$$
$$= b_{2,0} + b_{2,1}.$$

Thus by an iteration involving separate pairwise sums and differences, the coefficients are built up in a sequence of simple combinations. Each of these operations can be viewed as a linear filtering, as will be considered in Chapter 6 and the overall operation can be constructed using a cascade of filters and subsampling operations, a form of *subband filtering*. The interested reader can pursue the subject of subband filtering in Woods [37].

3.10   ⋆ Two-Dimensional Inversion

The basic ideas of Fourier inversion can be extended to two-dimensional transforms in the obvious way, e.g., for the continuous parameter infinite duration case we have the Fourier transform pair

$$G(f_X, f_Y) = \int_{-\infty}^{\infty} \int_{-\infty}^{\infty} g(x,y) e^{-i2\pi(f_X x + f_Y y)} \, dx \, dy; \; f_X \in \mathcal{R}, f_Y \in \mathcal{R},$$

(3.106)

$$g(x,y) = \int_{-\infty}^{\infty} \int_{-\infty}^{\infty} G(f_X, f_Y) e^{+i2\pi(f_X x + f_Y y)} \, df_X \, df_Y; \; x \in \mathcal{R}, y \in \mathcal{R},$$

(3.107)

provided that the signal is sufficiently well behaved.

Analogous to the one-dimensional case, the Fourier transform can be thought of as decomposing the signal $g(x,y)$ into elementary functions of the form $e^{+i2\pi(f_X x + f_Y y)}$, that is, two dimensional complex exponentials. To be more concrete, consider the real part of the elementary functions; that is, consider the signals $\cos(2\pi(f_X x + f_Y y))$. For fixed y, this signal is periodic in x with period $1/f_X$. For fixed x it is periodic in y with period $1/f_Y$. The signal will have 0 phase (mod 2π) when the argument satisfies

$$f_X x + f_Y y = k; \; k \in \mathcal{Z}.$$

These formulas define straight lines in the $X - Y$ plane.

2D signals that are separable in rectangular coordinates are easily inverted; the inverse is simply the product of the two one dimensional inverses for each coordinate. Signals that are separable in polar coordinates take a bit more effort, but it can be shown that we can invert the *zero-order Hankel transform* or the *Fourier-Bessel transform*

$$G(\rho, \phi) = 2\pi \int_{0}^{\infty} r g_R(r) J_0(2\pi r \rho) \, dr = G(\rho)$$

by

$$g_R(r) = 2\pi \int_{0}^{\infty} \rho G(\rho) J_0(2\pi r \rho) \, d\rho$$

at all points of continuity of g_R. Thus the transform and inverse transform operations are identical in the circularly symmetric case.

3.11 Problems

3.1. Given a signal $g = \{g(t); t \in \mathcal{T}\}$, what is $\mathcal{F}(\mathcal{F}(g))$? What is $\mathcal{F}(\mathcal{F}(\mathcal{F}(g)))$?

3.2. Prove the two-sided discrete time finite duration Fourier inversion formula (3.18).

3.3. Recall that the DCT of a signal $\{g(k,j); \ k = 0,1,\ldots,N-1; \ j = 0,1,\ldots,N-1\}$ is defined by

$$G(l,m) = \frac{2}{N}C(l)C(m) \sum_{k=0}^{N-1}\sum_{j=0}^{N-1} g(k,j) \cos\frac{(2k+1)l\pi}{2N} \cos\frac{(2j+1)m\pi}{2N},$$

$$(3.108)$$

where

$$C(n) = \begin{cases} \frac{1}{\sqrt{2}} & \text{if } n = 0 \\ 1 & \text{otherwise.} \end{cases}$$

Show that the inverse DCT is given by

$$g(k,j) = \frac{2}{N} \sum_{l=0}^{N-1}\sum_{m=0}^{N-1} C(l)C(m)G(l,m) \cos\frac{(2k+1)l\pi}{2N} \cos\frac{(2j+1)m\pi}{2N}.$$

$$(3.109)$$

Warning: This problem takes some hacking, but it is an important result for engineering practice.

3.4. Find a Fourier series for the two-sided discrete time signal $\{r^{-|n|}; \ n = -N, \cdots, 0, N\}$ for the cases $|r| < 1$, $|r| = 1$, and $r > 1$. Compare the result with the Fourier series of the one-sided discrete time geometric signal of (3.23). Write the Fourier series for the periodic extensions of both signals (period N for the one-sided signal and period $2N+1$ for the two sided signal) and sketch the two periodic signals.

3.5. Find a Fourier series for the discrete time infinite duration signal defined by $g_n = n \bmod 10$, $n \in \mathcal{Z}$. Is the Fourier series accurate for all integer n? Compare the result to the Fourier series for the continuous time ramp function $g(t) = t \bmod 10$, $t \in \mathcal{R}$.

3.6. Define the "roundoff" function $q(x)$ which maps real numbers x into the nearest integer, that is, $q(x) = n$ if $n - 1/2 < x \leq n + 1/2$. Define the roundoff error by $\epsilon(x) = q(x) - x$. Find a Fourier series in x for $\epsilon(x)$.

3.7. What signal has Fourier transform $e^{-|f|}$ for all real f?

3.8. Suppose that G is the infinite duration CT Fourier transform of a signal g. Then the definition at $f = 0$ gives the formula

$$\int_{-\infty}^{\infty} g(t)\, dt = G(0). \tag{3.110}$$

(a) Find an analogous result for the finite duration CT Fourier transform.

(b) Repeat for the finite duration DT Fourier transform.

(c) Suppose now we have an infinite duration time signal h defined by $h(t) = G(t)$ for all real t. In words, we are now looking at the function G as a time signal instead of a spectrum. What is the Fourier transform H of h (in terms of g)?

(d) Use the previous part to find the dual of Eq. (3.110), that is, relate $\int_{-\infty}^{\infty} G(f)\, df$ to g in a simple way for an infinite duration CT signal g.

(e) Use the previous part to evaluate the integral

$$\int_{\infty}^{\infty} \mathrm{sinc}(t)\, dt.$$

(To appreciate this shortcut you might try to evaluate this integral by straightforward calculus.)

3.9. What is the Fourier transform of the finite duration, continuous time signal

$$g(t) = \frac{\sin(2\pi t(\frac{5}{2}))}{\sin(\pi t)};\ t \in [-\frac{1}{2}, \frac{1}{2})?$$

Find a Fourier series representation for the periodic extension (having period 1) of this signal.

3.10. What infinite duration discrete time signal $\{g_n;\ n \in \mathcal{Z}\}$ has Fourier transform $\{\sqcap(4f);\ f \in [-\frac{1}{2}, \frac{1}{2})\}$?

3.11. Define the finite duration continuous time signal $g = \{g(t);\ t \in [-T/2, T/2)\}$ by $g(t) = A$ for $-T/4 \le t \le T/4$ and $g(t) = 0$ for $|t| > T/4$. Find the Fourier transform $G(f)$. Is the inverse Fourier transform of $G(f)$ equal to $g(t)$?

Now let $\hat{g}(t)$ be the infinite duration continuous time signal formed by zero-filling $g(t)$. Again find the Fourier transform and inverse Fourier transform. (Note that the Fourier transform has the same functional form in both cases, but the frequency domain of definition is different. The inverse Fourier transforms, however, are quite different.)

3.12. What is the Fourier transform of the continuous time finite duration signal $g = t; t \in [-\frac{1}{2}, \frac{1}{2})$? Find an exponential Fourier series representation for g. Find a trigonometric Fourier series representation for g.

3.13. Find a trigonometric Fourier series representation for the infinite duration continuous time periodic signal $\tilde{g}(t) = (t - 1/2)\text{mod}1 - 1/2$. (First sketch the waveform.) For what values of t is the Fourier series *not* accurate. Repeat for the signal $\tilde{g}(t - 1/2)$.

3.14. Prove the Fourier transform pair relationship of (3.84)–(3.85) by direct substitution.

3.15. Show how the a_n and b_n are related to $g(t)$ in Eq. 3.83. Express the a_n and b_n in terms of the c_n of the exponential Fourier series.

3.16. *Orthogonal Expansions*

A collection of signals $\{\phi_i(t); t \in [0, T]\}$, $i = -N, \cdots, 0, 1, 2, \cdots, N$ are said to be *orthonormal* on $[0, T]$ if

$$\int_0^T \phi_i(t)\phi_j^*(t)dt = \delta_{i-j}, \qquad (3.111)$$

that is, the integral is 1 if the functions are the same and 0 otherwise.

(a) Suppose that you are told that a real-valued signal g is given by

$$g(t) = \sum_{k=-N}^{N} b_k \phi_k(t).$$

How do you find the b_k from $g(t)$ and the $\phi_k(t)$? Evaluate the energy

$$\mathcal{E}_g = \int_0^T |g(t)|^2 dt$$

in terms of the b_i. (This is an example of Parseval's theorem.)

(b) Are the functions $A\sin(2\pi kt/T)$; $k = 1, 2, \ldots, N$ orthonormal on $[0, T]$?

(c) Suppose that we have an orthonormal set of functions $\{\phi_k(t); t \in [0, T]\}$, $k \in \mathcal{Z}$ and that we have an arbitrary signal $g(t), t \in [0, T]$. We want to construct an approximation $p(t)$ to $g(t)$ of the form

$$p(t) = \sum_{n=-N}^{N} c_n \phi_n(t).$$

Define the error signal resulting from this approximation by

$$e(t) = g(t) - p(t).$$

We wish to determine the best possible choice of the coefficients c_n. Toward this end define the coefficients

$$a_n = \int_0^T g(t)\phi_n^*(t)dt$$

and the approximation

$$q(t) = \sum_{n=-N}^{N} a_n\phi_n(t).$$

Consider the *mean squared error*

$$\frac{1}{T}\int_0^T |e(t)|^2 dt = \frac{1}{T}\int_0^T |g(t) - p(t)|^2 \, dt.$$

By adding and subtracting $q(t)$ inside of the brackets argue that

$$\frac{1}{T}\int_0^T |e(t)|^2 dt = \frac{1}{T}\int_0^T |g(t) - q(t)|^2 \, dt + \frac{1}{T}\int_0^T |q(t) - p(t)|^2 \, dt$$

$$- 2\frac{1}{T}\Re\left[\int_0^T [g(t) - q(t)][q(t) - p(t)]^* \, dt\right]$$

Show that the rightmost integral is 0 and use this to conclude that

$$\frac{1}{T}\int_0^T |e(t)|^2 dt \geq \frac{1}{T}\int_0^T |g(t) - q(t)|^2 dt,$$

that is, the a_n are the optimum coefficients in the sense of minimizing the mean squared error.

3.17. What signal has Fourier transform $\wedge(f)$; $f \in \mathcal{R}$?

3.18. Suppose that you have a finite duration, discrete time signal $\{g_n; n = 0, 1, \ldots, N-1\}$ and you have found its DFT $\{G_k = G(k/N); k = 0, 1, \ldots, N-1\}$. Just to be difficult, a local curmudgeon demands a formula for the Fourier transform for *arbitrary* real f, that is,

$$G(f) = \sum_{n=0}^{N-1} g_n e^{-i2\pi fn}, \; f \in \mathcal{R}.$$

Find an expression for $G(f)$ in terms of the G_k which is valid for arbitrary real f.

The point of this problem is that knowing the DFT for only the finite collection of frequencies is enough to compute the formula giving the transform for all real f. This is a trivial form of "sampling theorem" since $G(f)$ is recovered from its "samples" $G(k/N)$.

3.19. Suppose that you are told that a signal $g(t)$; $t \in \mathcal{R}$ is time-limited to $[-\Delta/2, \Delta/2]$ and that its spectrum $G(f)$ has the property that $G(k/\Delta) = 0$ for all nonzero integers k and $G(0) \neq 0$. What is $g(t)$? Suppose that you only know that g is time-limited to $[-\Delta, \Delta]$, but it still has the property that $G(k/\Delta) = 0$ for all nonzero integers k. Give examples of at least two nontrivial signals that meet this condition.

3.20. Suppose that $g = \{e^{-t}; \ t \in [0,T)\}$ is a finite duration continuous time signal. ($T > 0$ is a fixed parameter.)

   (a) Find the Fourier transform G of g.

   (b) Find a Fourier series representation for g.

   (c) Find the Fourier transform G_T of a zero filled extension g_T of g, i.e., $g_T(t) = g(t)$ for $t \in [0,T)$ and $g_T(t) = 0$ for all other real t. Sketch g_T.

   (d) Find a Fourier series for the periodic extension $\tilde{g}$ of g, i.e., $\tilde{g}(t) = g(t)$ for $t \in [0,T)$ and $\tilde{g}$ is periodic with period T. Sketch $\tilde{g}$.

   (e) Now suppose that T becomes very large. Let G_1 denote the CTFT of the signal $g_1 = \{e^{-t}u_{-1}(t); t \in \mathcal{R}\}$. Describe the differences and similarities between G_1 and G_T. Is it true that $G_1(f) = \lim_{T \to \infty} G_T(f)$ in some sense? (You are not required to come up with a mathematical proof, but you are expected to look at the limit and comment on why it might or might not converge to the guessed limit.)

3.21. Consider the matrix notation for the DFT and its inverse: $\mathbf{G} = \mathbf{W}g$ and $\mathbf{g} = \mathbf{W}^{-1}\mathbf{G}$. Define the energy in the signal and its transform by

$$\mathcal{E}_g = \|g\|^2 = \mathbf{g}^*\mathbf{g} = \sum_{n=0}^{N-1} |g(n)|^2$$

and

$$\mathcal{E}_G = \|G\|^2 = \mathbf{G}^*\mathbf{G} = \sum_{n=0}^{N-1} |G(n/N)|^2,$$

respectively. Show that

$$||g||^2 = N^{-1}||G||^2.$$

This is Parseval's equality for the DFT.

Hint: Use the matrix form of the transform and recall the relation of $\mathbf{W}^{-1}$ and $\mathbf{W}^*$.

3.22. This problem treats a special case of an important application of Fourier (and other transforms) called *transform coding*. Suppose that a signal $g = (g_0, \cdots, g_{N-1})$, represented as a column vector $\mathbf{g} = (g_0, \cdots, g_{N-1})^t$ (we use vector notation here for convenience), is to be reconstructed by the following sequence of operations. First we take its Fourier transform $\mathbf{G} = \mathbf{Wg}$, where the matrix $\mathbf{W}$ is given by (2.6). This transform is then approximated by another vector $\hat{\mathbf{G}}$. For example, $\hat{\mathbf{G}}$ might be formed by quantizing or digitizing $\mathbf{G}$ in order to store it in a digital medium or transmit it over a digital channel. We then use $\hat{\mathbf{G}}$ to reconstruct an approximation to $\mathbf{g}$ by inverse transforming to form $\hat{\mathbf{g}} = \mathbf{W}^{-1}\hat{\mathbf{G}}$. Suppose that we define the *mean squared error* or MSE of the overall reconstruction using the notation of Problem 3.21 as

$$\epsilon_g^2 = N^{-1}||g - \hat{g}||^2 = N^{-1}\sum_{n=0}^{N-1}|g_n - \hat{g}_n|^2. \tag{3.112}$$

We can similarly define the mean squared error in the Fourier domain by

$$\epsilon_G^2 = N^{-1}||G - \hat{G}||^2 = N^{-1}\sum_{n=0}^{N-1}|G(\frac{n}{N}) - \hat{G}(\frac{n}{N})|^2. \tag{3.113}$$

Show that the MSE in the original time domain is proportional to that in the frequency domain and find the constant of proportionality.

Hint: See Problem 3.21.

Now suppose that we wish to "compress" the image by throwing away some of the transform coefficients. In other words, instead of keeping all N floating point numbers describing the $G(\frac{n}{N})$; $n = 0, 1, \ldots, N-1$, we only keep $M < N$ of these coefficients and assume all the remaining $N - M$ coefficients are 0 for purposes of reconstruction. Assuming a fixed number of bytes, say m, for representing each floating point number on a digital computer, we have reduced the storage requirement for the signal from Nm bytes to Mm bytes, achieving a *compression ratio* of $N : M$. Obviously this comes at a cost as the setting of some coefficients to zero causes error in the reconstruction and one has to know in advance which of the coefficients are to be kept. If the reconstruction error is small, it will not be perceivable or will not damage the intended use of the image.

Suppose that one is allowed to optimally choose for a particular image which of the $M < N$ are the best coefficients to keep and what is the resulting mean squared error? Note that in this case one would need to specify which coefficients are nonzero as well as the values of the M nonzero coefficients. Thus in this case the "compression ratio" $N : M$ alone does not really indicate the information needed to reconstruct an approximate image.

The two dimensional version of transform coding is the most popular technique for compressing image data and it is an integral part of several international standards, including the international ISO/ITU-T JPEG (Joint Photographic Expert Group) standard. For a discussion of the standard and further details on how the floating point numbers are quantized and coded, see, e.g., [34, 26].

You might wonder why one would wish to do the approximating in the frequency domain rather than in the original time domain. There are many reasons, two of which can be easily described. First, the Fourier transform of most interesting data sources tends to concentrate the energy in the lower frequencies, which means often many of the higher frequency coefficients can be reproduced very coarsely or simply thrown away without much damage. This effectively reduces the dimensionality of the problem and provides some compression. Second, both the eye and ear seem to be sensitive to signal behavior in the Fourier domain, which means that doing a good job approximating the original signal in the Fourier domain will result in a reproduction that looks or sounds good.

3.23. Consider two finite duration signals: $h = \{h(t) = 6 - t; \ t \in [0,6)\}$ and $g = \{g(t) = \cos(\pi t/2), t \in [0,4)\}$. Let $\overline{h} = \{\overline{h}(t); \ t \in \mathcal{R}\}$ be the zero-filled extension of h, i.e.,

$$\overline{h}(t) = \begin{cases} 6 - t & t \in [0,6) \\ 0 & \text{otherwise} \end{cases}.$$

Let $\tilde{g}$ be the periodic extension of g, and let $\tilde{h}$ be the periodic extension of h. Define $r = \tilde{g} + \tilde{h}$.

(a) Find the Fourier transform of $\overline{h}$.

(b) Find the Fourier series coefficients $c_h(k)$ for the signal h.

(c) Find the Fourier series coefficients $c_g(k)$ for the signal g.

(d) What is the period of $r(t)$?

(e) Express the Fourier series coefficients $c_r(k)$ for the periodic signal $r(t)$ in terms of the coefficients $c_h(k)$ and $c_g(k)$ that you found in parts (b) and (c).

3.24. Suppose that U is a unitary matrix which defines a transform $G = Wg$. Suppose that V is yet another unitary matrix which also can be used to define a transform. Is the matrix product VU also unitary? If so, then how do you invert the transform UVg?

Chapter 4

Basic Properties

In this chapter the fundamental properties of Fourier transforms are derived. These properties are useful in manipulating, evaluating, verifying, and applying Fourier transforms.

4.1 Linearity

Recall that a linear combination of two signals g and h is a signal of the form $ag + bh = \{ag(t) + bh(t); t \in \mathcal{T}\}$. The most important elementary property of Fourier transforms is given by the following theorem.

Theorem 4.1 *The Fourier transform is linear; that is, given two signals g and h and two complex numbers a and b, then*

$$\mathcal{F}(ag + bh) = a\mathcal{F}(g) + b\mathcal{F}(h). \tag{4.1}$$

The theorem follows immediately from the fact that the Fourier transform is defined by a sum or integral and that sums and integrals have the linearity property. We have already seen that the DFT is linear by expressing it in matrix form. Linearity can also easily be proved directly from the definitions in this case:

$$
\begin{aligned}
\sum_{n=0}^{N-1} \left(ag_n + bh_n \right) e^{-i2\pi fn} &= a \sum_{n=0}^{N-1} g_n e^{-i2\pi fn} + b \sum_{n=0}^{N-1} h_n e^{-i2\pi fn} \\
&= aG(f) + bH(f)
\end{aligned}
$$

as claimed.

The linearity property is also sometimes called the *superposition property*.

Recall that the Fourier transform operation can be considered as an example of a system mapping an input signal (the original time domain signal) into an output signal (the Fourier transform or spectrum of the time domain signal, a frequency domain signal). Viewed in this way, the theorem simply states that the system described by the Fourier transform operation is a linear system.

The idea of linearity for more general systems is fundamental to many applications and so we here point out the general definition and interpret the linearity property of Fourier transforms as a special case.

The basic linearity property implies a similar result for any finite collection of signals; that is,

$$\mathcal{F}(\sum_{n=1}^{K} a_n g^{(n)}) = \sum_{n=1}^{K} a_n \mathcal{F}(g^{(n)}). \tag{4.2}$$

This result is quite useful in computing Fourier transforms: if a signal is the sum of a finite number of signals for which the transforms are known, then the transform of the sum is the sum of the transforms.

An often useful property which does not immediately follow from linearity, but which can be proved under certain conditions, is called *countable linearity* or *extended linearity* wherein a property like the above one holds for an infinite summation as well as a finite summation; that is,

$$\mathcal{F}(\sum_{n=1}^{\infty} a_n g^{(n)}) = \sum_{n=1}^{\infty} a_n \mathcal{F}(g^{(n)}). \tag{4.3}$$

We usually assume that the Fourier transform has the countable linearity property, but be forewarned that this is true only when the signals $g^{(n)}$ and the sequence of weights a_n are sufficiently well behaved.

As an example, suppose that

$$g_n = a r^n u_{-1}(n) + b \rho^n u_{-1}(n); \; n \in \mathcal{Z},$$

where $|r| < 1$ and $|\rho| < 1$. Then from linearity the spectrum is immediately

$$G(f) = \frac{a}{1 - r e^{-i2\pi f}} + \frac{b}{1 - \rho e^{-i2\pi f}}; \; f \in [0, 1).$$

4.2   Shifts

In the previous section the effect on transforms of linear combinations of signals was considered. Next the effect on transforms of shifting or delaying a signal is treated.

Theorem 4.2 *The Shift Theorem*

Given a signal $g = \{g(t); t \in \mathcal{T}\}$ with Fourier transform $G = \mathcal{F}(g)$, suppose that for $\tau \in \mathcal{T}$, $g_\tau = \{g(t - \tau); \tau \in \mathcal{T}\}$ is the shifted signal (the shift is a cyclic shift in the finite duration case). Then

$$\mathcal{F}(g_\tau) = \{e^{-i2\pi f\tau} \mathcal{F}_f(g); \ f \in \mathcal{S}\}. \tag{4.4}$$

Thus a delay in the time domain produces a frequency dependent exponential multiplier in the frequency domain. To interpret this result, consider two sinusoids (or equivalently complex exponentials), one having a low frequency and the other a high frequency. Suppose both sinusoids are shifted by τ. The sinusoid with the long period may be shifted by some small fraction of its period, while simultaneously the sinusoid with the short period is shifted by a large number of periods. Some further thought shows that each sinusoidal component is shifted by a fraction of its period that increases linearly with its frequency. Hence a fixed time shift corresponds to a phase shift linearly proportional to frequency (as well as to τ). For a complex exponential, this linear phase shift is equivalent to multiplication by another complex exponential with exponent $-i2\pi f\tau$.

From the systems viewpoint, the theorem tells us how to find the transform of the system output in terms of the transform of the system input. Many of the results of this chapter will be of this general form.

The proof is similar in all cases, so we only prove two of them. First consider the DFT case of a signal $\{g_n; n \in \{0, 1, \ldots, N-1\}\}$. Here

$$
\begin{aligned}
\mathcal{F}_{\frac{k}{N}}(g_\tau) &= \sum_{n=0}^{N-1} g_{(n-\tau)\bmod N} e^{-i2\pi \frac{k}{N} n} \\
&= \sum_{n=0}^{N-1} g_{(n-\tau)\bmod N} e^{-i2\pi \frac{k}{N}(n-\tau)} e^{-i2\pi\tau \frac{k}{N}} \\
&= \sum_{n=0}^{N-1} g_{(n-\tau)\bmod N} e^{-i2\pi \frac{k}{N}((n-\tau)\bmod N)} e^{-i2\pi\tau \frac{k}{N}} \\
&= e^{-i2\pi\tau \frac{k}{N}} \sum_{n=0}^{N-1} g_n e^{-i2\pi \frac{k}{N} n} = e^{-i2\pi\tau \frac{k}{N}} \mathcal{F}_{\frac{k}{N}}(g); \ k \in \mathcal{Z}_N,
\end{aligned}
$$

proving the result. Note that we have used the fact that $e^{-i2\pi \frac{k}{N} n} = e^{-i2\pi \frac{k}{N}(n \bmod N)}$ which holds since for $n = KN + l$ with $0 \leq l \leq N - 1$ $e^{-i2\pi \frac{k}{N}(KN+l)} = e^{-i2\pi kK} e^{-i2\pi \frac{k}{N} l} = e^{-i2\pi \frac{k}{N} l}$.

Next consider the infinite duration CTFT. Here we simply change variables $\alpha = t - \tau$ to find

$$
\begin{aligned}
\mathcal{F}_f(g_\tau) &= \int_{-\infty}^{\infty} g(t-\tau)e^{-i2\pi f t}\, dt \\
&= \int_{-\infty}^{\infty} g(\alpha)e^{-i2\pi f(\alpha+\tau)}\, d\alpha = e^{-i2\pi f\tau}\int_{-\infty}^{\infty} g(\alpha)e^{-i2\pi f\alpha}\, d\alpha \\
&= e^{-i2\pi f\tau}\mathcal{F}_f(g).
\end{aligned}
$$

The finite duration CTFT and the infinite duration DTFT follow by similar methods.

As an example, consider the infinite duration continuous time pulse $p(t) = A$ for $t \in [0, T)$ and 0 otherwise. This pulse can be considered as a scaled and shifted box function

$$
p(t) = A\square_{T/2}(t - \frac{T}{2})
$$

and hence using linearity and the shift theorem

$$
\mathcal{F}_f(p) = Ae^{-i2\pi f\frac{T}{2}}T\operatorname{sinc}(Tf). \tag{4.5}
$$

4.3   Modulation

The modulation theorem treats the modulation of a signal by a complex exponential. It will be seen to be a dual result to the shift theorem; that is, it can be viewed as the shift theorem with the roles of time and frequency interchanged.

Suppose that $g = \{g(t); t \in \mathcal{T}\}$ is a signal with Fourier transform $G = \{G(f); f \in \mathcal{S}\}$. Consider the new signal $g_e(t) = g(t)e^{i2\pi f_0 t}; t \in \mathcal{T}$ where $f_0 \in \mathcal{S}$ is a fixed frequency (sometimes called the *carrier frequency*). The signal $g_e(t)$ is said to be formed by *modulating* the complex exponential $e^{i2\pi f_0 t}$, called the *carrier*, by the original signal $g(t)$. In general, modulating is the methodical alteration of one waveform, here the complex exponential, by another waveform, called the signal. When the signal and the carrier are simply multiplied together, the modulation is called *amplitude modulation* or *AM*. In general AM includes multiplication by a complex exponential or by sinusoids as in $g_c(t) = g(t)\cos(2\pi f_0 t); t \in \mathcal{T}$ and $g_s(t) = g(t)\sin(2\pi f_0 t); t \in \mathcal{T}$. Often AM is used in a strict sense to mean signals of the form $g_a(t) = A[1 + mg(t)]\cos(2\pi f_0 t)$ which contains a separate carrier term $A\cos(2\pi f_0 t)$. $g_a(t)$ is referred to as double sideband (DSB) or double sideband amplitude modulation (DSB-AM), while the simpler forms of g_c or g_s are called double sideband suppressed carrier

(DSB-SC). The parameter m is called the *modulation index* and sets the relative strengths of the signal and the carrier. Typically it is required that m and g are chosen so that $|mg(t)| < 1$ for all t.

Amplitude modulation without the carrier term, i.e., g_c or g_s and not g_a, are called *linear* modulation because the modulation is accomplished by a linear operation, albeit a time varying one.

The operation of modulation can be considered as a system in a mathematical sense: the original signal put into the system produces at the output a modulated version of the input signal. It is perhaps less obvious in this case than in the ideal delay case that the resulting system is linear (for the type of modulation considered — other forms of modulation can result in nonlinear systems).

Theorem 4.3 *The Modulation Theorem.*
  Given a signal $\{g(t); t \in \mathcal{T}\}$ *with spectrum* $\{G(f); f \in \mathcal{S}\}$, *then*

$$\{g(t)e^{i2\pi f_0 t}; t \in \mathcal{T}\} \quad \supset \quad \{G(f - f_0); f \in \mathcal{S}\}$$

$$\{g(t)\cos(2\pi f_0 t); t \in \mathcal{T}\} \quad \supset \quad \{\frac{1}{2}G(f - f_0) + \frac{1}{2}G(f + f_0); f \in \mathcal{S}\}$$

$$\{g(t)\sin(2\pi f_0 t); t \in \mathcal{T}\} \quad \supset \quad \{\frac{i}{2}G(f + f_0) - \frac{i}{2}G(f - f_0); f \in \mathcal{S}\}$$

where the difference $f - f_0$ *and sum* $f + f_0$ *in the frequency domain are treated like the shift in the time domain, that is, ordinary algebra in the case of* $\mathcal{S} = \mathcal{R}$ *or* $\{k/T; k \in \mathcal{Z}\}$ *and modulo the frequency domain in the case of* $[0, 1)$ *or* $\{k/N; k \in \mathcal{Z}_N\}$.

Thus modulating a signal by an exponential shifts the spectrum in the frequency domain. Modulation by a cosine with spectrum $G(f)$ causes replicas of $G(f)$ to be placed at plus and minus the carrier frequency. These replicas are sometimes called *sidebands*.

Proof: As usual we consider only two cases, leaving the others for an exercise. In the infinite duration continuous time case,

$$\begin{aligned} G_e(f) &= \int_{-\infty}^{\infty} (g(t)e^{i2\pi f_0 t})e^{-i2\pi f t}\, dt \\ &= \int_{-\infty}^{\infty} g(t)e^{-i2\pi(f - f_0)t}\, dt \\ &= G(f - f_0). \end{aligned}$$

The results for cosine and sine modulation then follow via Euler's relations. For the DFT we have that with a frequency $f_0 = k_0/N$

$$G_e\left(\frac{k}{N}\right) = \sum_{n=0}^{N-1} (g_n e^{i2\pi \frac{k_0}{N} n})e^{-i2\pi \frac{k}{N} n}$$

$$= \sum_{n=0}^{N-1} (g_n e^{-i2\pi(\frac{k-k_0}{N})n})$$

$$= G(\frac{k-k_0}{N}).$$

Thus, for example, the Fourier transform of $\sqcap(t)\cos(\pi t)$ is given by

$$\mathcal{F}_f(\{\sqcap(t)\cos(\pi t);\ t \in \mathcal{R}\}) = \frac{1}{2}(\text{sinc}(f-1/2)+\text{sinc}(f+1/2));\ f \in \mathcal{R}. \quad (4.6)$$

4.4 Parseval's Theorem

In this section we develop a result popularly known as Parseval's theorem, although Parseval only considered the special case of continuous time Fourier series. The result might be better described as Rayleigh's theorem after Lord Rayleigh, who proved the result for the continuous time Fourier transform. We will follow common use, however, and collect the four similar results for the four signal types under the common name of Parseval's theorem.

The *energy* of a continuous time infinite duration signal is defined by

$$\mathcal{E}_g = \int_{-\infty}^{\infty} |g(t)|^2\, dt$$

and it has the interpretation of being the energy dissipated in a one ohm resistor if g is considered to be a voltage. It can also be viewed as a measure of the size of a signal. In a similar manner we can define the energy of the appropriate Fourier transform as

$$\mathcal{E}_G = \int_{-\infty}^{\infty} |G(f)|^2\, df.$$

These two energies are easily related by substituting the definition of the transform, changing the order of integration, and using the inversion formula:

$$\begin{aligned}
\mathcal{E}_G &= \int_{-\infty}^{\infty} G(f)G^*(f)\, df \\
&= \int_{-\infty}^{\infty} G(f) \left(\int_{-\infty}^{\infty} g(t)e^{-i2\pi ft}\, dt \right)^*\, df \\
&= \int_{-\infty}^{\infty} g^*(t) \left(\int_{-\infty}^{\infty} G(f)e^{i2\pi ft}\, df \right)\, dt \\
&= \int_{-\infty}^{\infty} g^*(t)g(t)\, dt \\
&= \mathcal{E}_g;
\end{aligned}$$

proving that the energies in the two domains are the same for the continuous time infinite duration case.

The corresponding result for the DFT can be proved by the analogous string of equalities for discrete time finite duration or by matrix manipulation as in Problem 3.21. In that case the result can be expressed in terms of the energies defined by

$$\mathcal{E}_g = \sum_{n=0}^{N-1} |g(n)|^2$$

and

$$\mathcal{E}_G = \sum_{n=0}^{N-1} |G(n/N)|^2$$

as

$$\mathcal{E}_g = \frac{1}{N}\mathcal{E}_G.$$

The following theorem summarizes the general result and its specialization to the various signal types.

Theorem 4.4 *Parseval's Theorem*
Given a signal g with Fourier transform G, then $\mathcal{E}_g = C\mathcal{E}_G$, where the energy is defined as the integral or sum of the signal squared over its domain of definition and C is a constant that equals 1 for infinite duration signals and the inverse duration of the signal for finite duration signals. To be specific:

1. *If the signals are infinite duration continuous time signals, then*

$$\mathcal{E}_g = \int_{-\infty}^{\infty} |g(t)|^2\, dt = \int_{-\infty}^{\infty} |G(f)|^2\, df = \mathcal{E}_G.$$

2. *If the signals are finite duration continuous time signals, then*

$$\mathcal{E}_g = \int_0^T |g(t)|^2\, dt = \frac{1}{T} \sum_{n=-\infty}^{\infty} |G(\frac{n}{T})|^2 = \frac{1}{T}\mathcal{E}_G.$$

3. *If the signals are infinite duration discrete time signals, then*

$$\mathcal{E}_g = \sum_{n=-\infty}^{\infty} |g_n|^2 = \int_{-\frac{1}{2}}^{\frac{1}{2}} |G(f)|^2\, df = \mathcal{E}_G.$$

4. If the signals are finite duration discrete time signals, then

$$\mathcal{E}_g = \sum_{n=0}^{N-1} |g_n|^2 = \frac{1}{N} \sum_{n=0}^{N-1} |G(\frac{n}{N})|^2 = \frac{1}{N}\mathcal{E}_G.$$

As mentioned earlier, the first relation is better described as Rayleigh's theorem and the second relation is Parseval's theorem as originally developed. The second relation can be stated in its classical form as follows. If a continuous time signal g is periodic with period T, then the the signal has a Fourier series described by (3.81)-(3.82), where $c_k = G(\frac{k}{T})/T$. Thus the finite duration result becomes

$$\frac{1}{T} \int_0^T |g(t)|^2 \, dt = \sum_{n=-\infty}^{\infty} |c_k|^2. \tag{4.7}$$

The results extend in a straightforward way to integrals and sums of products of signals. Consider for example two continuous time infinite duration signals g and h with Fourier transforms G and H and consider the integral

$$< g,h >= \int_{-\infty}^{\infty} g(t)h^*(t) \, dt,$$

where we have used the inner product notation as an abbreviation. The inner product is also called the *scalar product* and is often denoted by (g, h).

Exactly as in the earlier case where $g = h$, we have that

$$
\begin{aligned}
< G,H > &= \int_{-\infty}^{\infty} G(f)H^*(f) \, df \\
&= \int_{-\infty}^{\infty} \left(\int_{-\infty}^{\infty} g(t)e^{-i2\pi ft} \, dt \right) H^*(f) \, df \\
&= \int_{-\infty}^{\infty} g(t) \left(\int_{-\infty}^{\infty} H^*(f)e^{-i2\pi ft} \, df \right) dt \\
&= \int_{-\infty}^{\infty} g(t) \left(\int_{-\infty}^{\infty} H(f)e^{i2\pi ft} \, df \right)^* dt \\
&= \int_{-\infty}^{\infty} g(t)h^*(t) \, dt \\
&= < g,h > .
\end{aligned}
$$

In a similar fashion we can define inner products for discrete time finite duration signals in the natural way as

$$< g,h >= \sum_{n=0}^{N-1} g(n)h^*(n)$$

and

$$< G, H >= \sum_{n=0}^{N-1} G(\frac{n}{N}) H^*(\frac{n}{N})$$

and derive by a similar argument that for the DFT case

$$< g, h >= N^{-1} < G, H > .$$

Repeating these arguments for the various signal types yields the following general form of Parseval's theorem, which also goes by other names such as Rayleigh's theorem, Plancherel's theorem, and the power theorem.

Theorem 4.5 *Parseval's Theorem: General Form*

Given two signals g and h with Fourier transforms G and H, respectively, then $< g, h >= C < G, H >$, where C is defined as in the previous theorem. In particular,

1. *If the signals are infinite duration continuous time signals, then*

$$< g, h >= \int_{-\infty}^{\infty} g(t) h^*(t) \, dt = \int_{-\infty}^{\infty} G(f) H^*(f) \, df =< G, H > .$$

2. *If the signals are finite duration continuous time signals, then*

$$< g, h >= \int_{0}^{T} g(t) h^*(t) \, dt = \frac{1}{T} \sum_{n=-\infty}^{\infty} G(\frac{n}{T}) H^*(\frac{n}{T}) = \frac{1}{T} < G, H > .$$

3. *If the signals are infinite duration discrete time signals, then*

$$< g, h >= \sum_{n=-\infty}^{\infty} g_n h_n^* = \int_{-\frac{1}{2}}^{\frac{1}{2}} G(f) H^*(f) \, df =< G, H > .$$

4. *If the signals are finite duration discrete time signals, then*

$$< g, h >= \sum_{n=0}^{N-1} g_n h_n^* = \frac{1}{N} \sum_{n=0}^{N-1} G(\frac{n}{N}) H^*(\frac{n}{N}) = \frac{1}{N} < G, H > .$$

We shall later see that Parseval's theorem is itself just a special case of the convolution theorem, but we do not defer its statement as it is a handy result to have without waiting for the additional ideas required for the more general result.

Parseval's theorem is extremely useful for evaluating integrals. For example, the integral

$$\int_{-\infty}^{\infty} \text{sinc}^2(t)\,dt$$

is difficult to evaluate using straightforward calculus. Since $\text{sinc}(t) \leftrightarrow \sqcap(f)$, where the double arrow was defined in (3.2) as denoting that the signal and spectrum are a Fourier transform pair, Parseval's Theorem can be applied to yield that

$$\int_{-\infty}^{\infty} \text{sinc}^2(t)\,dt = \int_{-\infty}^{\infty} \sqcap^2(f)\,df = \int_{-\infty}^{\infty} \sqcap(f)\,df = 1.$$

As an example of the general theorem observe that

$$
\begin{aligned}
\int_{-\infty}^{\infty} \text{sinc}^3(t)\,dt &= \int_{-\infty}^{\infty} \text{sinc}(t)\text{sinc}^2(t)\,dt \\
&= \int_{-\infty}^{\infty} \sqcap(f) \wedge (f)\,df \\
&= 2\int_{0}^{\frac{1}{2}} (1-f)\,df = \frac{3}{4}.
\end{aligned}
$$

4.5   The Sampling Theorem

We now turn to a powerful but simple application of Fourier series. We use the ideas of Fourier series representations of finite duration or periodic functions to obtain a representation for infinite duration continuous time band-limited signals. This provides a surprising connection between continuous time and discrete time signals formed by sampling the continuous time signals. This topic will be further explored later in the book, but the fundamental result is derived here.

Suppose that g is an infinite duration continuous time signal with a spectrum G having the property that $G(f) = 0$ for $|f| \geq W$; that is, the signal is band-limited. For the development we can choose any W for which this is true, but a particularly important example will be the smallest W for which the assumption is true, a value which we shall call $W_{\min}$ and to which we shall refer as the *bandwidth* of the signal. For the development that follows, any $W \geq W_{\min}$ will do.

Consider the truncated finite bandwidth spectrum $\hat{G} = \{G(f); f \in (-W, W)\}$. As with a finite duration signal, we can write a Fourier series for $G(f)$ in this region:

$$G(f) = \sum_{n=-\infty}^{\infty} c_n e^{-i2\pi \frac{f}{2W} n}; \quad f \in [-W, W] \tag{4.8}$$

where

$$c_n = \frac{1}{2W} \int_{-W}^{W} \hat{G}(f) e^{i2\pi \frac{f}{2W} n} \, df = \frac{1}{2W} \int_{-W}^{W} G(f) e^{i2\pi \frac{f}{2W} n} \, df; \quad n \in \mathcal{Z}. \quad (4.9)$$

Note that the only thing unusual in this derivation is the interchange of signs in the exponentials and the fact that we have formed a Fourier series for a finite bandwidth spectrum instead of a Fourier series for a finite duration signal. It is this interchange of roles for time and frequency that suggests the corresponding changes in the signs of the exponentials.

Since $G(f)$ is assumed to be zero outside of $(-W, W)$, we can rewrite the coefficients as

$$
\begin{aligned}
c_n &= \frac{1}{2W} \int_{-W}^{W} G(f) e^{i2\pi \frac{f}{2W} n} \, df \\
&= \frac{1}{2W} \int_{-\infty}^{\infty} G(f) e^{i2\pi \frac{f}{2W} n} \, df \\
&= \frac{1}{2W} g\left(\frac{n}{2W}\right); \quad n \in \mathcal{Z}.
\end{aligned}
$$

Thus we have the following formula expressing the spectrum of a band-limited signal in terms of the samples of the signal itself:

$$G(f) = \begin{cases} \sum_{n=-\infty}^{\infty} \frac{g\left(\frac{n}{2W}\right)}{2W} e^{-i2\pi \frac{f}{2W} n}; & f \in (-W, W) \\ 0 & \text{otherwise} \end{cases} \quad (4.10)$$

This formula yields an interesting observation. Suppose that we define the discrete time signal $\gamma = \{\gamma_n; \, n \in \mathcal{Z}\}$ by the samples of g, i.e.,

$$\gamma_n = g(nT_s); \quad n \in \mathcal{Z}, \quad (4.11)$$

where $T_s = 1/2W \leq 1/2W_{\min}$ is the *sampling period*, the time between consecutive samples. Then the DTFT of the sampled signal γ_n is

$$
\begin{aligned}
\Gamma(f) &= \sum_{n=-\infty}^{\infty} \gamma_n e^{-i2\pi nf} \\
&= \sum_{n=-\infty}^{\infty} g(nT_s) e^{-i2\pi nf}, \quad f \in \left[-\frac{1}{2}, \frac{1}{2}\right). \quad (4.12)
\end{aligned}
$$

Comparing (4.10) and (4.12) yields

$$G(f) = \begin{cases} T_s \Gamma(fT_s); & f \in \left[-\frac{1}{2T_s}, \frac{1}{2T_s}\right] \\ 0 & \text{else} \end{cases} \quad (4.13)$$

We will see in the next section how to generalize this relation between the transform of a continuous time signal to the DTFT of a sampled version of the same signal to the case where $T_s > 1/2W_{\min}$, i.e., the signal is not bandlimited or it is bandlimited but the sampling period is too large for the above analysis to hold.

Knowing the spectrum in terms of the samples of the signal means that we can take an inverse Fourier transform and find the original signal *in terms of its samples!* In other words, knowing $g(n/2W)$ for all $n \in \mathcal{Z}$ determines the entire continuous time signal. Taking the inverse Fourier transform we have that

$$
\begin{aligned}
g(t) &= \int_{-\infty}^{\infty} G(f) e^{i2\pi f t}\, df \\
&= \int_{-W}^{W} \left(\sum_{n=-\infty}^{\infty} \frac{g(\frac{n}{2W})}{2W} e^{-i2\pi \frac{f}{2W} n} \right) e^{i2\pi f t}\, df \\
&= \sum_{n=-\infty}^{\infty} g(\frac{n}{2W}) \frac{1}{2W} \int_{-W}^{W} e^{i2\pi f(t-\frac{n}{2W})}\, df \\
&= \sum_{n=-\infty}^{\infty} g(\frac{n}{2W}) \operatorname{sinc}\left[2W(t - \frac{n}{2W}) \right].
\end{aligned}
$$

Summarizing the above, provided W is chosen so that $W \geq W_{\min}$ and hence so that $G(f) = 0$ when $|f| \geq W$, then the resulting bandlimited signal can be written in terms of $T_s = 1/2W$ as

$$
g(t) = \sum_{n=-\infty}^{\infty} g(nT_s) \operatorname{sinc}\left[\frac{t}{T_s} - n \right]. \tag{4.14}
$$

Alternatively, we can express the formula in terms of the *sampling frequency* $f_s = 1/T_s$, the number of samples per second, as

$$
g(t) = \sum_{n=-\infty}^{\infty} g(\frac{n}{f_s}) \operatorname{sinc}\left[f_s t - n \right], \tag{4.15}
$$

provided only that $f_s \geq 2W_{\min}$, a number known as the *Nyquist frequency* or *Nyquist rate*.

This is not a Fourier series because sinc functions and not exponentials appear, but it can be shown to be an orthogonal expansion since the sinc functions can be shown to be orthogonal on $\mathcal{R}$. Because of its fundamental importance in applications of Fourier analysis, we summarize this result formally.

Theorem 4.6 *The Whittaker-Shannon-Kotelnikov Sampling Theorem*

Suppose that $g = \{g(t); \ t \in \mathcal{R}\}$ *is a continuous time infinite duration signal that is bandlimited to* $(-W, W)$; *that is, its Fourier transform* $G(f)$ *has the property that*

$$G(f) = 0; \ f \notin (-W, W).$$

If the signal is sampled with sampling period T_s *and sampling frequency* $f_s = 1/T_s$ *and if* $f_s \geq 2W$, *then the original signal can be perfectly recovered from its samples as*

$$g(t) = \sum_{n=-\infty}^{\infty} g(nT_s) \operatorname{sinc} [f_s t - n]. \qquad (4.16)$$

The sampling theorem states that provided a bandlimited signal is sampled fast enough (at least twice its maximum frequency in Hz), then the signal can be recovered perfectly from its samples from the sampling expansion.

We shall later see a more engineering motivated derivation of the same result using the ideal sampling train and the properties of convolution. The previous derivation points out, however, that the result is easily demonstrated directly using Fourier series.

The basic idea can also be applied to obtain the dual result for time limited signals. Suppose that a signal g has the property that $g(t) = 0$ for $|t| \geq T$. Then it can be shown that

$$G(f) = \sum_{n=-\infty}^{\infty} G(\frac{n}{2T}) \operatorname{sinc} \left[2T(f - \frac{n}{2T}) \right]. \qquad (4.17)$$

This is sometimes called sampling in the frequency domain.

4.6   The DTFT of a Sampled Signal

The idea of sampling raises an interesting question: We have defined a continuous time Fourier transform for the original signal, and we could apply a discrete time Fourier transform to the sampled signal. How do these two Fourier transforms relate to each other? The answer to this question will provide a means of resolving another question that was implicit in the previous section. What happens if you sample a signal too slowly, i.e., below the Nyquist rate?

Suppose that $\{g(t); \ t \in \mathcal{R}\}$ and its Fourier transform $\{G(f); \ f \in \mathcal{R}\}$ are both absolutely integrable and continuous functions. Fix a positive number T_s and define a discrete time signal $\gamma = \{\gamma_n; \ n \in \mathcal{Z}\}$ by

$$\gamma_n = g(nT_s); \ n \in \mathcal{Z};$$

that is, γ is the sampled version of g. Unlike the previous section, no assumptions are made to the effect that g is bandlimited so there is no guarantee that the sampling theorem holds or that g can be reconstructed from γ. The immediate question is the following. How does the DTFT Γ of the sampled signal, defined by

$$\Gamma(f) = \sum_{k=-\infty}^{\infty} \gamma_k e^{-i2\pi f k}; f \in [-\frac{1}{2}, \frac{1}{2}) \tag{4.18}$$

compare to the CTFT G of the original signal? We establish this important result in two ways, each of which is of interest in its own right. The first is the more direct and the shorter. The second provides an application of Fourier series and yields an important side result, the Poisson summation formula.

First Approach

Since the unknown $\Gamma(f)$ is the Fourier transform of the discrete time signal $\gamma = \{\gamma_n = g(nT); \ n \in \mathcal{Z}\}$, γ must be the inverse DTFT of Γ:

$$\gamma_n = \int_{-\frac{1}{2}}^{\frac{1}{2}} e^{i2\pi f n} \Gamma(f) \, df. \tag{4.19}$$

We also have that

$$
\begin{aligned}
\gamma_n &= g(nT) \\
&= \int_{-\infty}^{\infty} G(f) e^{i2\pi f n T_s} \, df \\
&= \sum_{k=-\infty}^{\infty} \int_{\frac{(k-1/2)}{T_s}}^{\frac{(k+1/2)}{T_s}} G(f) e^{i2\pi f n T_s} \, df
\end{aligned}
\tag{4.20}
$$

where we have broken up the integral into an infinite sum of integrals over disjoint intervals of length $1/T_s$. Each of these integrals becomes with a change of variables $f' = fT_s + k$

$$
\begin{aligned}
\int_{\frac{(k-1/2)}{T_s}}^{\frac{(k+1/2)}{T_s}} G(f) e^{i2\pi f n T_s} \, df &= \int_{-\frac{1}{2}}^{\frac{1}{2}} G(\frac{f'-k}{T_s}) e^{i2\pi (f'-k)n} \, df' \\
&= \int_{-\frac{1}{2}}^{\frac{1}{2}} G(\frac{f'-k}{T_s}) e^{i2\pi f' n} \, df'
\end{aligned}
\tag{4.21}
$$

so that interchanging the sum and integral in (4.20) yields the formula

$$\gamma_n = \int_{-\frac{1}{2}}^{\frac{1}{2}} e^{i2\pi fn} \left[\sum_{k=-\infty}^{\infty} G(\frac{f-k}{T_s}) \right] df. \qquad (4.22)$$

Comparison with (4.19) identifies the term in brackets as the DTFT of γ; that is,

$$\Gamma(f) = \sum_{k=-\infty}^{\infty} G(\frac{f-k}{T_s}). \qquad (4.23)$$

Second Approach

We establish this relation in an indirect manner, but one which yields an interesting and well known side result. Let $\tilde{\Gamma}$ denote the periodic extension of Γ with period 1; that is,

$$\tilde{\Gamma}(f) = \sum_{k=-\infty}^{\infty} \gamma_k e^{-i2\pi fk}; f \in \mathcal{R}. \qquad (4.24)$$

In much of the literature the same notation is used for Γ and $\tilde{\Gamma}$ with the domain of definition left to context, but we will distinguish them.

Consider the following function of frequency formed by adding an infinite number of scaled and shifted versions of G:

$$\alpha(f) = \frac{1}{T_s} \sum_{k=-\infty}^{\infty} G(\frac{f-k}{T_s}); \ f \in \mathcal{R}. \qquad (4.25)$$

This function is a continuous parameter function (f is a continuous argument) and it is periodic in f with period 1 since adding any integer to f simply results in a reindexing of the infinite sum and does not change the value of the sum. Hence we can expand $\alpha(f)$ in a Fourier series in f, e.g.,

$$\alpha(f) = \sum_{k=-\infty}^{\infty} c_k e^{-i2\pi fk}, \qquad (4.26)$$

where

$$c_k = \int_{-1/2}^{1/2} \alpha(f) e^{i2\pi fk} df.$$

Before evaluating these coefficients, note the similarity of (4.24) and (4.26). In fact we will demonstrate that $c_k = \gamma_k = g(kT_s)$, thereby showing that $\tilde{\Gamma}(f) = \alpha(f)$ and hence $\Gamma(f) = \alpha(f)$ for $\alpha \in [-1/2, 1/2)$ since the two continuous functions have the same Fourier series, which will provide the desired formula relating Γ and G.

To evaluate the coefficients c_k we use the usual trick of substitution followed by a change of order of integration and summation.

$$
\begin{aligned}
c_k &= \int_{-\frac{1}{2}}^{\frac{1}{2}} \frac{1}{T_s} \sum_{n=-\infty}^{\infty} G(\frac{f-n}{T_s}) e^{i2\pi kf}\, df \\
&= \frac{1}{T_s} \sum_{n=-\infty}^{\infty} \int_{-\frac{1}{2}}^{\frac{1}{2}} G(\frac{f-n}{T_s}) e^{i2\pi kf}\, df.
\end{aligned}
\tag{4.27}
$$

Changing variables $f' = (f-n)/T_s$ in the rightmost integrals we have

$$
\begin{aligned}
c_k &= \frac{1}{T_s} \sum_{n=-\infty}^{\infty} \int_{-\frac{1}{2T_s}-\frac{n}{T_s}}^{\frac{1}{2T_s}-\frac{n}{T_s}} G(f') e^{i2\pi k(T_s f'+n)}\, T_s df' \\
&= \sum_{n=-\infty}^{\infty} \int_{-\frac{1}{2T_s}-\frac{n}{T_s}}^{\frac{1}{2T_s}-\frac{n}{T_s}} G(f') e^{i2\pi kT_s f'}\, df' \\
&= \int_{-\infty}^{\infty} G(f') e^{i2\pi kT_s f'}\, df' \\
&= g(kT_s).
\end{aligned}
\tag{4.28}
$$

The infinite sum of integrals over a disjoint collection of intervals together constituting the entire real line resulted in a single integral in the penultimate step above.

We have thus proved that the DTFT of the discrete time signal γ formed by sampling the continuous time signal g with sampling period T_s is

$$
\Gamma(f) = \frac{1}{T_s} \sum_{k=-\infty}^{\infty} G(\frac{f-k}{T_s});\, f \in [-\frac{1}{2}, \frac{1}{2}),
\tag{4.29}
$$

where G is the CTFT of the continuous time signal g. Before interpreting this result, however, we point out a side result that is important in its own right. The above development proved the following result, which is a version of the Poisson summation formula. We state the result here for later reference.

Theorem 4.7 *The Poisson Summation Formula*
Given an infinite duration signal $\{g(t);\, t \in \mathcal{R}\}$ with absolutely integrable continuous Fourier transform $G(f)$, then for any $T > 0$

$$
\frac{1}{T} \sum_{k=-\infty}^{\infty} G(\frac{f-k}{T}) = \sum_{n=-\infty}^{\infty} g(nT) e^{-i2\pi nf};\, f \in \mathcal{R}.
\tag{4.30}
$$

In particular, if $f = 0$, then

$$\sum_{k=-\infty}^{\infty} \frac{G(\frac{k}{T})}{T} = \sum_{n=-\infty}^{\infty} g(nT). \tag{4.31}$$

Returning to the consideration of the DTFT of a sampled process, we formally state (4.29) in context as a theorem and then interpret it in light of the sampling theorem.

Theorem 4.8 *Given an infinite duration signal $\{g(t); t \in \mathcal{R}\}$ with absolutely integrable continuous time Fourier transform $G(f)$, then the discrete time Fourier transform of the signal $\{\gamma_n = g(nT_s); n \in \mathcal{Z}\}$ is given by*

$$\Gamma(f) = \frac{1}{T_s} \sum_{k=-\infty}^{\infty} G(\frac{f - k}{T_s}); \ f \in [-\frac{1}{2}, \frac{1}{2}). \tag{4.32}$$

Thus the DTFT of a sampled signal is the sum of frequency scaled shifted replicas of the spectrum of the original continuous time signal.

As an example, consider the simple Fourier transform of a simple continuous time signal in Figure 4.1. The DTFT of the sampled signal will

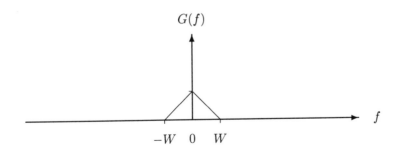

$G(f)$

$-W \quad 0 \quad W$

Figure 4.1: Original CTFT $G(f)$

be the sum of shifted replicas of $G(f/T_s)$, the original spectrum with the argument "stretched." We depict this basic waveform by simply relabeling the time axis as in Figure 4.2.

If the sampling period T_s is chosen so that $1/T_s \geq 2W$ or $1/2 \geq WT_s$, then the DTFT of the sampled signal is given by (4.29) and the individual terms in the sum do not overlap, yielding the picture of Figure 4.3 with separate "islands" for each term in the sum. Only one term, the $k = 0$ term, will be nonzero in the frequency region $[-1/2, 1/2]$. In this case,

$$\Gamma(f) = G(f/T_s)/T_s; \ f \in (-1/2, 1/2) \tag{4.33}$$

$G(f/T_s)$

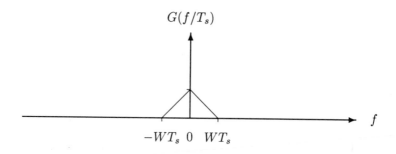

$$-WT_s \quad 0 \quad WT_s$$

Figure 4.2: Stretched Original CTFT $G(f/T_s)$

$\tilde{\Gamma}(f)$

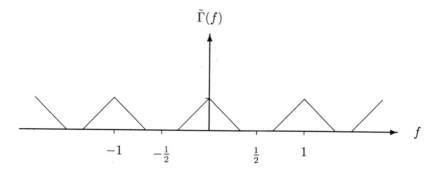

$$-1 \qquad -\tfrac{1}{2} \qquad\qquad \tfrac{1}{2} \qquad 1$$

Figure 4.3: DTFT of Sampled Signal: $1/T_s > 2W$

and the DTFT and CTFT are simply frequency and amplitude scaled versions of each other and the continuous time signal g can be recovered from the discrete time signal γ by inverting

$$G(f) = \begin{cases} T_s\Gamma(T_sf) & f \in [-\tfrac{1}{2T_s}, \tfrac{1}{2T_s}] \\ 0 & \text{else} \end{cases}, \qquad (4.34)$$

which is the same as (4.13) and provides another proof of the sampling theorem! If g is not bandlimited, however, the separate scaled images of G in the sum giving $\Gamma(f)$ will overlap as depicted in Figure 4.4 so that taking the sum to form the spectrum will yield a distorted version in $(-1/2, 1/2)$, as shown in Figure 4.5. This prevents recovery of G and hence of g in general. This phenomenon is known as *aliasing*.

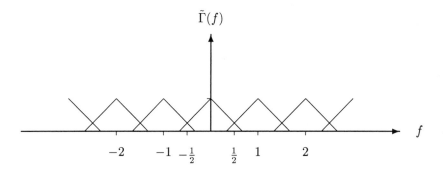

Figure 4.4: DTFT of Sampled Signal is sum of overlapping islands when $1/T_s < 2W$

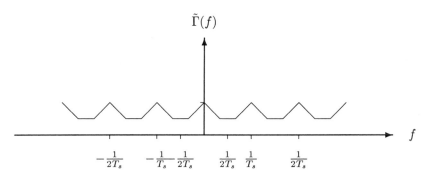

Figure 4.5: DTFT of Sampled Signal: $1/T_s < 2W$

The Poisson Summation

The Poisson summation formula of Theorem 4.7 is primarily important here because of its basic role in deriving the DTFT of a sampled signal. It also provides some interesting equalities for sums in the two domains and is thereby useful in evaluating some apparently complicated infinite summations. As an example, consider the Gaussian signal $g(t) = e^{-\pi t^2}$; $t \in \mathcal{R}$ which has as Fourier transform $G(f) = e^{-\pi f^2}$; $f \in \mathcal{R}$. Then for any $T > 0$

$$\sum_{n=-\infty}^{\infty} e^{-\pi(nT)^2} = \sum_{k=-\infty}^{\infty} \frac{e^{-\pi(\frac{k}{T})^2}}{T}. \tag{4.35}$$

This equality is surprising and not trivial to prove.

The Poisson summation formula is also seen in the dual form with the sum of replicas in the time domain instead of the frequency domain. The following version can be proved in a similar manner to the previous or by using duality.

Theorem 4.9 *The Poisson Summation Formula: Second Form*

Given an infinite duration, absolutely integrable, and continuous signal $\{g(t);\ t \in \mathcal{R}\}$ *with Fourier transform* $G(f)$, *then for any* $T > 0$

$$\sum_{n=-\infty}^{\infty} g(t-nT) = \sum_{k=-\infty}^{\infty} \frac{G(\frac{k}{T})}{T} e^{i2\pi \frac{k}{T} t}. \tag{4.36}$$

Setting $t = 0$ yields the special case of (4.31).

4.7 ⋆ Pulse Amplitude Modulation (PAM)

In the previous section the DTFT of a sampled continuous waveform was derived. We now reverse the process and see what happens when a discrete time signal γ with a known DTFT Γ is used to construct a continuous time signal by modulating a pulse sequence. We begin without any assumptions regarding the origin of the discrete time signal. Later the case of sampling will be revisited and the effect of bandlimiting again considered.

Given the discrete time signal γ, define a continuous time signal $r = \{r(t);\ t \in \mathcal{R}\}$ by forming a PAM signal

$$r(t) = \sum_{n=-\infty}^{\infty} \gamma_n p(t - nT_s);\ t \in \mathcal{R}, \tag{4.37}$$

where the signal (pulse) $p = \{p(t);\ t \in \mathcal{R}\}$ has a Fourier transform $P(f)$. In the ideal sampling expansion, the pulses would be sinc functions and the transforms box functions. A more realistic pulse might be

$$p(t) = \begin{cases} 1/T_s & t \in [0, T_s) \\ 0 & \text{otherwise} \end{cases} \tag{4.38}$$

which would correspond to the overall system being an idealized sample and hold circuit. What is the transform of r and how does it relate to those of Γ and P? We have that

$$R(f) \quad = \quad \int_{-\infty}^{\infty} r(t) e^{-i2\pi ft}\, dt$$

$$= \int_{-\infty}^{\infty} \sum_{n=-\infty}^{\infty} \gamma_n p(t - nT_s) e^{-i2\pi ft} \, dt$$

$$= \sum_{n=-\infty}^{\infty} \gamma_n \int_{-\infty}^{\infty} p(t - nT_s) e^{-i2\pi ft} \, dt$$

$$= \sum_{n=-\infty}^{\infty} \gamma_n P(f) e^{-i2\pi nfT_s}, \tag{4.39}$$

where the last step used the shift theorem for continuous time Fourier transforms. Pulling $P(f)$ out of the sum we are left with

$$R(f) = P(f) \sum_{n=-\infty}^{\infty} \gamma_n e^{-i2\pi nfT_s}$$

$$= P(f)\tilde{\Gamma}(fT_s), \ f \in \mathcal{R}, \tag{4.40}$$

where $\tilde{\Gamma}$ is the periodic extension of Γ.

Now return to the case where γ is formed by sampling a continuous time signal g with transform G. From (4.29) we then have that the Fourier transform of the reconstructed signal will be

$$R(f) = P(f)\frac{1}{T_s} \sum_{k=-\infty}^{\infty} G(f - \frac{k}{T_s}). \tag{4.41}$$

An important aspect of this formula is that if $P(f)$ is itself bandlimited, then so is $R(f)$. For example, if we choose the sinc pulse of the idealized sampling expansion, $p(t) = \text{sinc}(t/T_s)$, then $P(f)$ is a box function which is zero for $|f| \geq 1/2T_s$ and hence $R(f)$ is similarly bandlimited. If g is bandlimited to the same band, e.g., if g is bandlimited to $[-W, W]$ and the sampling frequency $f_s = 1/T_s$ exceeds the Nyquist frequency $2W$, then (4.41) simplifies to $R(f) = G(f)$ and we have once again proved the sampling theorem since the PAM signal constructed by modulating sincs with the samples has the same transform as the original signal.

Suppose on the other hand that one uses the rectangular pulse of (4.38), then

$$P(f) = \text{sinc}(\frac{fT_s}{2}) e^{-i2\pi fT_s/4}$$

which is not bandlimited, and hence neither is $R(f)$.

4.8 The Stretch Theorem

We have seen several specific examples of scaling a time or frequency domain variable. We now develop the general cases. Unfortunately, these results differ in important ways for the different signal types and hence we are forced to consider the cases separately.

Theorem 4.10 *Continuous Time Stretch Theorem*

Given a continuous time, infinite duration signal $g = \{g(t); t \in \mathcal{R}\}$ with Fourier transform $G = \{G(f); f \in \mathcal{R}\}$ and a real-valued nonzero constant a, then

$$\{g(at); t \in \mathcal{R}\} \leftrightarrow \{\frac{1}{|a|}G(\frac{f}{a}); f \in \mathcal{R}\}. \tag{4.42}$$

Proof: We here consider only the case $a > 0$. The case of negative a is left as an exersise. If a is strictly positive, just change variables $\tau = at$ to obtain

$$\int_{-\infty}^{\infty} g(at)e^{-i2\pi tf}\,dt = \int_{-\infty}^{\infty} g(\tau)e^{-i2\pi\frac{\tau}{a}f}\frac{d\tau}{a} = \frac{1}{a}G(\frac{f}{a}).$$

As an example of the theorem, the Fourier transform of $g(t) = \sqcap(2t)$ is immediately seen to be $\frac{1}{2}\text{sinc}(\frac{f}{2})$.

The stretch theorem can be applied to a finite duration continuous time signal, but the stretching changes the domain of definition of the signal, e.g., from $[0,T)$ to $[0,T/a)$ if $a > 0$. (For this discussion we confine attention to the case of positive a.) If $g = \{g(t); t \in [0,T)\}$ has a Fourier transform $\{G(\frac{k}{T}); k \in \mathcal{Z}\}$, then the stretched signal $g_a = \{g(at); t \in [0,T/a)\}$ will have a spectrum $G_a = \{\frac{1}{|a|}G(\frac{1}{a}\frac{k}{T/a}); k \in \mathcal{Z}\} = \{\frac{1}{|a|}G(\frac{k}{T}); k \in \mathcal{Z}\}$, that is, the multiplier inside G is cancelled by a corresponding stretch in the length of the time domain. This apparent oddity is in fact consistent with the infinite duration stretch theorem, as can be seen in two ways. First, we have seen that the spectrum of any finite duration signal of duration D has the form $H = \{H(k/D); k \in \mathcal{Z}\}$. Thus since the duration of G_a is T/a, we have that $G_a = \{G_a(k/(T/a)) = G_a(ak/T); k \in \mathcal{Z}\}$. Comparing this with the previous formula for G_a shows that $G_a(\frac{k}{T/a}) = \frac{1}{|a|}G(\frac{k}{T})$. Setting $f = k/(T/a)$ this is

$$G_a(f) = \frac{1}{|a|}G(\frac{f}{a}),$$

exactly as in the infinite duration case.

4.9 ⋆ Downsampling

Stretching the time domain variable has a much different behavior in discrete time. Suppose that $\{g_n; n \in \mathcal{Z}\}$ is a discrete time signal. If analogous to the continuous time case we try to form a new signal $\{g_{an}; n \in \mathcal{Z}\}$ we run into an immediate problem: if an is not an integer, then g_{an} is not defined and the new signal does not make any sense. In order to have g_{an} be a reasonable signal we must do one of two things. Either we must restrict an to be an integer for all integers n so that g_{an} is well-defined or we must extend the definition of our original signal to cover a wider index set; that is, define the values of g_{an} for all of those an which are not integers. We focus on the first approach and later return to the second.

If an is required to be an integer for all integers n, then a must itself be an integer (or else an would not be an integer for the case $n = 1$). Hence let's rename a as M and consider the new discrete time infinite duration process $g^{[M]} = \{g_{Mn}; n \in \mathcal{Z}\}$. Observe that the new process is formed by taking every Mth sample from the original process. Hence stretching in discrete time is called *downsampling*. Unlike the continuous time case, we have not just stretched the original signal, we have thrown away several of its values and hence have likely caused more distortion than a mere compression of the time scale. Alternatively, stretching a continuous time signal by a could be undone by restretching using $1/a$; nothing was irretrievably lost. On the other hand, stretching in the discrete time case has destroyed data.

The questions are now: Is there an analogous stretching theorem for the discrete time case and are there conditions under which one can recover the original signal from the downsampled signal? We shall see that these questions are interrelated and are resolved by a discrete time version of the sampling theorem.

As previously, let g denote the original signal and let $g^{[M]}$ denote the downsampled signal. Let G and $G^{[M]}$ denote the corresponding Fourier transforms. We have that

$$G^{[M]}(f) = \sum_{n=-\infty}^{\infty} g_{nM} e^{-i2\pi fn}$$

which in general bears no simple relationship to the spectrum

$$G(f) = \sum_{n=-\infty}^{\infty} g_n e^{-i2\pi fn}.$$

Unlike the continuous time case we cannot simply change variables.

The original signal g can be obtained by inverting G as

$$g_n = \int_{-\frac{1}{2}}^{\frac{1}{2}} G(f) e^{i2\pi fn}\, df$$

and hence we can find the downsampled signal by simply plugging in the appropriate values of time:

$$g_{nM} = \int_{-\frac{1}{2}}^{\frac{1}{2}} G(f)e^{i2\pi fnM} \, df.$$

Here we can change variables $\sigma = fM$ to write

$$g_{nM} = \int_{-\frac{M}{2}}^{\frac{M}{2}} \frac{G(\frac{\sigma}{M})}{M} e^{i2\pi\sigma n} \, d\sigma. \tag{4.43}$$

This formula can be compared with the inverse Fourier transform representation for the downsampled process:

$$g_{nM} = \int_{-\frac{1}{2}}^{\frac{1}{2}} G^{[M]}(f)e^{i2\pi fn} \, df. \tag{4.44}$$

It is tempting to identify $G^{[M]}(f)$ with $G(f/M)/M$, but this inference cannot in general be made because of the differing regions of integration. Suppose, however, that $G(f/M) = 0$ for $|f| \geq 1/2$; that is, that

$$G(f) = 0 \text{ for } |f| \geq \frac{1}{2M}.$$

This is a requirement that the discrete time signal be *bandlimited* analogous to the definition for continuous time signals. If we make this requirement, then (4.43) becomes

$$g_{nM} = \int_{-\frac{1}{2}}^{\frac{1}{2}} \frac{G(\frac{f}{M})}{M} e^{i2\pi fn} \, df \tag{4.45}$$

and this formula combined with (4.44) and the uniqueness of Fourier transforms proves the following discrete time stretch theorem.

Theorem 4.11 *Discrete Time Stretch Theorem (Downsampling Theorem)*
Given a discrete time, infinite duration signal $g = \{g_n; n \in \mathcal{Z}\}$ with Fourier transform $G = \{G(f); f \in [-\frac{1}{2}, \frac{1}{2})\}$. If there is an integer L such that $G(f) = 0$ for $|f| \geq 1/2L$, then for any positive integer $M \leq L$

$$\{g_{nM}; n \in \mathcal{Z}\} \leftrightarrow \{\frac{1}{M}G(\frac{f}{M}); f \in [-\frac{1}{2}, \frac{1}{2})\}. \tag{4.46}$$

The theorem is illustrated for $M = 2$ in Figure 4.6.

The bandlimited condition and the implied result easily provide a discrete time sampling theorem.

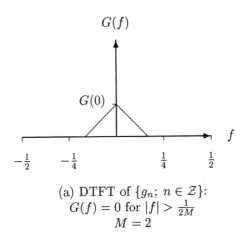

(a) DTFT of $\{g_n; \ n \in \mathcal{Z}\}$:
$$G(f) = 0 \text{ for } |f| > \tfrac{1}{2M}$$
$$M = 2$$

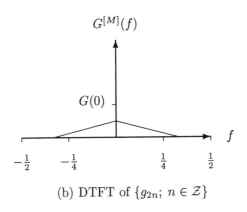

(b) DTFT of $\{g_{2n}; \ n \in \mathcal{Z}\}$

Figure 4.6: DTFT of Downsampled Signal

Theorem 4.12 *The Discrete Time Sampling Theorem*

Suppose that $g = \{g_n; n \in \mathcal{Z}\}$ is a discrete time, infinite duration signal with Fourier transform $G = \{G(f); f \in [-\frac{1}{2}, \frac{1}{2})\}$. Then if there is an integer L such that $G(f) = 0$ for $|f| \geq 1/2L$, then for any $M \leq L$

$$g_n = \sum_{k=-\infty}^{\infty} g_{kM} \operatorname{sinc}(\frac{n}{M} - k) \qquad (4.47)$$

and hence the original signal can be recovered from its downsampled version.

Proof: We have, using the bandlimited property and making a change of variables, that

$$
\begin{aligned}
g_n &= \int_{-\frac{1}{2}}^{\frac{1}{2}} G(\sigma) e^{i2\pi\sigma n} \, d\sigma = \int_{-\frac{1}{2M}}^{\frac{1}{2M}} G(\sigma) e^{i2\pi\sigma n} \, d\sigma \\
&= \int_{-\frac{1}{2}}^{\frac{1}{2}} \frac{G(\frac{f}{M})}{M} e^{i2\pi\frac{f}{M}n} \, df \\
&= \int_{-\frac{1}{2}}^{\frac{1}{2}} G^{[M]}(f) e^{i2\pi f \frac{n}{M}} \, df \\
&= \int_{-\frac{1}{2}}^{\frac{1}{2}} \sum_{k=-\infty}^{\infty} g_{kM} e^{-i2\pi fk} e^{i2\pi f \frac{n}{M}} \, df \\
&= \sum_{k=-\infty}^{\infty} g_{kM} \int_{-\frac{1}{2}}^{\frac{1}{2}} e^{i2\pi f(\frac{n}{M} - k)} \, df \\
&= \sum_{k=-\infty}^{\infty} g_{kM} \operatorname{sinc}(\frac{n}{M} - k).
\end{aligned}
$$

The theorem can also be proved in a manner much like the continuous time sampling theorem. One begins by expanding the bandlimited transform $G(f)$ on the region $[-1/2M, 1/2M)$ into a Fourier series, the coefficients of which are in terms of the samples. An inverse DTFT then yields the theorem.

4.10   ⋆ Upsampling

As previously mentioned, another method of stretching the time variable in discrete time is to permit a scaling constant that is not an integer, but to extend the definition of the signal to the new time indices. The most common example of this is to take the scale constant to be of the form $a = 1/M$, where M is an integer. Here we begin with a signal $g = \{g_n; n \in \mathcal{Z}\}$ and form a new signal, say $h = \{h_n; n \in \mathcal{Z}\}$ where $h_n = g_{n/M}$ for those values of n for which n/M is an integer and where we have to define h_n for other values of n. One common, simple definition is to make $h_n = 0$ for all n that are not multiples of M. This has the effect of producing a signal that has every Mth sample equal to a sample from g_n with 0s in between. This is the reverse process of downsampling and is called *upsampling*. An alternative (and perhaps more sensible) approach would be to interpolate between the values. We consider zero filling, however, because of its simplicity. Interpolation could be accomplished by subsequent smoothing using a linear filter.

If we upsample and fill with zeros, then observe that the Fourier transform of h is

$$
\begin{aligned}
H(f) &= \sum_{n=-\infty}^{\infty} h_n e^{-i2\pi fn} \\
&= \sum_{n=-\infty}^{\infty} h_{nM} e^{-i2\pi fnM} \\
&= \sum_{n=-\infty}^{\infty} g_n e^{-i2\pi fnM}; \ f \in [-\frac{1}{2}, \frac{1}{2}).
\end{aligned}
$$

The latter formula resembles $G(f)$, the Fourier transform of g, with a scale change of the frequency range. Strictly speaking, $G(f)$ is only defined for $f \in [-\frac{1}{2}, \frac{1}{2})$, but the right-hand side of the above equation has frequencies fM which vary from $[-\frac{M}{2}, \frac{M}{2})$. Thus $H(f)$ looks like M periodic replicas of a compressed $G(f)$. In particular, if $\tilde{G}(f)$ is the periodic extension of $G(f)$, then

$$
H(f) = \tilde{G}(Mf); \ f \in [-\frac{1}{2}, \frac{1}{2}).
$$

The case for $M = 2$ is depicted in Figure 4.7.

4.11   The Derivative and Difference Theorems

Suppose that $g = \{g(t); t \in \mathcal{R}\}$ is an infinite duration continuous time signal that is everywhere differentiable and suppose that $g' = \{g'(t); t \in \mathcal{R}\}$ is the signal defined by

$$
g'(t) = \frac{dg(t)}{dt}.
$$

$G(f)$

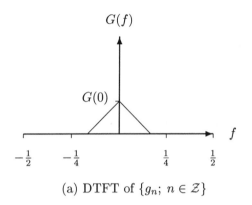

(a) DTFT of $\{g_n;\ n \in \mathcal{Z}\}$

$H(f)$

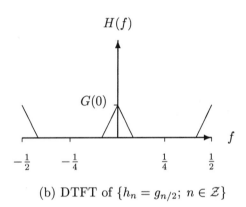

(b) DTFT of $\{h_n = g_{n/2};\ n \in \mathcal{Z}\}$

Figure 4.7: DTFT of Upsampled Signal

Let $G(f)$ denote the Fourier transform of g. Then if the signal is nice enough for the Fourier inversion formula to hold and for us to be able to interchange the order of differentiation and integration:

$$
\begin{aligned}
g'(t) &= \frac{d}{dt} g(t) \\
&= \frac{d}{dt} \left(\int_{-\infty}^{\infty} G(f) e^{i2\pi ft}\, df \right)
\end{aligned}
$$

$$= \int_{-\infty}^{\infty} G(f) \frac{d}{dt} e^{i2\pi ft} \, df$$

$$= \int_{-\infty}^{\infty} G(f)(i2\pi f) e^{i2\pi ft} \, df.$$

From the inversion formula we can identify $i2\pi f G(f)$ as the Fourier transform of g', yielding the following result.

Theorem 4.13 *The Derivative Theorem*

Given a continuous time infinite duration everywhere differentiable signal g with Fourier transform $G(f)$, then

$$\mathcal{F}_f(g') = i2\pi f G(f); \ f \in \mathcal{R}. \tag{4.48}$$

Similarly, if g is n times differentiable and if $g^{(n)}$ is the signal defined by $g^{(n)}(t) = \frac{d^n g(t)}{dt^n}$, then

$$\mathcal{F}_f(g^{(n)}) = (i2\pi f)^n G(f); \ f \in \mathcal{R}. \tag{4.49}$$

As an example of the derivative theorem, consider the infinite duration continuous time Gaussian pulse $g = \{e^{-\pi t^2}; \ t \in \mathcal{R}\}$. The theorem implies that the CTFT of the signal $g'(t) = -2\pi t e^{-\pi t^2}$ is $i2\pi f e^{-\pi f^2}$.

The analogous discrete time result follows by replacing the derivative with a difference, sometimes called a discrete time derivative. If $g = \{g_n; \ n \in \mathcal{Z}\}$ is an infinite duration discrete time signal, define the difference signal g' by

$$g'_n = g_n - g_{n-1}.$$

In this case we have

$$
\begin{aligned}
\mathcal{F}_f(g') &= \sum_{n=-\infty}^{\infty} g_n e^{-i2\pi fn} - \sum_{n=-\infty}^{\infty} g_{n-1} e^{-i2\pi fn} \\
&= \sum_{n=-\infty}^{\infty} g_n e^{-i2\pi fn} - \sum_{n=-\infty}^{\infty} g_n e^{-i2\pi f(n+1)} \\
&= (1 - e^{-i2\pi f}) \sum_{n=-\infty}^{\infty} g_n e^{-i2\pi fn}.
\end{aligned}
$$

This yields the discrete time equivalent of the derivative theorem.

Theorem 4.14 *The Difference Theorem*

Given a discrete time infinite duration signal g with Fourier transform $G(f)$, then

$$\mathcal{F}_f(g') = (1 - e^{-i2\pi f})G(f); \; f \in [-\frac{1}{2}, \frac{1}{2}). \qquad (4.50)$$

Similarly, if $g^{(k)}$ is the kth order difference of the signal g, then

$$\mathcal{F}_f(g^{(k)}) = (1 - e^{-i2\pi f})^k G(f); \; f \in [-\frac{1}{2}, \frac{1}{2}). \qquad (4.51)$$

Note that the nth order difference of a signal can be defined iteratively by

$$g_n^{(k)} = g_n^{(k-1)} - g_{n-1}^{(k-1)},$$

e.g.,

$$g_n^{(2)} = g_n^{(1)} - g_{n-1}^{(1)} = g_n - 2g_{n-1} + g_{n-2},$$

and so on.

The derivative and difference theorems are similar, but the dependence on f in the extra term is different. In particular, the multiplier in the discrete time case is periodic in f with period 1 (as it should be).

Both results give further transform pairs considering the dual results. In the continuous time case we have that the Fourier transform of the signal $tg(t)$ is

$$\mathcal{F}_f(\{tg(t); t \in \mathcal{R}\}) = \frac{i}{2\pi}G'(f); \; f \in \mathcal{R}.$$

This can be proved directly by differentiating the formula for $G(f)$ with respect to f which results in a multiplier of $-i2\pi t$. The discrete time equivalent is left as an exercise.

4.12   Moment Generating

This section collects a group of results often called *moment theorems* because they show how moments or averages in one domain can be evaluated in the other domain. These results are useful as a tool for evaluating integrals of a particular kind and for studying the relation between "pulse width" and "bandwidth" of a signal. The results are also useful simply because they provide practice in manipulating Fourier sums and integrals.

Because of the importance of Fourier transforms in finding moments, a variation of the Fourier transform used in probability theory is often called the *moment generating function*. Such variations may differ from the ordinary Fourier transform in the sign in the exponential and in the nature of the independent variable, but the basic ideas of evaluating moments from transforms remain the same.

We focus on moments in the time domain, but duality can be invoked to find similar formulas for frequency domain moments. We will consider both continuous and discrete time, but we focus on the infinite duration case for simplicity. It is often convenient to normalize moments, as we shall do later.

We begin with the general definition of moments (unnormalized) and we will then consider several special cases. Given a signal g, define for any nonnegative integer n the nth order moment

$$M_g^{(n)} = \begin{cases} \int\limits_{-\infty}^{\infty} t^n g(t)\, dt & \text{continuous time} \\[2ex] \sum\limits_{k=-\infty}^{\infty} k^n g_k & \text{discrete time} \end{cases}$$

If g is a probability density function or probability mass function (a nonnegative signal that integrates or sums to one, respectively) then the nth order moment is the usual statistical nth order moment, the mean or average of the nth power of the independent variable weighted by the signal.

The simplest example of a moment is the 0th order moment or *area* of the signal:

$$M_g^{(0)} = \begin{cases} \int\limits_{-\infty}^{\infty} g(t)\, dt & \text{continuous time} \\[2ex] \sum\limits_{k=-\infty}^{\infty} g_k & \text{discrete time} \end{cases}$$

This is easily recognized in both cases as being $G(0)$, where $G(f)$ is the Fourier transform of g. Thus for both continuous and discrete time,

$$M_g^{(0)} = G(0). \tag{4.52}$$

The dual of this result is similarly found to be

$$M_G^{(0)} = \begin{cases} \int\limits_{-\infty}^{\infty} G(f)\, df & \text{continuous time} \\[2ex] \int\limits_{-\frac{1}{2}}^{\frac{1}{2}} G(f)\, df & \text{discrete time} \end{cases} = g(0). \tag{4.53}$$

Eq. (4.52) is called the *area property*.

Note the difference of appearance of the two results: both continuous time and discrete time signals have continuous spectra and hence both spectral moments are integrals. Similar results could be obtained for the finite duration counterparts, but that is left as an exercise.

As an example of the use of the area property consider the continuous time signal $g(t) = J_0(2\pi t)$, a zeroth order ordinary Bessel function. From the transform tables $G(f) = \frac{\Pi(f/2)}{\pi\sqrt{1-f^2}}$. Thus from the area property

$$\int_{-\infty}^{\infty} J_0(2\pi t)\, dt = G(0) = \frac{1}{\pi}.$$

Next consider the first moment. We treat the continuous time case; the discrete time case follows in exactly the same way with sums replacing integrals. Observe that if we differentiate the spectrum with respect to f we have

$$G'(f) = \frac{d}{df}\left(\int_{-\infty}^{\infty} g(t)e^{-i2\pi ft}\, dt\right) = (-i2\pi)\int_{-\infty}^{\infty} tg(t)e^{-i2\pi ft}\, dt.$$

(This is the dual of the derivative theorem.) Thus choosing $f = 0$ we have that

$$\int_{-\infty}^{\infty} tg(t)\, dt = \frac{i}{2\pi}G'(0). \tag{4.54}$$

The above formula of course holds only if the derivative and integral can be interchanged. This is the case if the absolute moment exists,

$$\int_{-\infty}^{\infty} |tg(t)|\, dt < \infty.$$

In general (continuous or discrete time) we have the following:

$$M_g^{(1)} = \frac{i}{2\pi}G'(0). \tag{4.55}$$

This result is called the *first moment property*.

The importance of this result is that it is often easier to differentiate than to integrate. Thus if one knows the spectrum, it is easier to take a single derivative than to integrate to find a moment.

As an example of the first moment property, consider the continuous time signal $g(t) = \Pi(t - 3)$ and find $\int_{-\infty}^{\infty} tg(t)\, dt$. We have that $G(f) = \text{sinc}(f)e^{-i6\pi f}$ and hence

$$\frac{dG(f)}{df} = -6\pi i\, \text{sinc}(f)e^{-6i\pi f} + e^{-6i\pi f}\frac{d}{df}\,\text{sinc}(f).$$

At $f = 0$, $\frac{d\,\text{sinc}(f)}{df} = 0$ (any even function has zero slope at origin if the derivative is well-defined) and therefore $G'(0) = -i6\pi$ and hence

$$\int_{-\infty}^{\infty} tg(t)\, dt = \frac{iG'(0)}{2\pi} = \frac{6\pi}{2\pi} = 3.$$

This same procedure works for any moment: Differentiate the spectrum n times and evaluate the result at $f = 0$ to obtain the nth moment times some constant. We now summarize these results:

Theorem 4.15 *Moment Theorem*

Suppose that g is an infinite duration signal which satisfies the following condition:

$$\int_{-\infty}^{\infty} |t^n g(t)|\, dt \quad < \quad \infty; \quad continuous\ time \qquad (4.56)$$

$$\sum_{k=-\infty}^{\infty} |k^n g_k| \quad < \quad \infty; \quad discrete\ time; \qquad (4.57)$$

then

$$M_g^{(n)} = (\frac{i}{2\pi})^n G^{(n)}(0), \qquad (4.58)$$

where

$$G^{(n)}(0) = \frac{d^n}{df^n} G(f)|_{f=0}.$$

As an example of a proof, in the infinite duration continuous time case we have that

$$G^{(n)}(f) \quad = \quad \frac{d^n}{df^n} \int_{-\infty}^{\infty} g(t) e^{-i2\pi ft}\, dt$$

$$= \quad \int_{-\infty}^{\infty} (-2\pi it)^n g(t) e^{-i2\pi ft}\, dt$$

$$= \quad (-2\pi i)^n \int_{-\infty}^{\infty} t^n g(t) e^{-i2\pi ft}\, dt.$$

Setting $f = 0$ then yields the theorem. Not setting $f = 0$, however, leaves one with another result of interest, which we state formally.

Theorem 4.16 *Given a continuous time infinite duration signal $g(t)$ with spectrum $G(f)$,*

$$\mathcal{F}_f(\{t^n g(t); \ t \in \mathcal{R}\}) = (\frac{i}{2\pi})^n G^{(n)}(f). \tag{4.59}$$

Similar results can be proved for other signal types.

In the special case $n = 2$ the second moment is called the *moment of inertia* and the moment theorem reduces in the continuous time case to

$$\int_{-\infty}^{\infty} t^2 g(t) \, dt = -\frac{G''(0)}{4\pi^2},$$

called the *second moment property*. This result has an interesting implication. Suppose that the signal is nonnegative and hence can be thought of as a density (say of mass or probability) so that the moment of inertia can be thought of as a measure of the spread of the signal. In other words, if the second moment is small the signal is clumped around the origin. If it is large, there is significant "mass" away from the origin. The above second moment property implies that a low moment of inertia corresponds to a spectrum with a low negative second derivative which means a small curvature or relatively flat behavior around the origin. Correspondingly, a large moment of inertia means that the spectrum has a large curvature at the origin and hence is very "peaky". Thus a signal with a peak at the origin produces a spectrum that is very flat at the origin and a signal that is very flat produces a spectrum that is very steep. This apparent tradeoff between steepness in one domain and flatness in another will be explored in more depth later.

As an example that provides a simple evaluation of an important integral, consider the continuous time signal $g(t) = e^{-\pi t^2}$ which has spectrum $G(f) = e^{-\pi f^2}$. We have that

$$G'(f) = -2\pi f e^{-\pi f^2}$$

$$G''(f) = -2\pi e^{-\pi f^2}(1 - 2\pi f^2)$$

which implies from the moment theorem that

$$\int_{-\infty}^{\infty} e^{-\pi t^2} \, dt = G(0) = 1,$$

$$\int_{-\infty}^{\infty} t e^{-\pi t^2} \, dt = \frac{i}{2\pi} G'(0) = 0,$$

(since the integrand is odd) and

$$\int_{-\infty}^{\infty} t^2 e^{-\pi t^2}\, dt = -\frac{1}{4\pi^2} G''(0) = \frac{1}{2\pi}.$$

As a second example and an object lesson in caution when dealing with moments, consider the continuous time signal $g(t) = 2/(1 + (2\pi t)^2)$ and its transform $G(f) = e^{-|f|}$. The 0th moment is

$$\int_{-\infty}^{\infty} g(t)\, dt = G(0) = 1.$$

If one tries to find the first moment, however, one cannot use the moment theorem because G' is not continuous at 0! In particular, $G'(0^+) = -1$ and $G'(0^-) = +1$. The problem here is that the integrand in

$$\int_{-\infty}^{\infty} tg(t)\, dt = \int_{-\infty}^{\infty} \frac{2t}{1 + (2\pi t)^2}\, dt$$

falls as $1/t$ for large t, which is not integrable. In other words, the first moment blows up. The integral does exist in a Cauchy sense (it is 0). In fact, this signal corresponds to the so-called Cauchy distribution in probability theory. Note that it violates the sufficient condition for the moment theorem to hold, i.e., $|tg(t)|$ is not integrable.

As a final example pointing out a more serious peril, consider the signal $\operatorname{sinc}(t)$. Its spectrum is $\sqcap(f)$ which is infinitely differentiable at the origin and the derivatives are all 0. Thus one would suspect that the moments are all 0 (except for the area). This is easily seen to not be the case for the second moment, however, by direct integration. Integration by parts shows that in fact

$$\int_{-T}^{T} t^2 \operatorname{sinc}(t)\, dt = \sin T - T \cos T$$

does not converge as $T \to \infty$ and hence the second moment does not exist. The problem is that the conditions for validity of the theorem are violated since the second absolute moment does not exist. Thus existence of the derivatives is not sufficient to ensure that the formula makes sense. In order to apply the formula, one needs to at least argue or demonstrate by other means that the desired moments exist.

⋆ Normalized Moments

Normalizing moments allows us to bring out more clearly some of their basic properties. If the signal is nonnegative, the normalized signal can be thought of as a probability or mass density function in the continuous time case and a probability mass function or point mass function in the discrete time case. The normalized moments are obtained by replacing the weighting by the signal g in the sum or integral by a signal having unit area; that is, the weighting is $g(t)/\int_{-\infty}^{\infty} g(\alpha)\, d\alpha$ in the continuous time case and $g_n/\sum_{k=-\infty}^{\infty} g_k$ in the discrete time case. Clearly this normalization makes sense only when the signal has nonzero area. Normalizing in this way is equivalent to dividing the nth moment by the 0th order moment (the area of the signal). To be specific, define the normalized moments $< t^n >_g$ by

$$< t^n >_g = \frac{M_g^{(n)}}{M_g^{(0)}}. \tag{4.60}$$

For both continuous and discrete time,

$$< t^0 >_g = 1.$$

The normalized first moment is called the *centroid* or *mean* of the signal. In the continuous time case this is given by

$$< t >_g = \frac{\int_{-\infty}^{\infty} t g(t)\, dt}{\int_{-\infty}^{\infty} g(t)\, dt} = \frac{i}{2\pi} \frac{G'(0)}{G(0)},$$

where the moment theorem has been used. The second moment is called the *mean squared abscissa* and it is given in the continuous time case by

$$< t^2 >_g = \frac{\int_{-\infty}^{\infty} t^2 g(t)\, dt}{\int_{-\infty}^{\infty} g(t)\, dt} = -\frac{1}{4\pi^2} \frac{G''(0)}{G(0)}.$$

In addition to normalizing moments, they are often *centralized* in the sense of removing the area or mean before taking the power. The principal example is the *variance* of a signal, defined by

$$\sigma_g^2 = < (t - < t >_g)^2 >_g .$$

The variance or its square root, the standard deviation, is often used as a measure of the spread or width of a signal. If the signal is unimodal, then the "hump" in the signal will be wide (narrow) if the variance is large (small). This interpretation must be made with care, however, as the variance may not be a good measure of the physical width of a signal. It can be negative, for example, if the signal is not required to be nonnegative.

When computing the variance, it is usually easier to use the fact that

$$\sigma_g^2 = < t^2 >_g - < t >_g^2 \tag{4.61}$$

which can be proved by expanding and manipulating the definition.

We will see later in this chapter that any real even signal will have a real and even spectrum. Since an even function has 0 derivative at time 0, this means that $< t >_g = 0$ for any real even signal. If the centroid is 0, then the variance and the mean squared abscissa are equal.

4.13 Bandwidth and Pulse Width

We have several times referred to the notions of the width of a signal or spectrum. While the width of a signal such as $\sqcap(t)$ or $\wedge(t)$ is obvious, a meaningful definition of width for an arbitrary non-time-limited signal (or an arbitrary non-band-limited spectrum) is not obvious. Intuitively, the width of a signal or spectrum should measure the amount of time or frequency required to contain most of the signal or spectrum. When considering the second order moment property, we observed that there is a tradeoff between width in the two domains: A narrow (broad) signal corresponds to a broad (narrow) spectrum. This observation can be further quantified in the special case of the rectangle function. From the continuous time stretch (or similarity) theorem, the transform of $\sqcap(t/T)$ is $T \operatorname{sinc}(Tf)$. The box has a width of T by any reasonable definition. The width of the sinc function is at least indicated by the difference between the first two zeros, which occur at $-1/T$ and $1/T$. Thus if we increase (decrease) the width of the time signal, the spectrum decreases (increases) in proportion. The object of this section is to obtain a general result along these lines which provides a useful quantitative notion of time width and band-width and which generalizes the observation that the two widths are inversely proportional: wide (narrow) signals yield narrow (wide) spectra.

We have already encountered in Section 4.5 applications where the notion of width is important: band-limited spectra are required for the sampling theorem and time-limited signals can be expanded in a Fourier series over the region where they are nonzero. Both of these results required absolute time or band limitation, but a weaker notion of time-limited or band-limited should allow these results to at least provide useful approximations. Other applications abound. If a narrow signal has a wide spectrum, then a quantitative measure of the signal width is necessary to determine the bandwidth of a communications channel necessary to pass the signal without distortion. An important attribute of an antenna is its beamwidth, a measure of its angular region of maximum sensitivity. Resolution in radar is determined by pulse width.

As one might guess, there is not a single definition of width. We shall consider a few of the simplest and most commonly encountered definitions along with some properties. We will introduce the definitions for signals, but they have obvious counterparts for spectra. The entire section concentrates on the case of continuous time infinite duration signals so that indices in both time and frequency are continuous. The notion of width for discrete time or discrete frequency is of much less interest.

Equivalent Width

The simplest notion of the width of a signal is its *equivalent width* defined as the width of a rectangle signal with the same area and the same maximum height as the given signal. The area of the rectangle $\sqcap(t/T)$ is T. Given a signal g with maximum height g_{max} and area $\int_{-\infty}^{\infty} g(t)\, dt = G(0)$ (using the moment property), then we define the *equivalent width* W_g so that the rectangular pulse $g_{max} \sqcap (t/W_g)$ has the same area as g; that is, $g_{max} W_g = G(0)$. Thus

$$W_g = \frac{G(0)}{g_{max}}.$$

In the special but important case where $g(t)$ attains its maximum at the origin, this becomes

$$W_g = \frac{G(0)}{g(0)}.$$

An obvious drawback to the above definition arises when a signal has zero area and hence the width is zero. For example, the signal $\sqcap(t - 1/2) - \sqcap(t + 1/2)$ is assigned a zero width when its actual width should be 2. Another shortcoming is that the definition makes no sense for an idealized pulse like the impulse or Dirac delta function.

The equivalent width is usually easy to find. The equivalent width of $\sqcap(t)$ is 1, as are the equivalent widths of $\wedge(t)$ and $\text{sinc}(t)$. These signals have very different physical widths, but their equal areas result in equal equivalent widths.

When used in the time domain, the equivalent width is often called the *equivalent pulse width*. We shall also use the term equivalent time width. The same idea can be used in the frequency domain to define *equivalent bandwidth*:

$$W_G = \frac{\int_{-\infty}^{\infty} G(f)\, df}{G_{max}} = \frac{g(0)}{G_{max}}.$$

In the common special case where $G(f)$ attains its maximum value at the origin, this becomes

$$W_G = \frac{g(0)}{G(0)}.$$

Note that even if $G(f)$ is complex, its area will be real if $g(t)$ is real. Note also that if the signal and spectra achieve their maxima at the origin, then

$$W_g W_G = 1$$

and hence the width in one domain is indeed inversely proportional to the width in the other domain.

Magnitude Equivalent Width

In the communications systems literature, equivalent bandwidth is usually defined not in terms of the area under the spectrum, but in terms of the area under the magnitude spectrum. To avoid confusion, we will call this the *magnitude equivalent bandwidth* and denote it by

$$B_{eq} = \frac{\int_{-\infty}^{\infty} |G(f)| \, df}{G_{max}}.$$

We again assume that $G_{max} = G(0)$, which is reasonable for some baseband pulses. Defining the pulse width as before we still have a simple relation between pulse and bandwidth, except that now it is an inequality instead of an equality. In particular, since for all t

$$
\begin{aligned}
|g(t)| &= |\int_{-\infty}^{\infty} G(f) e^{i2\pi ft} \, df| \\
&\leq \int_{-\infty}^{\infty} |G(f) e^{i2\pi ft}| \, df \\
&= \int_{-\infty}^{\infty} |G(f)| \, df \\
&= B_{eq} G_{max},
\end{aligned}
$$

then choosing t so that $g(t) = g_{max}$ we have that

$$B_{eq} \geq \frac{g_{max}}{G_{max}} = \frac{1}{W_g}.$$

4.14 Symmetry Properties

The remaining simple properties have to do with the symmetry properties of Fourier transforms. These properties are useful for checking the correctness of Fourier transforms of signals since symmetry properties of the signals imply corresponding symmetries in the transform. They are also helpful for occasionally suggesting shortcuts to computation.

We here focus on the infinite duration CTFT; similar properties are easily seen for the infinite duration DTFT.

The Fourier transforms of even and odd signals have special symmetries. The following result shows that all signals can be decomposed into even and odd parts and the signal is a linear combination of those parts.

Theorem 4.17 *Any signal $g(t)$ can be uniquely decomposed into an even part and an odd part; that is,*

$$g(t) = g_e(t) + g_o(t)$$

where $g_e(t)$ is even and $g_o(t)$ is odd.

To get such a representation just define

$$g_e(t) = \frac{1}{2}(g(t) + g(-t)) \tag{4.62}$$

$$g_o(t) = \frac{1}{2}(g(t) - g(-t)) \tag{4.63}$$

By construction the two functions are even and odd and their sum is $g(t)$. The representation is unique, since if it were not, there would be another even function $e(t)$ and odd function $o(t)$ with $g(t) = e(t) + o(t)$. But this would mean that

$$e(t) + o(t) = g_e(t) + g_o(t)$$

and hence

$$e(t) - g_e(t) = g_o(t) - o(t).$$

Since the left-hand side is even and the right-hand side is odd, this is only possible if both sides are everywhere 0; that is, if $e(t) = g_e(t)$ and $o(t) = g_o(t)$.

Remarks

1. The choices of $g_e(t)$ and $g_o(t)$ depend on the time origin, e.g., $\cos t$ is even while $\cos(t - \pi/2)$ is odd.

2. $\int_{-\infty}^{\infty} g_o(t)\, dt = 0$, at least in the Cauchy principal value sense. It is true in the general improper integral sense if $g_o(t)$ is absolutely integrable. (This problem is pointed out by the function

$$\text{sgn}(t) = \begin{cases} 1 & t > 0 \\ 0 & t = 0 \\ -1 & t < 0 \end{cases} \tag{4.64}$$

which has a 0 integral in the Cauchy principal value sense but which is not integrable in the usual improper Riemann integral sense. The function $g(t) = 1/t$ is similarly unpleasant.)

3. If $e_1(t)$ and $e_2(t)$ are even functions and $o_1(t)$ and $o_2(t)$ are odd functions, then $e_1(t) \pm e_2(t)$ is even, $o_1(t) \pm o_2(t)$ is odd, $e_1(t)e_2(t)$ is even, $o_1(t)o_2(t)$ is even, and $e_1(t)o_2(t)$ is odd. The proof is left as an exercise.

4. All of the ideas and results for even and odd signals can be applied to infinite duration two-sided discrete time signals.

We can now consider the Fourier transforms of even and odd functions. Again recall that $g(t) = g_e(t) + g_o(t)$, where the even and odd parts in general can be complex. Then

$$
\begin{aligned}
G(f) &= \int_{-\infty}^{\infty} g(t)e^{-i2\pi ft}dt \\
&= \int_{-\infty}^{\infty} (g_e(t) + g_o(t))\,(\cos(2\pi ft) - i\sin(2\pi ft))\;dt \\
&= \int_{-\infty}^{\infty} g_e(t)\cos(2\pi ft)dt - i\int_{-\infty}^{\infty} g_e(t)\sin(2\pi ft)\;dt + \\
&\quad \int_{-\infty}^{\infty} g_o(t)\cos(2\pi ft)\;dt - i\int_{-\infty}^{\infty} g_o(t)\sin(2\pi ft)\;dt.
\end{aligned}
$$

Since the second and third terms in the final expression are the integrals of odd functions, they are zero and hence

$$
\begin{aligned}
G(f) &= \int_{-\infty}^{\infty} g_e(t)\cos(2\pi ft)\;dt - i\int_{-\infty}^{\infty} g_o(t)\sin(2\pi ft)\;dt \\
&= G_e(f) + G_o(f), \qquad\qquad\qquad\qquad\qquad (4.65)
\end{aligned}
$$

where $G_e(f)$ is the cosine transform of the even part of $g(t)$ and $G_o(f)$ is $-i$ times the sine transform of the odd part. (Recall that the cosine and sine transforms may have normalization constants for convenience.) Note that if $g(t)$ is an even (odd) function of t, then $G(f)$ is an even (odd) function of f.

As an interesting special case, suppose that $g(t)$ is a real-valued signal and hence that $g_e(t)$ and $g_o(t)$ are also both real. Then the real and imaginary parts of the spectrum are immediately identifiable as

$$
\Re(G(f)) = \int_{-\infty}^{\infty} g_e(t)\cos(2\pi ft)\;dt \qquad\qquad (4.66)
$$

$$
\Im(G(f)) = -\int_{-\infty}^{\infty} g_o(t)\sin(2\pi ft)\;dt. \qquad\qquad (4.67)
$$

Observe that the real part of $G(f)$ is even in f and the imaginary part of $G(f)$ is odd in f. Observe also that if $g(t)$ is real and even (odd) in t, then $G(f)$ is real (imaginary) and even (odd) in f.

If $g(t)$ is real valued we further have that

$$\begin{aligned} G(-f) &= \Re(G(-f)) + i\Im(G(-f)) \\ &= \Re(G(f)) - i\Im(G(f)) \\ &= G^*(f), \end{aligned}$$

which implies that *the Fourier transform of a real-valued signal is Hermitian.*

By a similar analysis we can show that if the signal $g(t)$ is purely imaginary and hence $g_e(t)$ and $g_o(t)$ are imaginary, then

$$\Re(G(f)) = -i \int_{-\infty}^{\infty} g_o(t) \sin(2\pi t f)\, dt$$

is odd in f, and

$$\Im(G(f)) = -i \int_{-\infty}^{\infty} g_e(t) \cos(2\pi t f)\, dt$$

is even in f, and

$$G(-f) = -G^*(f); \tag{4.68}$$

that is, the spectrum is *anti-Hermitian.*

We can summarize the symmetry properties for a general complex signal $g(t)$ as follows:

$$g(t) = g_e(t) + g_o(t) = e_R(t) + ie_I(t) + o_R(t) + io_I(t) \tag{4.69}$$

$$G(f) = G_e(f) + G_o(f) = E_R(f) + iE_I(f) + O_R(f) + iO_I(f) \tag{4.70}$$

where

$$\begin{aligned} g(t) &\supset G(f) \\ g_e(t) &\supset G_e(f) \\ g_o(t) &\supset G_o(f) \\ e_R(t) &\supset E_R(f) \\ e_I(t) &\supset E_I(f) \\ o_R(t) &\supset iO_I(f) \\ o_I(t) &\supset -iO_R(f) \end{aligned} \tag{4.71}$$

These formulas can be useful in computing Fourier transforms of complicated signals in terms of simpler parts. They also provide a quick check on the symmetry properties of Fourier transforms. For example, the transform of a real and even signal must be real and even and the transform of an odd and real signal must be odd and imaginary.

4.15   Problems

4.1. What is the DFT of the signal

$$g_n = \sin(2\pi 3\frac{n}{8}) - \sin(2\pi\frac{n}{4}); \; n = 0, 1, \ldots, 7?$$

4.2. Prove the shift theorem for infinite duration discrete time signals and finite duration continuous time signals.

4.3. Let $g = \{g_n; \; n \in \mathcal{Z}\}$, where $\mathcal{Z}$ is the set of all integers, be defined by

$$g_n = \begin{cases} (-1)^n 3^{-n} \cos(\frac{\pi n}{9}) & n = 0, 1, 2, \cdots \\ 0 & \text{otherwise} \end{cases}$$

Find the Fourier transform G of g.

4.4. Prove that the Fourier transform of the infinite duration continuous time signal $\{g(at - b); \; t \in \mathcal{R}\}$ is $\frac{1}{|a|}G(f/a)e^{-i2\pi fb/a}$, where G is the Fourier transform of g.

4.5. Find the Fourier transform of the following continous time infinite duration signal: $g(t) = e^{-|t-3|}; \; t \in \mathcal{R}$. Repeat for the discrete time case (now $t \in \mathcal{Z}$).

4.6. Suppose that $g(t)$ and $G(f)$ are infinite duration continuous time Fourier transform pairs. What is the transform of $\cos^2(2\pi f_0 t)g(t)$?

4.7. What is the Fourier transform of $\{\operatorname{sinc}(t)\cos(2\pi t); \; t \in \mathcal{R}\}$?

4.8. State and prove the following properties for the two-dimensional finite duration discrete time Fourier transform (the two-dimensional DFT): Linearity, the shift theorem, and the modulation theorem.

4.9. Given a discrete time, infinite duration signal $g_n = r^n$ for $n \geq 0$ and $g_n = 0$ for $n < 0$ with $|r| < 1$, suppose that we form a new signal h_n which is equal to g_n whenever n is a multiple of 10 and is 0 otherwise: $h_n = g_n$ for $n = \ldots, -20, -10, 0, 10, 20, \ldots$. What is the Fourier transform $H(f)$ of h_n?

It was pointed out that the true meaning of "decimation" is not to produce the sequence h_n (i.e., the dead centurions in the Roman use of the word), but to produce the remaining or live centurions. Thus the correct meaning of "decimating" the sequence g_n is to produce the sequence f_n for which

$$f_n = \begin{cases} g_n & \text{if } n \text{ is not a multiple of 10} \\ 0 & \text{if } n \text{ is a multiple of 10} \end{cases}$$

Find the Fourier transform $F(f)$ of f_n.

4.10. Suppose you have two discrete time signals $\{h_n; n = 0, 1, \ldots, N-1\}$ and $\{g_n; n = 0, 1, \ldots, N-1\}$ with DFTs $H(k/N)$ and $G(k/N)$, $k = 0, 1, \ldots, N-1$, respectively. Form a new signal $\{w_n; n = 0, 1, \ldots, 2N-1\}$ by "multiplexing" these two signals to form $h_0, g_0, h_1, g_1, \ldots, h_{N-1}, g_{N-1}$; that is,

$$w_n = \begin{cases} h_{\frac{n}{2}} & \text{if } n \text{ is zero or even} \\ g_{\frac{n-1}{2}} & \text{if } n \text{ is odd} \end{cases}$$

What is the DFT $W(l/2N)$; $l = 0, 1, \ldots, 2N-1$ in terms of H and G?

4.11. Prove Parseval's equality for energy for the continuous time finite duration case.

4.12. Prove Parseval's equality for energy for the discrete time infinite duration case.

4.13. Evaluate the integral

$$\int_{-\infty}^{\infty} \text{sinc}[2B(t - \frac{n}{2B})] \, \text{sinc}[2B(t - \frac{m}{2B})] \, dt \qquad (4.72)$$

for all integers n, m. What does this say about the signals $\text{sinc}(2B(t - \frac{n}{2B}))$ for integer n?

4.14. This problem considers a simplified model of "oversampling" techniques used in CD player audio reconstruction.

A continuous time infinite duration signal $g = \{g(t); t \in \mathcal{R}\}$ is bandlimited to ± 22 kHz. It is sampled at $f_0 = 44$ kHz to form a discrete time signal $\hat{g} = \{\hat{g}_n; n \in \mathcal{Z}\}$, where $\hat{g}_n = g(n/f_0)$.

A new discrete time signal, $\hat{h} = \{\hat{h}_n; n \in \mathcal{Z}\}$ is formed by repeating each value of $\hat{g}$ four times; that is,

$$\hat{h}_{4n} = \hat{h}_{4n+1} = \hat{h}_{4n+2} = \hat{h}_{4n+3} = \hat{g}_n; \quad n \in \mathcal{Z}.$$

Note that this step can be expressed in terms of the upsampling operation: If we define the signal $r = \{r_n; \ n \in \mathcal{Z}\}$ by

$$r_n = \begin{cases} \hat{g}_{n/4} & \text{if } n = 0, \pm 4, \pm 8, \cdots \\ 0 & \text{if } n \text{ is not a multiple of 4} \end{cases}$$

then

$$\hat{h}_n = r_n + r_{n-1} + r_{n-2} + r_{n-3}.$$

Lastly, a continuous time signal $h = \{h(t); \ t \in \mathcal{R}\}$ is formed from $\hat{h}$ using a sampling expansion:

$$h(t) = \sum_{n=-\infty}^{\infty} \hat{h}_n \operatorname{sinc}(176000t - n).$$

(a) Find $H(f)$; $f \in \mathcal{R}$ in terms of $G(f)$; $f \in \mathcal{R}$.

(b) If $g(t) = \cos 2000\pi t + \cos 4000\pi t$, what frequencies have nonzero values of $H(f)$?

(c) If $h(t)$ is then filtered to ± 22 kHz by an ideal low pass filter (i.e., its Fourier transform is multiplied by a unit magnitude box filter having this bandwidth), is the result an exact reproduction of $g(t)$? Why or why not?

4.15. Suppose that a discrete time signal $g = \{g_n; \ n \in \mathcal{Z}\}$ is not band limited so that it cannot be recovered from its downsampled version $h = \{g_{2n}; \ n \in \mathcal{Z}\}$. Suppose that we also form a second downsampled signal $r = \{g_{2n+1}; \ n \in \mathcal{Z}\}$ which contains all of the samples in g that are not in h. Find a formula for G in terms of H and R. Thus if we form all possible distinct downsampled versions of a signal and compute their transforms, the transforms can be combined to find the transform of the original signal. The construction strongly resembles that used for the FFT.

4.16. Develop an analog of the discrete time sampling theorem for the DFT, that is, for finite duration discrete time signals.

4.17. The DFT of the sequence $\{g_0, g_1, \ldots, g_{N-1}\}$ is $\{G_0, G_1, \ldots, G_{N-1}\}$. What is the DFT of the sequence $\{g_0, 0, g_1, 0, \ldots, g_{N-1}, 0\}$?

4.18. If $g = \{g_n; \ n = 0, 1, \ldots, N-1\}$ has Fourier transform

$$G(f) = \sum_{n=0}^{N-1} g_n e^{-i2\pi fn},$$

find an expression for

$$\sum_{n=0}^{N-1} n g_n$$

in terms of $G(f)$. (*Hint:* Stick with the Fourier transform here, not the DFT, i.e., do not restrict the range of f at the start. This problem makes the point that sometimes it is useful to consider a more general frequency domain than that required for inversion.)

4.19. Suppose you have an infinite-duration discrete-time signal $x = \{x_k; \ k \in \mathcal{Z}\}$ with Fourier transform $X = \{X(f); \ f \in [-1/2, 1, 2)\}$. A second discrete-time signal $y = \{y_k; \ k \in \mathcal{Z}\}$ is defined in terms of x by

$$y_k = x_k + x_{k-1} + x_{k-2} + x_{k-3} + x_{k-4} + x_{k-5} + x_{k-6} + x_{k-7}.$$

(This is an example of what we will later call a discrete-time linear filter, but you do not need any linear filtering theory to do this problem.)

(a) Find the Fourier transform Y of y for the case where

$$x_k = e^{-k} u_{-1}(k); \ k \in \mathcal{Z}$$

where $u_{-1}(k)$ is the unit step function defined as 1 for $k \geq 0$ and 0 otherwise.

(b) Form the truncated signal $\hat{y} = \{\hat{y}_k; \ k \in \mathcal{Z}_{16}\}$ defined by $\hat{y}_n = y_n$ for $n \in \mathcal{Z}_{16}$. Find the Fourier transform $\hat{Y}$ of $\hat{y}$ for the case where x is defined by

$$x_k = \sin(2\pi \frac{3k}{16}) \ k \in \mathcal{Z}.$$

4.20. Suppose that $g = \{g(n); n \in \mathcal{Z}\}$ is a discrete time signal with Fourier transform $G = \{G(f); \ f \in [0, 1)\}$, that is not necessarily bandlimited. For a fixed integer M define the subsampled signal $\gamma = \{\gamma(n) = g(nM); \ n \in \mathcal{Z}\}$. Find the DTFT Γ of γ.

Hint: Analogous to the continuous time case, consider the function

$$\alpha(f) = \sum_{m=0}^{M-1} G(\frac{f - m}{M}),$$

where the shift inside G is cyclic on the frequency domain, e.g., using $[0, 1)$ as the frequency domain,

$$G(\frac{f - m}{M}) = G([\frac{f - m}{M}] \bmod 1).$$

Expand α in a Fourier series on $[0, 1)$ and compare it to the definition of γ.

4.21. Suppose $g = \{g(t); \ t \in \mathcal{R}\}$ where

$$g(t) = \mathrm{sinc}^2\left(\frac{t}{4}\right).$$

(a) What is $G(f)$? (Give an explicit expression.) Make a sketch labeling points of interest in frequency and amplitude.

(b) The signal g is now sampled at the Nyquist rate and a discrete time sequence h_n is formed with the samples: $h_n = g(nT_s)$. What is the transform of the sampled sequence? (Give an explicit expression for $H(f)$ for the given g.) Make a labeled sketch.

(c) h_n is then upsampled by 5 to form

$$v_n = \begin{cases} h_{n/5} & \text{if } n \text{ is an integer multiple of 5} \\ 0 & \text{otherwise} \end{cases}$$

Give an explicit expression and a labeled sketch of the transform $V(f)$ of the signal v_n.

(d) Suppose we want to obtain the sequence w_n which is formed by sampling the continuous time signal g with a sampling period $T_s = 1.2$. How can this be done given the sequence v_n alone? Make a block diagram of your procedure.

(e) Now the system you've designed in part (d) is altered so that h_n is upsampled by 2 instead of 5. Can the resulting sequence at the output be used to reconstruct the signal g? Why or why not?

4.22. Consider the 9 point sequence $g = \{120120121\}$. Define the Fourier transform (DFT) of g to be $G = \{G(k/9); \ k = 0, 1, \dots, 8\}$.

Let $\tilde{g}$ be the periodic extension of g. Define two new sequences: $h_n = r^n \tilde{g}_n u_{-1}(n), |r| < 1$; and $v_n = h_{3n+1}$.

(a) What is $G(1/3)$?

(b) What is $V(f)$?

(c) What is $H(f)$?

4.23. Sketch and find the Fourier transform of the infinite duration discrete time signal $g_n = 2\square_4(n) - \square_2(n)$. What is the inverse Fourier transform of the resulting spectrum?

4.24. Given a continuous time infinite duration signal $\{p(t); t \in \mathcal{R}\}$ with Fourier transform $P(f)$, find the Fourier transform of the signal $g(t) = p(T - t)$, where T is fixed. Specialize your result for the cases $p(t) = \sqcap(t)$ and $p(t) = \wedge(t)$ and $T = 1$ (e.g., what is the transform of $\sqcap(1 - t)$).

4.25. What is the Fourier transform of the signal $g = \{g_n = (1/2)^n + (1/3)^{n-1}; n = 0, 1, \ldots\}$.

4.26. Suppose that you have a discrete time signal $g = \{g(n); n \in \mathcal{Z}\}$ with a Fourier transform G that is bandlimited:

$$G(f) = 0; \text{ for } \frac{1}{2} \geq |f| \geq \frac{1}{8}.$$

(a) Find the Fourier transform H of the signal h defined by

$$h(n) = g(n) \cos(\frac{\pi n}{4}); \ n \in \mathcal{Z}.$$

(b) Derive the Fourier transform of the upsampled signal $g_{1/3}$ defined by

$$g_{1/3}(n) = \begin{cases} g(\frac{n}{3}) & \text{if } n \text{ is an integer multiple of } 3 \\ 0 & \text{otherwise} \end{cases}$$

(c) Find the Fourier transform of the signal w defined by

$$w_n = g_{1/3}(n) - g(n).$$

Compare this signal and its Fourier transform with h and its Fourier transform.

4.27. An infinite duration continuous time signal $g = \{g(t); t \in \mathcal{R}\}$ is band-limited to $(-W, W)$, i.e., its Fourier transform G satisfies

$$G(f) = 0 \text{ for } |f| \geq W.$$

We showed that we can expand G in a Fourier series on $[-W, W)$ for this case. Use this expansion to find an expression for

$$\int_{-\infty}^{\infty} |G(f)|^2 \, df$$

in terms of the samples $g(nT)$.

4.28. A continuous time signal $g = \{g(t); \ t \in \mathcal{R}\}$ has a spectrum $G = \{G(f); f \in \mathcal{R}\}$ defined by

$$G(f) = \begin{cases} \frac{1}{2} & f \in [-\frac{1}{2}, \frac{1}{2}] \\ 1 & f \in [-1, -\frac{1}{2}) \text{ or } f \in (\frac{1}{2}, 1] \\ 0 & \text{otherwise} \end{cases}$$

(a) Find g.

(b) Write a sampling expansion for g using a sampling period of T and state for which T the formula is valid.

For the remainder of this problem assume that T meets this condition.

(c) Find the DTFT Γ for the sampled sequence $\gamma = \{\gamma_n = g(nT); \ n \in \mathcal{Z}\}$.

(d) Evaluate the sum

$$\sum_{n=-\infty}^{\infty} g(nT).$$

(e) Prove or disprove

$$\sum_{n=-\infty}^{\infty} g^2(nT) = \int_{-\infty}^{\infty} g^2(t)\, dt$$

Keep in mind that in Parts (c)–(e) it is assumed that T meets the condition of (b) necessary to ensure the validity of the sampling theorem.

4.29. Suppose that a bandlimited continuous time signal $y(t)$ is sampled faster than the Nyquist rate to produce a discrete time signal $x(n) = y(nT)$. Relate the energy of the discrete time signal, $\sum |x(n)|^2$, to that of the continuous time process, $\int |y(t)|^2\, dt$.

4.30. An infinite duration continuous time signal $g = \{g(t); t \in \mathcal{R}\}$ is band-limited to $(-1/2, 1/2)$, i.e., its Fourier transform G satisfies

$$G(f) = 0 \text{ for } |f| \geq \frac{1}{2}.$$

A discrete time signal $\hat{g} = \{\hat{g}_n; \ n \in \mathcal{Z}\}$ is then defined by sampling as $\hat{g}_n = g(n)$. A new continous time signal $h = \{h(t); t \in \mathcal{R}\}$ is then defined by

$$h(t) = \sum_{n=-\infty}^{\infty} \hat{g}_n p(t-n),$$

where $p = \{p(t); t \in \mathcal{R}\}$ is a continous time signal (not necessarily a pulse or time limited) with Fourier transform P.

Find an expression for the Fourier transform H of h in terms of G and P. Specialize your answer to the cases where $p(t) = \mathrm{sinc}(t)$ and $p(t) = \square_{1/2}(t)$. Both of these results are forms of PAM, but you should find that h is trivially related to g in one case and has a simple relation in the other. Note that neither of these p are really physical pulses, but both can be approximated by physical pulses.

4.31. What is the Fourier transform of $g = \{te^{-\pi t^2}; \ t \in \mathcal{R}\}$?

4.32. Define the signal $g = \{g(t); \ t \in \mathcal{R}\}$ by

$$g(t) = \begin{cases} 1 + e^{-|t|} & |t| < \frac{1}{2} \\ e^{-|t|} & |t| \geq \frac{1}{2}. \end{cases}$$

(a) Find the Fourier transform G of g.
    Write a Fourier integral formula for g in terms of G. Does the formula hold for all t?

(b) What signal h has Fourier transform $\{H(f); \ f \in \mathcal{R}\}$ given by
$$H(f) = 2G(f)\cos(8\pi f)?$$
    Provide a labeled sketch of h.

(c) Define the truncated finite duration signal $\hat{g} = \{\hat{g}(t); t \in [-1/2, 1/2)\}$ where $\hat{g}(t) = g(t)$ for $t \in [-1/2, 1/2)$. Find the Fourier transform $\hat{G}$ of $\hat{g}$ and write a Fourier series representation for $\hat{g}$. Does the Fourier series give $\hat{g}$ for all $t \in [-1/2, 1/2)$?

4.33. (a) Define the discrete time, finite duration signal $g = \{g_0, g_1, g_2, g_3, g_4, g_5\}$ by
$$g = \{+1, -1, +1, -1, +1, -1\}$$
and define the signal h by
$$h = \{+1, -1, +1, +1, -1, +1\}.$$
Find the DFTs G and H of g and h, respectively. Compare and contrast G and H. (Remark on any similar or distinct properties.)

(b) Define the continuous time finite duration pulse $p = \{p(t); \ t \in [0,6)\}$ by
$$p(t) = \begin{cases} 1 & 0 \leq t < 1 \\ 0 & \text{otherwise} \end{cases}.$$
Find the Fourier transform P of p.

(c) Let g, h, and p be as above. Define the continuous time finite duration signals $\hat{g} = \{\hat{g}(t); \ t \in [0,6)\}$ and $\hat{h} = \{\hat{h}(t); \ t \in [0,6)\}$ by
$$\hat{g}(t) = \sum_{n=0}^{5} g_n p(t - n); \ t \in [0,6)$$

$$\hat{h}(t) = \sum_{n=0}^{5} h_n p(t - n); \ t \in [0, 6).$$

Find the Fourier transforms $\hat{G}$ and $\hat{H}$ of $\hat{g}$ and $\hat{h}$. How do these transforms differ from each other? How do they differ from the results of the previous part?

Hint: First try to find the required transforms in terms of P without plugging in the details from (b).

4.34. Consider a finite duration real-valued continuous time signal $g = \{g(t); \ t \in [0, 1)\}$ with Fourier transform $G = \{G(f); \ f \in \mathcal{Z}\}$.

(a) Find an expression for the energy

$$\mathcal{E}_g = \int_0^1 |g(t)|^2 \ dt$$

in terms of G.

(b) Suppose now that

$$G(f) = r^{-|f|}; \ f \in \mathcal{Z}, \tag{4.73}$$

where $r > 0$ is a real parameter. What is g? (Your final answer should be a closed form, not an infinite sum.) What did you have to assume about r to get this answer?

(c) Given g as in part (b) of this problem, evaluate

$$\int_0^1 g(t) \ dt .$$

(d) Given g as in part (b) of this problem, evaluate

$$\sum_{k=-\infty}^{\infty} G(k).$$

(e) Given g as in part (b) of this problem, evaluate the energy

$$\mathcal{E}_g = \int_0^1 |g(t)|^2 \ dt .$$

4.35. State and prove the moment theorem for finite duration discrete time signals.

4.36. State and prove the moment theorem for finite duration continuous time signals.

4.37. For the function $g(t) = \wedge(t)\cos(\pi t)$; $t \in \mathcal{R}$, find

   (a) The Fourier transform of g.
   (b) The equivalent width of g.

4.38. For the function $g(t) = e^{-|t|}$; $t \in \mathcal{R}$, find

   (a) the area of g.
   (b) The first moment of g.
   (c) The second moment of g.
   (d) The equivalent width of g.

4.39. Suppose that

$$g(t) = g_e(t) + g_o(t)$$
$$f(t) = f_e(t) + f_o(t)$$

where the subscripts e and o denote even and odd parts, respectively.

   (a) Find expressions for

$$\int_{-\infty}^{\infty} |g(t)|^2 \, dt$$

   and

$$\int_{-\infty}^{\infty} g(t)g(-t) \, dt$$

   in terms of integrals involving g_e and g_o.

   (b) If $h(t) = g(t)f(t)$, find expressions for h_o and h_e in terms of the even and odd parts of f and g.

4.40. Find the odd and even parts of the following signals ($\mathcal{T} = \mathcal{R}$):

   (a) e^{it}
   (b) $e^{-it}H(t)$ (Where $H(t)$ is the Heaviside step function.)
   (c) $|t|\sin(t - \pi/4)$
   (d) $e^{i\pi \sin(t)}$

4.41. Find the even and odd parts of the continuous time infinite duration signals

(a) $g(t) = |t|^{\frac{1}{3}} \cos(t - \frac{\pi}{8})$

(b) $g(t) = e^{i\pi(t-t_0)^2}$

4.42. Suppose that we decompose an infinite duration continuous time signal $g(t)$ into its odd and even parts: $g(t) = g_e(t) + g_o(t)$ and suppose that $g(t) \leftrightarrow G(f)$ form a Fourier transform pair. What is the Fourier transform of $G(t)$ in terms of g_e and g_o?

4.43. Is it true that the magnitude spectrum $|G(f)|$ of a real signal must be even?

4.44. Match the signals in the first list with their corresponding DFT's in the second. The DFT's are rounded to one decimal place.

Hint: Very little computation is required here!

A $(1,2,3,4,3,4,3,2)$
B $(0,2,3,4,0,-4,-3,-2)$
C $(i,3i,2i,4i,3i,4i,2i,3i)$
D $(1,2,2,3,3,4,4,3)$
E $(0,3,2,5i,0,-5i,-2,-3)$
1 $(22,-3.4+3.4i,-2,-.6-.6i,-2,-.6+.6i,-2,-3.4-3.4i)$
2 $(22i,-3.4i,0,-.6i,-.6i,-.6i,0,-3.4i)$
3 $(22,-4.8,-2,.8,-2,.8,-2,-4.8)$
4 $(0,7.1-8.2i,-10-6i,7.1-.2i,0,-7.1+.2i,10+6i,-7.1+8.2i)$
5 $(0,-14.5i,4i,-2.5i,0,2.5i,-4i,14.5i)$

4.45. Match the signals in the first list with their corresponding DFT's in the second. The DFT's are rounded to one decimal place.

Hint: Very little computation is required here: no programmable calculators allowed on this problem!

A	$(-1, -1, -3, 4, 1, 4, -3, -1)$
B	$(0, 1, -3, 4, 0, -4, 3, -1)$
C	$(j, -j, -4, 2, j, -2, 4, -j)$
D	$(6, -4, j, -2, 6, -2, -j, -4)$
E	$(0, j, 4j, -5j, 0, -5j, 4j, j)$
F	$(-2, 3, j, -2, 1, -2, -j, 3)$
G	$(2 + j, 0, 0, 0, -2 - j, 0, 0, 0)$
1	$(0, 8.5j, -8j, -8.5j, 16j, -8.5j, -8j, 8.5j)$
2	$(0, -1.1j, 6j, -13.1j, 0, 13.1j, -6j, 1.1j)$
3	$(0, -9.1, 6, 5.1, -12, 5.1, 6, -9.1)$
4	$(0, 4 + 2j, 0, 4 + 2j, 0, 4 + 2j, 0, 4 + 2j)$
5	$(0, 3.8j, 6j, -9.4j, 4j, 12.2j, -2j, -6.6j)$
6	$(1, 6.1, -1, -12.1, -3, -8.1, -1, 2.1)$
7	$(0, -0.8, 12, 0.8, 24, 4.8, 12, -4.8)$

4.46. Table 4.1 has two lists of functions. For each of the functions on the left, show which functions are Fourier transform pairs by means of an arrow drawn between that function and a single function on the right (as illustrated in the top case).

4.47. What can you say about the Fourier transform of a signal that is

    (a) real and even?

    (b) real and odd?

    (c) imaginary and even?

    (d) complex and even?

    (e) even?

    (f) odd?

$$\{g(t); t \in \mathcal{R}\} \qquad \leftrightarrow \qquad \{G(f); f \in \mathcal{R}\}$$
$$\{g_n; n \in \mathcal{Z}\} \qquad \leftrightarrow \qquad \{G(f); f \in [0,1)\}$$
$$\{g(-t); t \in \mathcal{R}\} \qquad\qquad \{G^*(f); f \in \mathcal{R}\}$$
$$\{g^*(t); t \in \mathcal{R}\} \qquad\qquad \{G(-f); f \in \mathcal{R}\}$$
$$\{g_{-n}; n \in \mathcal{Z}\} \qquad\qquad \{G^*(-f); f \in \mathcal{R}\}$$
$$\{g_n^*; n \in \mathcal{Z}\} \qquad\qquad \{G^*(f); f \in [0,1)\}$$
$$\{G(t); t \in \mathcal{R}\} \qquad\qquad \{G(-f \bmod 1); f \in [0,1)\}$$
$$\{G(-t); t \in \mathcal{R}\} \qquad\qquad \{G^*(-f \bmod 1); f \in [0,1)\}$$
$$\{G^*(-t \bmod 1); t \in [0,1)\} \qquad\qquad \{g(f); f \in \mathcal{R}\}$$
$$\{G(-t \bmod 1); t \in [0,1)\} \qquad\qquad \{g(-f); f \in \mathcal{R}\}$$
$$\{g^*(f); f \in \mathcal{R}\}$$
$$\{g^*(-f); f \in \mathcal{R}\}$$
$$\{g_k; k \in \mathcal{Z}\}$$
$$\{g_{-k}; k \in \mathcal{Z}\}$$
$$\{g_k^*; k \in \mathcal{Z}\}$$
$$\{g_{-k}^*; k \in \mathcal{Z}\}$$

Table 4.1: Symmetry Properties

Chapter 5

Generalized Transforms and Functions

In Chapter 2 the basic definitions of Fourier transforms were introduced along with sufficient conditions for the transforms to exist. In this chapter we extend the Fourier transform definitions to include some important signals violating the sufficient conditions seen thus far. We shall see that in some cases the extension is straightforward and simply replaces the basic definitions by a limiting form. In other cases the definitions cannot be patched up so easily and we need to introduce more general ideas of functions or signals in order to construct useful Fourier transform pairs. The emphasis in this chapter is on continuous time, since that is where most of the difficulties arise.

5.1 Limiting Transforms

In this section we treat by example a class of processes for which the Fourier transform defined previously does not exist, yet a simple but natural trick allows us to define a meaningful Fourier transform that behaves in the desired way. The trick is to express the signal as a limit of better behaved signals which have Fourier transforms in the original sense. If the transforms of these signals converge, then the limiting transform is a reasonable definition of the transform of the original signal. A primary example of this technique is the infinite duration CTFT of the signum signal

$$g(t) = \text{sgn}(t) = \begin{cases} 1 & t > 0 \\ 0 & t = 0 \\ -1 & t < 0 \end{cases}. \tag{5.1}$$

This signal clearly violates the absolute integrability criterion since the integral of its absolute magnitude is infinite. Can a meaningful transform be defined? One approach is to consider a sequence of better behaved signals that converge to $g(t)$. If the corresponding sequence of Fourier transforms also converges to something, then that something is a candidate for the Fourier transform of $g(t)$ (in a generalized sense). One candidate sequence is

$$g_k(t) = e^{-\frac{|t|}{k}} \operatorname{sgn}(t) = \begin{cases} e^{-\frac{t}{k}} & t > 0 \\ 0 & t = 0 \\ -e^{\frac{t}{k}} & t < 0 \end{cases} ; \ k = 0, 1, \cdots. \qquad (5.2)$$

The signal sequence is depicted in Figure 5.1 for $k = 1, 10, 100$ along with the step function. These signals are absolutely integrable and piecewise

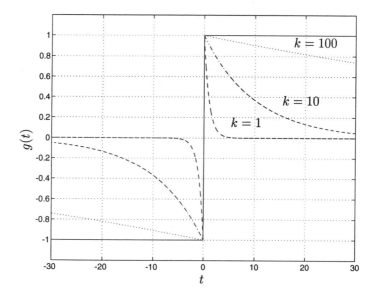

Figure 5.1: Function Sequence: The solid line is the sgn signal.

smooth and

$$\lim_{k \to \infty} g_k(t) = g(t). \qquad (5.3)$$

The CTFT of $g_k(t)$ is

$$G_k(f) = \frac{-i4\pi f}{k^{-2} + (2\pi f)^2} \qquad (5.4)$$

(do the integral for practice). We could then *define* the Fourier transform of sgn(t) to be

$$G(f) = \lim_{k \to \infty} G_k(f) = \begin{cases} -\frac{i}{\pi f} & f \neq 0 \\ 0 & f = 0 \end{cases}. \tag{5.5}$$

This approach has some obvious shortcomings: We may be able to find a $g_k(t)$ sequence that converges to $g(t)$, yet the Fourier transform sequence $G_k(f)$ may not converge for that sequence. To make matters worse, there may not be any such sequence of $g_k(t)$ which converges to $g(t)$ and yields a convergent sequence of Fourier transforms. Hence this trick may be useful for some specific signals, but it is limited in its utility.

The trick works in a similar fashion for DTFT evaluation. As an exercise try finding the Fourier transform of the discrete time analog to the signum function:

$$g(n) = \operatorname{sgn}(n) = \begin{cases} 1 & n > 0 \\ 0 & n = 0 \\ -1 & n < 0 \end{cases}. \tag{5.6}$$

(Just use the same sequence of limiting functions used in the continuous time case, but now use them only for discrete time.) You should be able to show that $G(0) = 0$ and for $f \neq 0$

$$G(f) = \frac{1 - e^{i2\pi f}}{1 - \cos 2\pi f} - 1 = \frac{-i \sin(2\pi f)}{1 - \cos(2\pi f)}. \tag{5.7}$$

In the discrete-time case it is occasionally useful to modify the definition of the signum function slightly to

$$\operatorname{sgn}_0(n) = \begin{cases} 1 & n \geq 0 \\ -1 & n < 0, \end{cases} \tag{5.8}$$

where the subscript 0 is intended to emphasize that this signum is different at the origin. Linearity and the previous result imply that the transform of the modified signum is $G(0) = 1$ and for $f \neq 0$

$$\frac{1 - e^{i2\pi f}}{1 - \cos 2\pi f}.$$

What continuous time infinite duration signal has Fourier transform $G(f) = i\operatorname{sgn}(f)$?

5.2   Periodic Signals and Fourier Series

In this section we modify one of the Fourier transforms already introduced to obtain the most important generalization of Fourier transforms to a class of signals for which the basic definitions fail. In particular, we show that the transforms of finite duration signals can be used to define useful transforms for periodic infinite duration signals, signals which are not absolutely summable or integrable and hence signals for which the usual definitions cannot be used. We begin with the simpler case of discrete time signals.

Recall that an infinite duration discrete time signal $g = \{g_n; \ n \in \mathcal{Z}\}$ is periodic with period N if

$$g_{n+N} = g_n \tag{5.9}$$

for all integers n. Recall also that if we truncate such a signal to produce a finite duration discrete time signal

$$\hat{g} = \{\hat{g}_n; \ n \in \mathcal{Z}_N\} = \{g_n; \ n \in \mathcal{Z}_N\}$$

and then we replicate $\hat{g}$ to form the periodic extension $\tilde{g} = \{\hat{g}_{n \bmod N}; \ n \in \mathcal{Z}\}$, then $\tilde{g} = g$. In other words, the periodic extension of one period of a periodic function is exactly the original periodic function. Intuitively this just says that we should be able to describe the transform of an infinite duration periodic function in terms of the transform of the finite duration signal consisting of one period of the infinite duration signal; that is, the Fourier transform of the periodic signal should be "the same" as the DFT of one period of the signal. We explore how this idea can be made precise.

Given a DTID periodic signal g that is not identically 0, then the signal is clearly not absolutely summable and hence violates the sufficient conditions for the existence of the DTFT. Recall from Chapter 3 however, that if we truncate g to form $\hat{g}$ and then form the DFT

$$\hat{G}(f) = \sum_{n=0}^{N-1} \hat{g}_n e^{-i2\pi f n} = \sum_{n=0}^{N-1} g_n e^{-i2\pi f n}, \tag{5.10}$$

then a Fourier series representation for the infinite duration signal is given from (3.28) as

$$g_n = \frac{1}{N} \sum_{k=0}^{N-1} \hat{G}(\frac{k}{N}) e^{i2\pi \frac{k}{N} n}; \ \text{all} \ n \in \mathcal{Z}. \tag{5.11}$$

Thus we have a Fourier-type decomposition of the original infinite duration signal into a weighted combination of exponentials, but unlike the basic DTFT formula, the signal is given by a *sum* of weighted exponentials rather than by an *integral* of weighted exponentials (e.g., the inverse transform is a sum rather than an integral). Unlike the DTFT, we have only a *discrete spectrum* since only a finite number of frequencies are required in this representation.

Exactly the same idea works for a continuous time infinite duration periodic signal g. Suppose that $g(t)$ has period T, that is, $g(t+T) = g(t)$ for all $t \in \mathcal{R}$. Then the finite duration CTFT of the signal $\hat{g} = \{g(t); t \in [0, T)\}$ is given by

$$\hat{G}(f) = \int_0^T g(t)e^{-i2\pi ft}dt \qquad (5.12)$$

and a Fourier series representation is then given by

$$g(t) = \sum_{n=-\infty}^{\infty} \frac{\hat{G}(\frac{n}{T})}{T}e^{i2\pi \frac{n}{T}t}. \qquad (5.13)$$

Again this formula resembles an inverse Fourier transform for g, but it is a sum and not an integral, that is, it does not have the usual form of an inverse CTFT.

For reasons of consistency and uniformity of notation, it would be nice to have the basic transform and its inverse for periodic infinite duration signals more closely resemble those of ordinary infinite duration signals. To see how this might be accomplished, we focus for the moment on a fundamental example, the complex-valued discrete time periodic signal e defined by

$$e_n = e^{i2\pi f_0 n}; \ n \in \mathcal{Z},$$

where $f_0 = m/N$ for some integer m. This signal has period N (and no smaller period if m/N is in lowest terms). For the moment the frequency f_0 will be considered as a fixed parameter. From the linearity of the Fourier transform, knowing the DTFT of any single exponential signal would then imply the DTFT for any discrete time periodic signal because of the representation of (5.11) of *any* such signal as a finite sum of weighted exponentials! Observe that the discrete time exponential is unchanged if we replace m by $m + MN$ for any integer M. In other words, all that matters is m mod N.

The DFT of one period of the signal is

$$\hat{E}(f) = \begin{cases} 1 & f = f_0 = \frac{m}{N} \\ 0 & f = \frac{l}{N}, l = 0, \ldots, N-1, l \neq m \end{cases}$$

The ordinary DTFT of the signal defined by

$$E(f) = \sum_{n=-\infty}^{\infty} e_n e^{-i2\pi fn}$$

does not exist (because the limiting sum does not converge). Observe also that e_n is clearly neither absolutely summable nor does it have finite energy since

$$\sum_{n=-\infty}^{\infty} |e_n| = \sum_{n=-\infty}^{\infty} 1 = \infty.$$

Suppose for the moment that the transform did exist and was equal to some function of f which we call for the moment $E(f)$. What properties should $E(f)$ have? Ideally we should be able to use the DTFT inversion formula on $E(f)$ to recover the original signal e_n, that is, we would like $E(f)$ to solve the integral equation

$$\int_{-\frac{1}{2}}^{\frac{1}{2}} E(f) e^{i2\pi f n} \, df = e^{i2\pi f_0 n}, \quad n \in \mathcal{Z}, \tag{5.14}$$

in which case we would know that e_n was indeed the inverse DTFT of $E(f)$.

It will simplify things a bit if we consider the frequency domain to be $[0, 1)$ instead of $[-\frac{1}{2}, \frac{1}{2})$ for the current discussion and hence we seek an $E(f)$ that solves

$$\int_{0}^{1} E(f) e^{i2\pi f n} \, df = e^{i2\pi f_0 n}, \quad n \in \mathcal{Z}. \tag{5.15}$$

What $E(f)$ will do this; that is, what $E(f)$ is such that integrating $E(f)$ times an exponential $e^{i2\pi f n}$ will exactly produce the value of the exponential in the integrand with $f = f_0$ for all n? The answer is that no *ordinary* function $E(f)$ will accomplish this, but by using the idea of generalized functions we will be able to make rigorous something like (5.15). So for the moment we continue the fantasy of supposing that there is a function $E(f)$ for which (5.15) holds and we look at the implications of the formula. This will eventually lead up to a precise definition.

Before continuing it is convenient to introduce a special notation for $E(f)$, even though it has not yet been precisely defined. Intuitively we would like something which will have a unit area, i.e.,

$$\int_{0}^{1} E(f) \, df = 1,$$

which is concentrated in an infinitesimally small region around f_0. If $E(f)$ has these properties, then the integrand will be 0 for f not near f_0 and $E(f)e^{i2\pi f n} \approx E(f)e^{i2\pi f_0 n}$ for $f \approx f_0$ and hence

$$\int_{0}^{1} E(f) e^{i2\pi f n} \, df \approx e^{i2\pi f_0 n} \int_{0}^{1} E(f) \, df$$

$$\approx e^{i2\pi f_0 n},$$

as desired. Again we emphasize that no ordinary function behaves like this, but the idea of a very narrow, very tall pulse with unit area is useful for intuition (provided it is not taken too literally) and hence we give this hypothetical object the name of a *unit impulse function* or *Dirac delta function* at f_0, denote it by $\delta(f - f_0)$, and depict it graphically as in Figure 5.2.

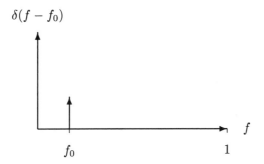

Figure 5.2: Graphical representation of a Dirac delta

Equation (5.15) is an example of what is called a *sifting* property. If $E(f) = \delta(f - f_0)$ is multiplied by a complex exponential and integrated, the resulting integration yields exactly the value of the complex exponential at the fixed frequency f_0 so that the combination of multiplication by $\delta(f - f_0)$ and integration exactly sifts out or samples one value of the complex exponential. We next show that if $\delta(f - f_0)$ sifts exponentials at f_0, it must also sift more general signals.

Suppose that $X(f); f \in [0, 1)$ is a frequency domain signal. Analogous to a continuous time finite duration signal, we can expand $X(f)$ as a Fourier series

$$X(f) = \sum_{n \in Z} x_n e^{-i2\pi n f}$$

where

$$x_n = \int_0^1 X(f) e^{i2\pi f n} \, df.$$

We have changed the signs in the exponentials because we have reversed the usual roles of time and frequency, i.e., we are writing a Fourier series for a frequency domain signal rather than a time domain signal. We assume for simplicity that $X(f)$ is continuous at f_0 (that is, $X(f_0 + \epsilon)$ and $X(f_0 - \epsilon)$ go to $X(f_0)$ as $\epsilon \to 0$) so that the Fourier series actually holds with equality at f_0. Consider the integral

$$
\begin{aligned}
\int_0^1 \delta(f - f_0) X(f)\, df &= \int_0^1 \delta(f - f_0)\Big(\sum_{n \in \mathcal{Z}} x_n e^{-i2\pi n f}\Big)\, df \\
&= \sum_{n \in \mathcal{Z}} x_n \int_0^1 \delta(f - f_0) e^{-i2\pi n f}\, df \\
&= \sum_{n \in \mathcal{Z}} x_n e^{-i2\pi n f_0} \\
&= X(f_0), \qquad\qquad\qquad (5.16)
\end{aligned}
$$

where we have used the property (5.15) that $\delta(f - f_0)$ sifts complex exponentials at f_0. The point is we have shown that if $\delta(f - f_0)$ sifts complex exponentials, it also sifts all other continuous frequency domain signals (assuming they are well behaved enough to have Fourier series).

We now summarize our hand-waving development to this point: If the signal $\{e^{i2\pi f_0 n};\ n \in \mathcal{Z}\}$ has a Fourier transform $\delta(f - f_0); f \in [0, 1)$, then this Fourier transform should satisfy the sifting property, i.e., for any suitably well behaved continuous frequency domain signal $G(f)$

$$
\int_0^1 \delta(f - f_0) G(f)\, df = G(f_0), \qquad\qquad (5.17)
$$

for $f_0 = m/N \in [0, 1)$.

When suitable theoretical machinery is introduced, the sifting property will in fact provide a rigorous definition of $\{\delta(f - f_0);\ f \in [0, 1)\}$, the Dirac delta function at f_0. There are two equivalent ways of making the sifting property rigorous. The first is operational: ordinary integrals are used in a limiting statement to obtain a result that looks like the sifting property. The second is essentially a redefinition of the integral itself using the concept of a *generalized function* or *distribution*. We shall see and use both methods.

The important fact to keep in mind when dealing with Dirac delta functions is that they only really make sense inside of an integral; one cannot hope that they will behave in a reasonable way when alone.

As in the modulation theorem, the difference between frequencies $f - f_0$ in (5.17), is handled like shifts in the time domain; that is, it is taken modulo the frequency domain of definition $\mathcal{S}$ so that $f - f_0 \in \mathcal{S}$.

The Dirac delta sifting property can be extended to non-continuous functions in the same way that Fourier series were so extended. If $G(f)$ has a jump discontinuity at f_0, then a reasonable requirement for a sifting property is

$$\int_0^1 \delta(f - f_0)G(f)\, df = \frac{G(f_0^+) + G(f_0^-)}{2}, \tag{5.18}$$

the midpoint of the upper and lower limits of $G(f)$ at f_0. This is the general form of the sifting property. Again observe that no ordinary function has this property.

Before making the Dirac delta rigorous, we return to the original question of finding a generalized DTFT for periodic signals and show how the sifting property provides a solution.

If we set $E(f) = \delta(f - \frac{k}{N})$, then (5.15) holds. Thus we could consider the DTFT of an exponential to be a Dirac delta function; that is, we would have the Fourier transform pair for $k \in \mathcal{Z}_N$

$$\{e^{i2\pi \frac{k}{N} n}; \ n \in \mathcal{Z}\} \leftrightarrow \{\delta(f - \frac{k}{N}); f \in [0, 1)\}, \tag{5.19}$$

where the frequency difference $f - k/N$ is here taken modulo 1, that is, $\delta(f - k/N) = \delta((f - k/N) \bmod 1)$.

If (5.19) were true, then (5.11) and linearity would imply that the DTFT G of a periodic discrete time signal g would be given by

$$G(f) = \sum_{k=0}^{N-1} \frac{\hat{G}(\frac{k}{N})}{N} \delta(f - \frac{k}{N}); \ f \in [0, 1). \tag{5.20}$$

Thus the spectrum of a periodic discrete function with period N can be considered to be a sequence of N Dirac deltas with areas weighted by the Fourier series coefficients for the periodic function, as depicted in Figure 5.3. The values $\hat{G}(k/N)$ which label the arrows are areas, not magnitudes.

It is important to differentiate between the two forms of Fourier transforms being used here: for a periodic signal g, $\hat{G}$ is the DFT of one period of g while G is the DTFT of the infinite duration signal g. Each of these can be inverted according to the usual rule for inverting a DFT or DTFT, thereby giving a weighted exponential form for g. Using the DFT inversion we have that

$$g_n = \frac{1}{N} \sum_{k=0}^{N-1} \hat{G}(\frac{k}{N}) e^{i2\pi \frac{k}{N} n}; \ \text{all } n \in \mathcal{Z}. \tag{5.21}$$

Using the DTFT inversion formula we have that

$$\begin{aligned} g_n &= \int_0^1 G(f) e^{i2\pi f n}\, df \\ &= \int_0^1 \sum_{k=0}^{N-1} \frac{\hat{G}(\frac{k}{N})}{N} \delta(f - \frac{k}{N}) e^{i2\pi f n}\, df \end{aligned}$$

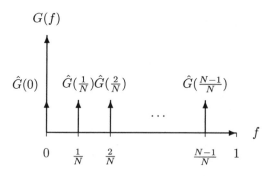

Figure 5.3: Graphical representation of DTFT of a periodic signal

which is the same as the previous equation because of the sifting properties of Dirac delta functions. Thus we can represent g_n either as a sum of weighted exponentials as in (5.21) (usually referred to as the Fourier series representation) or by an integral of weighted exponentials (the Fourier integral representation). Both forms are Fourier transforms, however. Which form is best? The Fourier series representation is probably the simplest to use when it suffices, but if one wants to consider both absolutely summable and periodic infinite duration signals together, then the integral representation using delta functions allows both signal types to be handled using the same notation.

To summarize the discussion thus far: given a discrete time periodic function g with period N, the following can be considered to be a Fourier transform pair:

$$G(f) = \sum_{k=0}^{N-1} \frac{\hat{G}(\frac{k}{N})}{N}\delta(f - \frac{k}{N}); f \in [0,1) \qquad (5.22)$$

$$g_n = \int_{-\frac{1}{2}}^{\frac{1}{2}} G(f)e^{i2\pi fn}\, df; \; n \in \mathcal{Z}, \qquad (5.23)$$

where

$$\hat{G}(\frac{k}{N}) = \sum_{n=0}^{N-1} g_n e^{-i2\pi \frac{k}{N}n}; k = 0, \ldots, N - 1. \qquad (5.24)$$

The same scenario works for continuous time periodic functions assuming the supposed properties of the Dirac delta function if we also assume that the Fourier transform is linear in a countable sense; that is, the transform of a countable sum of signals is the sum of the corresponding transforms. In this case the fundamental Fourier transform pair is that of the complex exponential:

$$\{e^{i2\pi f_0 t}; t \in \mathcal{R}\} \leftrightarrow \{\delta(f - f_0); f \in \mathcal{R}\} \qquad (5.25)$$

which allows us to write for a general continuous time periodic function g with period T the Fourier transform pair

$$G(f) = \sum_{k=-\infty}^{\infty} \frac{\hat{G}(\frac{k}{T})}{T} \delta(f - \frac{k}{T}); \; f \in \mathcal{R} \qquad (5.26)$$

$$g(t) = \int_{-\infty}^{\infty} G(f) e^{i2\pi f t} \, df; \; t \in \mathcal{R} \qquad (5.27)$$

where

$$\hat{G}(\frac{k}{T}) = \int_{0}^{T} g(t) e^{-i2\pi \frac{k}{T} t} dt; \; k \in \mathcal{Z}. \qquad (5.28)$$

Thus as in the discrete time case we can represent a continuous time periodic signal either as a sum (Fourier series) of weighted exponentials (5.13) or as an integral (5.27) (Fourier integral). Both representations have their uses. We next turn to the chore of making the idea of a Dirac delta more precise so as to justify the above development.

5.3 Generalized Functions

In this section we add rigor to the intuitive notion of a Dirac delta function using the theory of generalized functions. Recall that an (ordinary) function is just a mapping of one space into another. In our case we have been considering complex-valued functions defined on the real line, that is, functions of the form $g : \mathcal{R} \to \mathcal{C}$ which are mappings assigning a value $g(x) \in \mathcal{C}$ to every $x \in \mathcal{R}$. We here use x as a dummy variable since it could represent either time or frequency. Ordinary functions include things like $\sin x$, e^{-x^2}, $u_{-1}(x)$, etc. A *generalized function* or *distribution* is a mapping of functions into complex numbers according to certain properties. Given a function $g = \{g(x); x \in \mathcal{R}\}$, a distribution assigns a complex number $D(g)$ to g in a way satisfying the following rules:

1. (Linearity)

   Given two functions g_1 and g_2 and complex constants a_1 and a_2, then

   $$D(a_1 g_1 + a_2 g_2) = a_1 D(g_1) + a_2 D(g_2)$$

2. (Continuity)

If $\lim_{n\to\infty} g_n(x) = g(x)$ for all x, then also

$$\lim_{n\to\infty} D(g_n) = D(g)$$

A generalized function is called a *linear functional* or *linear operator* in functional analysis. The key idea to keep in mind is that a generalized function assigns a complex number to every function.

A common example of a generalized function is the following: Suppose that $h(x)$ is a fixed ordinary function and define the generalized function D by

$$D_h(g) = \int_{-\infty}^{\infty} g(x)h(x)dx,$$

that is, $D_h(g)$ assigns the value to g equal to the integral of the product of g with the fixed function h. The properties of integration then guarantee that D meets the required conditions to be a generalized function. Note that this generalized function has nothing strange about it; the above integral is an ordinary integral.

As a second example of a generalized function, consider the operator D_δ defined as follows:

$$D_\delta(g) = \begin{cases} g(0) & \text{if } g(t) \text{ is continuous} \\ & \text{at } t = 0 \\ \frac{g(0^+)+g(0^-)}{2} & \text{otherwise} \end{cases} \tag{5.29}$$

where $g(0^+)$ and $g(0^-)$ are the upper and lower limits of g at 0, respectively. We assume that $g(t)$ is sufficiently well behaved to ensure the existence of these limits, e.g., $g(t)$ is piecewise smooth.

We write this generalized function symbolically as

$$D_\delta(g) = \int_{-\infty}^{\infty} \delta(x)g(x)dx,$$

but this is only a symbol or alternative notation for D_δ, it is *NOT* an ordinary integral. The defined distribution is both linear and continuous. This is in fact the rigorous definition of a Dirac delta function. It is not really a function in the ordinary sense; it is a generalized function which assigns a real number to every function. Alternatively, it is not really $\delta(x)$ that we have defined, it is D_δ or an integral containing $\delta(x)$ that has been defined. Because of the linearity and continuity of a generalized function, we can often (but not always) deal with the above integral as if it were an ordinary integral.

The shifted Dirac delta $D_{\delta_{x_0}}$ is interpreted in a similar fashion. That is, when we write the integral

$$D_{\delta_{x_0}}(g) = \int_{-\infty}^{\infty} \delta(x - x_0)g(x)\, dx$$

what we really mean is that this is a generalized function defined by

$$D_{\delta_{x_0}}(g) = \frac{g(x_0^+) + g(x_0^-)}{2}.$$

The shifted Dirac delta can be related to the unshifted Dirac delta with a shifted argument. Define the shifted signal $g_{x_0}(x)$ by

$$g_{x_0}(x) = g(x + x_0).$$

Then it is easy to see that

$$D_{\delta}(g_{x_0}) = D_{\delta_{x_0}}(g).$$

This relationship becomes more familiar if we use the integral notation for the generalized function:

$$\int_{-\infty}^{\infty} \delta(x)g(x + x_0)\, dx = \int_{-\infty}^{\infty} \delta(x - x_0)g(x)\, dx. \qquad (5.30)$$

In this form it looks like an ordinary change of variables formula. If instead of a generalized function δ we had an ordinary function h, then

$$\int_{-\infty}^{\infty} h(x)g(x + x_0)\, dx = \int_{-\infty}^{\infty} h(x - x_0)g(x)\, dx$$

would follow immediately by changing the variable $x - x_0$ to x. A subtle but important point has been made here: although δ is not an ordinary function and the integrals in (5.30) are not ordinary integrals, under the integral sign it behaves like ordinary functions, at least in the special case of the simple change of variables by a shift. We shall see many more ways in which generalized functions inside integral signs behave as one would expect.

As a third example of a generalized function we consider an example which is fairly close to the usual physical description of a Dirac delta function. Suppose that we have a sequence of functions $h_n(x)$; $n = 1, 2, \ldots$ with the following properties:

1.

$$\int_{-\infty}^{\infty} h_n(x)\, dx = 1; \, n = 0, 1, 2, \ldots$$

2. For any function g

$$\lim_{n\to\infty} \int_{-\infty}^{\infty} h_n(x)g(x)\, dx = \begin{cases} g(0) & \text{if } g(t) \text{ is continuous at } t = 0 \\ \frac{g(0^+)+g(0^-)}{2} & \text{otherwise} \end{cases}$$

$$(5.31)$$

Define D_δ by

$$D_\delta(g) = \lim_{n\to\infty} \int_{-\infty}^{\infty} h_n(t)g(t)\, dt,$$

if the limit exists, where now the integrals are ordinary integrals. This provides an alternative description of the generalized function D_δ.

The limiting function description of the Dirac delta function is convenient for also describing the shifted delta function. If h_n is such a sequence, then an ordinary change of variables yields

$$\lim_{n\to\infty} \int_{-\infty}^{\infty} h_n(x - x_0)g(x)\, dx = \lim_{n\to\infty} \int_{-\infty}^{\infty} h_n(y)g(y + x_0)\, dy$$

$$= \frac{g(x_0^+) + g(x_0^-)}{2} \qquad (5.32)$$

and hence the shifted sequence satisfies the properties of a shifted Dirac delta. In fact, we can also prove (5.30) by defining the generalized function $D_{\delta_{x_0}}$ in terms of the limiting behavior of integrals of the shifted functions $h_n(x - x_0)$ in the above sense. This provides a very useful general approach to proving properties for the Dirac delta: find a sequence h_n describing the generalized function, prove the property for the members of the sequence using ordinary signals and integrals, then take the limit to get the implied property for the generalized function.

It is important to note that we are *NOT* making the claim that the delta function is itself the limit of the h_n, i.e., that "$\delta(t) = \lim_{n\to\infty} h_n(t)$." In fact, most sequences of functions $h_n(t)$ satisfying the required conditions will not have a finite limit at $t = 0$! In spite of this warning, it is sometimes useful to think of an impulse as a limit of functions satisfying these conditions. We next consider a few such sequences.

- $h_n(x) = n\square_{\frac{1}{2n}}(x)$. (See Figure 5.4.) Alternatively, one can use $n \sqcap (nx)$. This is the simplest such sequence, a rectangle with vanishing width and exploding height while preserving constant area. If $g(t)$ is continuous at $t = 0$, then the mean value theorem of calculus implies that

$$n \int_{-\frac{1}{2n}}^{\frac{1}{2n}} g(t)\, dt \approx n\frac{g(0)}{n}.$$

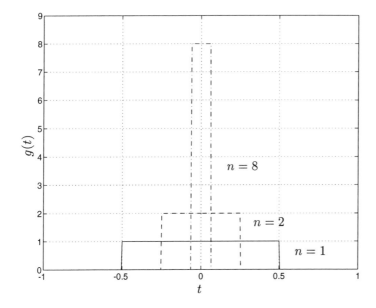

Figure 5.4: Impulse via Box Functions: $n\square_{\frac{1}{2n}}(t)$

Note that $h_n(x)$ has 0 as a limit as $n \to \infty$ for all x except the point $x = 0$. Although this sequence is simple, it has the disadvantage that its derivative does not exist at the edges.

- The sequence $h_n(x) = n\wedge(nx)$ also has the required properties, where $\wedge$ is the triangle function of (1.10). (See Figure 5.5.)

It is also not differentiable at its edges and at the origin. It is piecewise smooth. Again $h_n(x)$ has a limit (0) except for the point $x = 0$.

- The Dirichlet kernel $h_n(x) = D_n(x) = \frac{\sin[2\pi x(n+\frac{1}{2})]}{\sin(\pi t)}$. (See Figure 3.1.) Since this sequence is periodic with period 1, it is only useful for defining a Dirac delta when the domain of definition of x has width 1, as is the case here. It can be modified for any finite width interval of length T by scaling $\frac{1}{T}D_N(\frac{x}{T})$ as in (3.70). This sequence was used in the "proof" of the convergence of the finite duration CTFT and hence also in the proof of the Fourier series representation of a periodic function. Thus a rigorous proof that $D_n(x)$ indeed has the properties required to define the Dirac delta also thereby provides a rigorous proof of those Fourier inversion formulas.

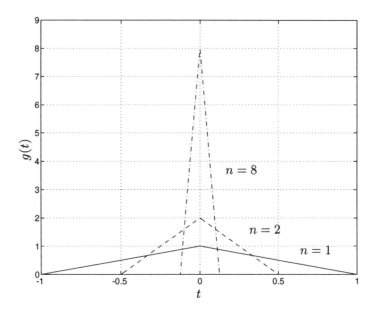

Figure 5.5: Impulse via Triangle Functions: $n \wedge (nt)$

- $h_n(x) = n \operatorname{sinc}(nx) = \sin(n\pi x)/\pi x$. (See Figure 3.2.) In this example the limit $\lim_{n \to \infty} h_n(x)$ does not exist for *any* x, yet the sequence is quite suitable for defining the Dirac delta. This is in fact the sequence of functions that arose in the "proof" of the Fourier integral theorem. That proof can now be viewed as accurate given the fact (which we do not prove) that this sequence indeed satisfies the required conditions of (5.29) to define the Dirac delta generalized function. Observe that these functions are everywhere differentiable.

As an aid to picturing the functions $n \operatorname{sinc}(nx)$, observe that

$$\lim_{x \to 0} \frac{\sin x}{x} = 1$$

and

$$\lim_{x \to 0} \frac{\sin(ax)}{x} = a.$$

Thus

$$\lim_{x \to 0} \frac{\sin(n\pi x)}{\pi x} = n.$$

These function sequences are useful for proving properties of Dirac delta functions. For example, consider the meaning of $\delta(ax)$, a delta function with an argument scaled by $a > 0$. Using the simplest limiting sequence we can argue that this should behave under the integral sign like

$$
\begin{aligned}
\int_{-\infty}^{\infty} \delta(ax)g(x)dx &= \lim_{n \to \infty} \int_{-\infty}^{\infty} n\square_{\frac{1}{2na}}(x)g(x)dx \\
&= \lim_{n \to \infty} \int_{-\infty}^{\infty} n\square_{\frac{1}{2na}}(\frac{y}{a})g(\frac{y}{a})\frac{dy}{a} \\
&= \lim_{n \to \infty} \int_{-\infty}^{\infty} n\square_{\frac{1}{2n}}(y)g(\frac{y}{a})\frac{dy}{a}
\end{aligned}
$$

by a simple change of variables. This, however, is just

$$
\frac{g(0)}{a} = \int_{-\infty}^{\infty} \frac{1}{a}\delta(x)g(x)\, dx.
$$

Identifying these two generalized functions (remember the above formulas hold for *all* continuous $g(x)$) implies that

$$
\delta(ax) = \frac{1}{a}\delta(x).
$$

A similar argument for negative a using the fact that now there is a sign change when the integration variable is changed gives the general result

$$
\delta(ax) = \frac{1}{|a|}\delta(x). \tag{5.33}
$$

A further generalization of this result is

$$
\delta(ax + b) = \frac{1}{|a|}\delta(x + \frac{b}{a}) \tag{5.34}
$$

for any $a \neq 0$. The proof is left as an exercise. (See Problem 6.) Other properties of Dirac δ functions are also developed in the exercises. One such property that is often useful is

$$
g(t)\delta(t - t_0) = g(t_0)\delta(t - t_0) \tag{5.35}
$$

if $g(t)$ is continuous at t_0. (See Problem 10.)

5.4 Fourier Transforms of Generalized Functions

With the definition of a generalized function in hand, we can now generalize the Fourier transform to include integral transforms of signals involving

Dirac delta functions by interpreting the integral as a generalized function. In this sense the formulas of the previous section are made precise. Thus, for example, in the continuous time case we have that

$$\mathcal{F}_f(\{\delta(t - t_0); t \in \mathcal{R}\}) = \int_{-\infty}^{\infty} \delta(t - t_0) e^{-i2\pi ft}\, dt$$

$$= e^{-i2\pi ft_0}; f \in \mathcal{R} \qquad (5.36)$$

using the sifting property. In particular, setting $t_0 = 0$ yields

$$\mathcal{F}(\{\delta(t); t \in \mathcal{R}\}) = 1; f \in \mathcal{R}; \qquad (5.37)$$

that is, the Fourier transform of a Dirac delta at the time origin is a constant. This result is the continuous time analog to the result that the DTFT of a Kronecker delta is a constant.

We have already argued that the Fourier transform of a complex exponential $\{e^{i2\pi f_0 t}; t \in \mathcal{R}\}$ should have the sifting property, that is, behave like the generalized function defining a Dirac delta. Thus we can also define

$$\mathcal{F}(\{e^{i2\pi tf_0}; t \in \mathcal{R}\}) = \{\delta(f - f_0); f \in \mathcal{R}\}; \qquad (5.38)$$

that is, the Fourier transform of a complex exponential is a delta function. Alternatively, if we consider a formal integral in a Cauchy sense we should have that

$$\mathcal{F}_f(\{e^{i2\pi f_0 t}; t \in \mathcal{R}\}) = \lim_{N \to \infty} \int_{-\frac{N}{2}}^{\frac{N}{2}} e^{-i2\pi t(f - f_0)}\, dt$$

$$= \lim_{N \to \infty} N \operatorname{sinc}(N(f - f_0)),$$

which "converges" to $\delta(f - f_0)$ in the sense that the function $N \operatorname{sinc}[N(f - f_0)]$ inside an integral has the sifting property defining the Dirac delta. (The formula makes no sense as an ordinary limit because the final limit does not exist.)

Again the special case of $f_0 = 0$ yields the relation

$$\mathcal{F}_f(\{1; t \in \mathcal{R}\}) = \delta(f); f \in \mathcal{R}; \qquad (5.39)$$

that is, the Fourier transform of a continuous time infinite duration signal that is a constant (a dc) is a delta function in frequency.

Note that these two above results are duals: Transforming a delta in one domain yields a complex exponential in the other domain.

In the discrete time case the intuition is slightly different and there is not the nice duality because there is no such thing as a Dirac delta in discrete time. The analogous idea is a Kronecker delta which is a perfectly well behaved quite ordinary function (whose transform happens to be a complex exponential). In the infinite duration discrete time case, one can, however, have Dirac deltas in the frequency domain and their inverse transform is a discrete time complex exponential since from the sifting property

$$\int_S \delta(f - f_0) e^{i2\pi f n} \, df = e^{i2\pi f_0 n}; \text{ if } f_0 \in S. \tag{5.40}$$

We have seen that the intuitive ideas of Dirac deltas can be made precise using generalized functions and that this provides a means of defining DTFTs and CTFTs for infinite duration periodic signals, even though these signals violate the sufficient conditions for the existence of ordinary DTFTs and CTFTs. Generalized functions also provide a means of carefully proving conjectured properties of Dirac delta functions either by limiting arguments or from the properties of distributions. We will occasionally have need to derive such properties. One should always be careful about treating delta functions as ordinary functions.

The previously derived properties of Fourier transforms extend to generalized transforms involving delta functions. For example, the differentiation theorem gives consistent results for some signals which are not strictly speaking differentiable. Consider, for example, the box function $\{\Box_T(t); t \in \mathcal{R}\}$. This function is not differentiable at $-T$ and $+T$ in the usual sense, but one can define the derivative as a generalized function as

$$\frac{d}{dt} \Box_T(t) = \delta(t + T) - \delta(t - T)$$

since if one integrates the generalized function on the right, one gets $\Box_T(t)$. Now the transform of $\Box_T(t)$ is $2T \operatorname{sinc}(2Tf)$ and hence the differentiation theorem implies that the transform of $\frac{d}{dt} \Box_T(t)$ should be $i2\pi f 2T \operatorname{sinc}(2Tf) = 2i \sin(2T\pi f)$, which is easily seen from the sifting property and Euler's relations to be the transform of $\delta(t + T) - \delta(t - T)$.

5.5 ⋆ Derivatives of Delta Functions

The Dirac delta is effectively defined by its behavior inside an integral and hence integrals of Dirac deltas make sense. Since integration and differentiation are inverse operations on ordinary functions, one might ask if one can define a derivative of a delta function. The answer is yes if we consider the new function as another generalized function. As a first approach, suppose that we consider the sequence of triangle functions of (1.10) that "converges" to the Dirac delta in the sense that (5.31) holds. This sequence of functions has a derivative defined by

$$h'_n(x) = \frac{d}{dx} n \wedge (nx) = \begin{cases} -n^2 & x \in (0, 1/n) \\ +n^2 & x \in (-1/n, 0) \\ 0 & |x| > 1/n \end{cases} . \qquad (5.41)$$

The derivative is not defined at the origin and at the points $\pm 1/n$ (because the left and right derivatives differ), but we are only interested in the behavior of this function under an integral sign and its differential behavior at these points does not matter. Suppose now that $g(x)$ is a well behaved function. In particular, suppose that it has a continuous derivative at the origin. For large n we will then have that

$$\int h'_n(x)g(x) \, dx \;=\; n^2 \int_{-1/n}^{0} g(x) \, dx - n^2 \int_0^{1/n} g(x) \, dx$$

$$\approx\; ng(-\frac{1}{2n}) - ng(\frac{1}{2n}),$$

where we have approximated each integral by the value of the function at the middle of the interval of integration times the length of the integral. As $n \to \infty$ this tends to the negative of the derivative, $dg(x)/dx$ evaluated at $x = 0$; that is, setting $\Delta x = 1/n$ we have that

$$\lim_{n\to\infty} [ng(\frac{1}{2n}) - ng(\frac{-1}{2n})] \;=\; \lim_{\Delta x \to 0} \frac{g(\frac{\Delta x}{2}) - g(\frac{-\Delta x}{2})}{\Delta x}$$

$$=\; \frac{dg(x)}{dx}|_{x=0} \overset{\Delta}{=} g'(0).$$

Thus

$$\lim_{n\to\infty} \int h'_n(x)g(x) \, dx = -g'(0). \qquad (5.42)$$

The intuition here is as follows: If the functions h_n "converge" to a Dirac delta, then their derivatives should "converge" to the derivative of a Dirac delta, say $\delta'(x)$, which should behave under the integral sign as above; that is, for any $g(x)$ that is differentiable at the origin,

$$\int \delta'(x)g(x) \, dx = -g'(0). \qquad (5.43)$$

The generalized function $\delta'(x)$ is called a *doublet*.

Analogous to the Dirac delta, we can consider the shifted doublet to obtain the formula

$$\int \delta'(t - x)g(x) \, dx = -g'(t) \qquad (5.44)$$

if $g(x)$ is differentiable at t. This type of integral is called a *convolution* and the result often described as showing that the convolution of a doublet with a signal produces the derivative of the signal at the location of the doublet.

The Fourier transform of the doublet is easily found to be

$$
\mathcal{F}_f(\{\delta'(t); \ t \in \mathcal{R}\}) \quad = \quad \int_{-\infty}^{\infty} \delta'(t) e^{-i2\pi f t} \, dt
$$

$$
= \quad -\frac{d}{dt} e^{-i2\pi f t}\big|_{t=0} = i2\pi f. \tag{5.45}
$$

In a similar manner higher order derivatives of the Dirac delta can be defined as generalized functions.

5.6   ⋆ The Generalized Function $\delta(g(t))$

Generalized functions provide a way to give meaning to a Dirac delta with an ordinary function as an argument, that is, a generalized function of the form $\delta(g(t))$. Suppose for the moment that $g(t)$ is a well-behaved function with a single zero at $t = t_0$; that is, $g(t_0) = 0$ and $g(t) \neq 0$ for $t \neq t_0$. If we think of the Dirac delta as a very tall and narrow pulse, then $\delta(g(t))$ will have a similar form and will be zero except for $t = t_0$, where $g(t) = 0$ and hence $\delta(g(t))$ has a jump. Hence a natural guess is that $\delta(g(t)) = A\delta(t-t_0)$, where the area A has to be determined. To make this precise we need to demonstrate that for any well-behaved function $r(t)$,

$$
\int_{-\infty}^{\infty} r(t)\delta(g(t)) \, dt = \int_{-\infty}^{\infty} r(t)A\delta(t - t_0) \, dt = Ar(t_0). \tag{5.46}
$$

We will sketch how this result is proved and find the constant A in the process.

Let $h_n(t) = n\Box_{\frac{1}{2n}}(t)$ denote the box sequence of functions converging to the Dirac delta under the integral sign and consider the limit defining the new generalized function:

$$
\int_{-\infty}^{\infty} r(t)\delta(g(t)) \, dt = \lim_{n \to \infty} \int_{-\infty}^{\infty} r(t)h_n(g(t)) \, dt. \tag{5.47}
$$

The function $h_n(g(t))$ is given by

$$
h_n(g(t)) = \begin{cases} n & -\frac{1}{2n} < g(t) < \frac{1}{2n} \\ 0 & \text{otherwise} \end{cases}.
$$

To see how this behaves in the limit as $n \to \infty$, we suppose that t is very near t_0 and expand $g(t)$ in a Taylor series around t_0 as

$$g(t) \approx g(t_0) + (t - t_0)g'(t_0) + \text{ higher order terms } \approx (t - t_0)g'(t_0).$$

With this approximation we have approximately for large n that

$$h_n(g(t)) = \begin{cases} n & -\frac{1}{2n} < (t - t_0)g'(t_0) < \frac{1}{2n} \\ 0 & \text{otherwise} \end{cases}.$$

Assume for the moment that $g'(t_0)$ is positive and not zero and we can rewrite the above equation as

$$h_n(g(t)) = \begin{cases} n & t_0 - \frac{1}{2ng'(t_0)} < t < t_0 + \frac{1}{2ng'(t_0)} \\ 0 & \text{otherwise} \end{cases}. \tag{5.48}$$

Thus $h_n(g(t))$ is a box function of height n and width $1/(ng'(t_0))$ and center at $t = t_0$. By an argument analogous to that used to demonstrate that the sequence h_n provides a limiting definition of the Dirac delta, multiplying $r(t)$ by this box function, integrating, and taking the limit will yield $r(t_0)/g'(t_0)$, giving (5.46) with $A = 1/g'(t_0)$.

If $g'(t_0)$ is negative, then $g'(t_0)$ in (5.48) is replaced by $|g'(t_0)|$ and the remainder of the argument proceeds with this substitution. If g has a finite collection of zeros, then a similar argument for each zero yields the following result.

If $g(t)$ is a well-behaved function with zeros at $t_1, t_2, \ldots, t_M$, then

$$\delta(g(t)) = \sum_{k=1}^{M} \frac{\delta(t - t_k)}{g'(t_k)}; \tag{5.49}$$

that is,

$$\int_{-\infty}^{\infty} r(t)\delta(g(t))\, dt = \sum_{k=1}^{M} \frac{r(t_k)}{g'(t_k)}. \tag{5.50}$$

5.7   Impulse Trains

Having defined impulse functions precisely using the idea of generalized functions and having used these ideas to describe the generalized Fourier transform pair consisting of delayed delta functions and shifted exponentials, we now develop an alternative Fourier representation of continuous time delta functions by applying Fourier series ideas to a single delta function considered as a finite duration signal. This provides a rather surprising formula for the delta function which can be proved both formally, pretending that the delta function is an ordinary signal, and carefully, using the ideas of generalized functions. This representation leads naturally to periodic impulse trains as infinite duration signals, the so-called ideal sampling function (or sampling waveform or impulse train). As some of the results may be counter-intuitive at first glance, it is helpful to first develop them in the simple context of a finite duration signal. The generalizations then come immediately using the periodic extension via Fourier series.

Suppose that we consider $\delta = \{\delta(t); -T/2 \le t < T/2\}$ as a finite duration signal which equals the Dirac delta function during $\mathcal{T} = [-T/2, T/2)$. Assume for the moment that we can treat this as an ordinary finite duration continuous time signal and derive its Fourier series; that is, we can represent $\delta(t)$ on $\mathcal{T}$ by a series of the form

$$\delta(t) = \sum_{n=-\infty}^{\infty} c_n e^{i2\pi \frac{t}{T} n}; \ t \in [-\frac{T}{2}, \frac{T}{2})$$

where the coefficients can be computed as

$$c_n = \frac{1}{T} \int_{-\frac{T}{2}}^{\frac{T}{2}} \delta(t) e^{-i2\pi \frac{t}{T} n} \, dt = \frac{1}{T}$$

using the defining property of delta functions. This yields the surprising representation for a Dirac delta:

$$\delta(t) = \sum_{n=-\infty}^{\infty} \frac{e^{i2\pi \frac{t}{T} n}}{T}; \ t \in [-\frac{T}{2}, \frac{T}{2}). \tag{5.51}$$

This suggests that an infinite sum of exponentials as above equals a Dirac delta, regardless of the value of T! This is not, however, a proof, because we cannot treat a delta function as an ordinary signal in this way. To verify (5.51) we must demonstrate that the right hand side is a generalized function with the necessary properties to be a Dirac delta. In particular, we must show that for any well behaved continuous function $g(t)$

$$g(0) = \int_{-\frac{T}{2}}^{\frac{T}{2}} g(t) \left(\sum_{n=-\infty}^{\infty} \frac{e^{i2\pi \frac{t}{T} n}}{T} \right) dt. \tag{5.52}$$

This still leaves a problem of rigor, however, as the infinite sum inside the right-hand integral may not exist; in fact it cannot exist if it is to equal a delta function. In order to make sense of the generalized function we wish to call $\sum_{n=-\infty}^{\infty} e^{i2\pi \frac{t}{T}n}/T$, we resort to the limiting definition of delta functions. Define the sequence of functions $h_k(t)$ by

$$h_k(t) = \sum_{n=-k}^{k} \frac{e^{i2\pi \frac{t}{T}n}}{T}. \tag{5.53}$$

We note in passing the strong resemblance of this formula to the Fourier transform of a discrete time rectangle, but we do not take advantage of this similarity. Instead observe that for all k these functions have integral

$$\int_{-\frac{T}{2}}^{\frac{T}{2}} h_k(t)\, dt = \int_{-\frac{T}{2}}^{\frac{T}{2}} \sum_{n=-k}^{k} \frac{e^{i2\pi \frac{t}{T}n}}{T}\, dt$$

$$= \sum_{n=-k}^{k} \int_{-\frac{T}{2}}^{\frac{T}{2}} \frac{e^{i2\pi \frac{t}{T}n}}{T}\, dt$$

$$= \sum_{n=-k}^{k} \delta_n = 1,$$

where δ_n is a Kronecker δ and we have used the fact that the integral of a complex exponential over one period is 0 unless the exponent is 0. Thus the h_k satisfy the first condition for defining an impulse as a limit. There is no problem with the interchange of integral and sum because the summation is finite. Next observe that if $g(t)$ is continuous at the origin,

$$\lim_{k\to\infty} \int_{-\frac{T}{2}}^{\frac{T}{2}} h_k(t)g(t)\, dt = \lim_{k\to\infty} \int_{-\frac{T}{2}}^{\frac{T}{2}} g(t) \sum_{n=-k}^{k} \frac{e^{i2\pi \frac{t}{T}n}}{T}\, dt$$

$$= \lim_{k\to\infty} \sum_{n=-k}^{k} \int_{-\frac{T}{2}}^{\frac{T}{2}} g(t)\frac{e^{i2\pi \frac{t}{T}n}}{T}\, dt$$

$$= \lim_{k\to\infty} \sum_{n=-k}^{k} \frac{G(-\frac{n}{T})}{T}$$

$$= \sum_{n=-\infty}^{\infty} \frac{G(\frac{n}{T})}{T}.$$

Does the last infinite sum above exist as implied? If the original signal $g(t)$ is well behaved in the sense that it has a Fourier transform G, then from the inversion formula for the finite duration continuous time Fourier transform we can write

$$g(0) = \sum_{n=-\infty}^{\infty} \frac{G(\frac{n}{T})}{T}, \tag{5.54}$$

that is, the claimed infinite sum exists and equals $g(0)$. But this implies that the sequence h_n also satisfies the second condition required of the limiting definition of a delta function:

$$\lim_{k \to \infty} \int_{-\frac{T}{2}}^{\frac{T}{2}} h_k(t)g(t)\, dt = g(0)$$

for continuous functions g. This shows that indeed the (5.51) is valid in a distribution sense.

We omit the details of what happens when the function $g(t)$ is not continuous at $t = 0$, but the proof can be completed by the stout of heart.

Eq. (5.51) provides a representation for a Dirac delta defined as a finite duration signal. As always with a Fourier series, however, we can consider the series to be defined for all time since it is is periodic with period T. Thus the Fourier series provides immediately the periodic extension of the original finite duration signal. In our case this consists of periodic replicas of a delta function. This argument leads to the formula for an impulse train

$$\Psi_T(t) = \sum_{n=-\infty}^{\infty} \delta(t - nT) = \sum_{n=-\infty}^{\infty} \frac{e^{i2\pi \frac{t}{T} n}}{T}; \quad -\infty < t < \infty. \tag{5.55}$$

Thus the Fourier series for a Dirac delta considered as a finite duration signal of duration T gives an infinite impulse train when considered as an infinite duration signal. The infinite impulse train is sometimes referred to as the "bed of nails" or "comb" function because of its symbolic appearance. It is also referred to as the *ideal sampling function* since using the properties of delta functions (especially (5.35)), we have that

$$g(t)\Psi_T(t) = \sum_{n=-\infty}^{\infty} g(nT)\delta(t - nT); \tag{5.56}$$

that is, multiplying the sampling function by an arbitrary signal yields a sequence of pulses with area proportional to the samples of the original signal. This representation will be useful when we later interpret the sampling theorem.

The same manipulations can be done in the frequency domain to obtain

$$\Psi_S(f) = \sum_{n=-\infty}^{\infty} \delta(f - nS) = \sum_{n=-\infty}^{\infty} \frac{1}{S} e^{i2\pi \frac{f}{S}n}.$$

The summation index can be negated without changing the sum to yield

$$\Psi_S(f) = \sum_{n=-\infty}^{\infty} \frac{1}{S} e^{-i2\pi \frac{f}{S}n}.$$

Recall that the Fourier transform of a periodic continuous time signal g has the form

$$G(f) = \sum_{n=-\infty}^{\infty} \frac{\hat{G}(\frac{n}{T})}{T} \delta(f - \frac{n}{T})$$

where

$$\hat{G}(f) = \int_0^T g(t) e^{-i2\pi ft}\, dt.$$

Thus we can now write

$$G(f) = \frac{1}{T}\Psi_{\frac{1}{T}}(f)\hat{G}(f); \tag{5.57}$$

that is, the spectrum of a periodic signal $g(t)$ is the *sampled* Fourier transform of the finite duration signal $\{g(t); t \in [0,T)\}$ consisting of a single period of the periodic waveform, when that transform is defined for all real f.

We have seen a Fourier series representation for the sampling function. An alternative Fourier representation for such a periodic signal is a Fourier transform. If we proceed formally this is found to be

$$\begin{aligned}
\mathcal{F}_f(\{\Psi_T(t); t \in \mathcal{R}\}) &= \int_{-\infty}^{\infty} \left(\sum_{n=-\infty}^{\infty} \delta(t - nT) \right) e^{-i2\pi ft}\, dt \\
&= \sum_{n=-\infty}^{\infty} \int_{-\infty}^{\infty} \delta(t - nT) e^{-i2\pi ft}\, dt \\
&= \sum_{n=-\infty}^{\infty} e^{-i2\pi fnT}
\end{aligned}$$

This exponential sum, however, is almost identical to the Fourier series that we have already seen for a sampling function, the only difference is that we have replaced the time variable t/T by the frequency variable fT. Thus we can conclude that

$$
\begin{aligned}
\mathcal{F}_f(\{\Psi_T(t); t \in \mathcal{R}\}) &= \sum_{n=-\infty}^{\infty} e^{-i2\pi fnT} \\
&= \frac{1}{T} \sum_{n=-\infty}^{\infty} \delta(f - \frac{n}{T}) \\
&= \frac{1}{T} \Psi_{\frac{1}{T}}(f);
\end{aligned} \tag{5.58}
$$

that is, the Fourier transform of the sampling function in time is another sampling function in frequency!

The sampling function provides an example in which we can use the properties of delta functions to demonstrate that generalized Fourier transforms can possess the same basic properties as ordinary Fourier transforms, as one would hope. The stretch theorem provides an example. Consider the periodic functions formed by replicating the finite duration signals. Let $\tilde{g} = \{\tilde{g}(t); t \in \mathcal{R}\}$ be the periodic extension of some signal g and $\tilde{g}_a$ the periodic extension of $g_a = \{g(at); t \in [0, T/a)\}$, with $a > 0$. Then $\tilde{g}$ has period T, while $\tilde{g}(at)$ has period T/a. The Fourier transform of $\tilde{g}(t)$ is

$$
\tilde{G}(f) = \sum_{k=-\infty}^{\infty} \frac{1}{T} G(\frac{k}{T}) \delta(f - \frac{k}{T})
$$

and the Fourier transform of $\tilde{g}(at)$ is, from the finite-duration argument,

$$
\sum_{k=-\infty}^{\infty} \frac{1}{T/a} \frac{1}{a} G(\frac{k}{T}) \delta(f - \frac{ka}{T}) = \sum_{k=-\infty}^{\infty} \frac{1}{T} G(\frac{k}{T}) \delta(f - \frac{ka}{T}).
$$

According to the infinite duration continuous time stretch theorem, however, this transform should be

$$
\frac{1}{a} \tilde{G}(\frac{f}{a}) = \frac{1}{a} \frac{1}{T} \sum_{k=-\infty}^{\infty} G(\frac{k}{T}) \delta(\frac{f}{a} - \frac{k}{T}),
$$

which from the stretch theorem for the Dirac delta function is

$$
\sum_{k=-\infty}^{\infty} \frac{1}{T} G(\frac{k}{T}) \delta(f - \frac{ka}{T}),
$$

which agrees with the previous result.

An important special case of the sampling function with a special name is the case with $T = 1$. Here the signal is called the *shah* function and is defined as

$$III(t) = \sum_{n=-\infty}^{\infty} \delta(t - n). \tag{5.59}$$

Bracewell [6] describes the history of the cyrillic letter III for the historically minded reader. Using the stretch formula for delta functions, (5.34), observe that

$$III(\frac{t}{T}) = \sum_{n=-\infty}^{\infty} \delta(\frac{t}{T} - n) = T \sum_{n=-\infty}^{\infty} \delta(t - nT); \tag{5.60}$$

that is, the sampling function is given by

$$\sum_{n=-\infty}^{\infty} \delta(\frac{t}{T} - n) = T^{-1}III(t/T). \tag{5.61}$$

Observe from the earlier results for sampling functions that III is its own Fourier transform; that is,

$$\mathcal{F}(\{III(t); t \in \mathcal{R}\}) = \{III(f); f \in \mathcal{R}\}. \tag{5.62}$$

Impulse Pairs

We close this section with two related generalized Fourier transforms. First consider the continuous time signal (actually, generalized function) defined by

$$II(t) = \frac{\delta(t + \frac{1}{2}) + \delta(t - \frac{1}{2})}{2}; t \in \mathcal{R}, \tag{5.63}$$

which is called an even impulse pair. Taking the generalized Fourier transform we have that

$$\begin{aligned} \mathcal{F}_f(II) &= \frac{1}{2} \int \delta(t + \frac{1}{2}) e^{-i2\pi ft}\, dt + \frac{1}{2} \int \delta(t - \frac{1}{2}) e^{-i2\pi ft}\, dt \\ &= \frac{e^{i\pi f} + e^{-i\pi f}}{2} = \cos(\pi f); f \in \mathcal{R}. \end{aligned} \tag{5.64}$$

Similarly we can define the odd impulse pair

$$I_I(t) = \frac{\delta(t + \frac{1}{2}) - \delta(t - \frac{1}{2})}{2}; t \in \mathcal{R}, \tag{5.65}$$

and find that

$$\mathcal{F}_f(I_I) = i \sin(\pi f); f \in \mathcal{R}. \tag{5.66}$$

5.8 Problems

5.1. Use the limiting transform method to find the Fourier transform of $\text{sgn}(t+3)$.

5.2. Suppose that we attempt to find the Fourier transform of $\text{sgn}(t)$ using the sequence $h_n(t)$ defined by

$$h_n(t) = \begin{cases} 1 & 0 < t \le n \\ -1 & -n \le t < 0 \\ 0 & \text{otherwise} \end{cases}$$

instead of the exponential sequence actually used. Does this approach yield the same result?

5.3. What is the (generalized) Fourier transform of the signal $g = \{g(t); \ t \in \mathcal{R}\}$ defined by $g(t) = 1/t$ for $t \ne 0$ and $g(0) = 0$?

5.4. We have seen that a discrete time infinite duration periodic signal g_n with period N can be expressed as a Fourier series

$$g_n = \sum_{k=-\infty}^{\infty} b_k e^{i2\pi \frac{k}{N} n}.$$

Suppose that we consider a more general but similar form of infinite duration discrete time signal. Consider the signal h_n given by

$$h_n = \sum_{k=-\infty}^{\infty} b_k e^{i2\pi \lambda_k n}$$

the difference being that we are now allowing frequencies which may not have the form $\frac{k}{N}$ of those in the Fourier series. A signal of this form is sometimes referred to as a generalized Fourier series or a *Bohr-Fourier series* after Harald Bohr who developed their theory [4]. Is a signal of this form periodic?

Hint: Consider a simple such signal $h_n = e^{i2\pi \lambda n}$ for which $2\pi \lambda$ is an irrational number.

Given such a sequence h_n how would you recover the values of b_n?

Hint: Consider the sums

$$\lim_{N \to \infty} \frac{1}{2N+1} \sum_{n=-N}^{N} g_n e^{-i2\pi \lambda_k n}$$

and assume that any limit interchanges are valid. This form of generalized Fourier analysis is useful for a class of signals known as *almost
periodic signals* and has been used extensively in studying quantization noise.

5.5. Find the DTFT of the signal $g_n = n \bmod N$; $n \in \mathcal{Z}$. Find the DTFT
of the periodic extension of the signal g given by

$$g_n = \begin{cases} 1 & n = 0, 1, \ldots, N/2 - 1 \\ 0 & n = N/2, \ldots, N - 1 \end{cases}$$

(N is even). Find the CTFT of the analogous continuous time signal,
the periodic extension of $g(t) = 1$ for $0 \le t < T/2$ and $g(t) = 0$ for
$T/2 \le t < T$. What are the duals to these transform pairs?

5.6. Prove (5.34), i.e., that if $a \neq 0$,

$$\delta(at + b) = \frac{1}{|a|}\delta(t + \frac{b}{a}).$$

5.7. What, if anything, does $\delta^2(t)$ mean?

5.8. What is the 2D transform of the signal $g(x, y) = \{\sin(2\pi ax)\sin(2\pi by);$
$x \in \mathcal{R}, y \in \mathcal{R}\}$?

5.9. Evaluate

$$\int_{-\infty}^{+\infty} \delta(-2x + 3) \wedge (\frac{x}{3}) \, dx.$$

5.10. Given an ordinary function h that is continuous at $t = 0$ and a Dirac
delta function (a distribution) $\delta(x)$, show that the product $h(x)\delta(x)$
can also be considered to be a distribution $D_{h\delta}$ with the property

$$D_{h\delta}(g) = h(0)g(0)$$

if g is continuous at the origin. This is symbolically written as

$$\int_{-\infty}^{\infty} h(x)\delta(x)g(x)dx = h(0)g(0).$$

This property is usually abbreviated as $h(x)\delta(x) = h(0)\delta(x)$. Extend
the property to prove (5.35), i.e., that $h(x)\delta(x - x_0) = h(x_0)\delta(x - x_0)$
if h is continuous at $x = x_0$.

5.11. Find the generalized Fourier transform $U(f)$ of the unit step function
$u_{-1}(t)$; $t \in \mathcal{R}$.

5.12. Are $t\delta'(t)$ and $-\delta(t)$ equal? (That is, are the corresponding distributions identical?)

5.13. What is the Fourier transform of

$$\sum_{k=0}^{N} ae^{-b_k(t-\tau_k)}u_{-1}(t),$$

where $b_k > 0$ for all k, the τ_k and a are fixed, and where $u_{-1}(t)$ is the unit step function?

5.14. Consider the impulse-like generalized function δ_1 defined by the sequence

$$h_n(x) = \frac{2n}{3} \wedge (n(x - \frac{1}{n})) + \frac{2n}{3} \wedge (n(x + \frac{1}{n})).$$

What is the value of

$$\int_{-\infty}^{\infty} g(x)\delta_1(x)\, dx$$

for $g(x)$ continuous at $x = 0$ and for $g(x)$ discontinuous at $x = 0$?

5.15. Determine the Fourier series coefficients and the Fourier transforms of the following periodic continuous time signals:

(a) $\sum_{n=-\infty}^{\infty} \wedge(t - 2n)$.

(b) $\sum_{n=-\infty}^{\infty} \wedge(t - 4n)$.

(c) $\sum_{n=-\infty}^{\infty} \wedge(\frac{4}{T}(t - nT))$.

(d) $\sum_{n=-\infty}^{\infty} e^{-\pi(t-n)^2}$. (Hint: Use the Poisson summation formula.)

(e) $|\cos(\pi t)|$.

5.16. Consider a continuous time finite duration signal $g = \{g(t);\ -1/2 \leq t < 1/2\}$ defined by

$$g(t) = \begin{cases} -1 & \text{if } \frac{1}{4} < t < \frac{1}{2} \\ 0 & \text{if } t = \pm\frac{1}{4} \text{ or } t = -\frac{1}{2} \\ 1 & \text{if } -\frac{1}{4} < t < \frac{1}{4} \\ -1 & \text{if } -\frac{1}{2} < t < -\frac{1}{4} \end{cases}.$$

(a) What is the Fourier transform $G(f)$ of $g(t)$?

(b) Find a Fourier series for $g(t)$. Does the series equal $g(t)$ exactly for all $-1/2 \le t < 1/2$?

(c) Suppose that $\{h_n; n \in \mathcal{Z}\}$ is a discrete time signal with Fourier transform $H(f) = g(f)$, where g is as above. What is h_n?

(d) Let $\tilde{g}(t)$ denote the periodic extension of $g(t)$ having period 1. Sketch $\tilde{g}(t)$ and write a Fourier series for it.

(e) Sketch the shifted signal $\{\tilde{g}(t - 1/4); t \in \mathcal{R}\}$ and find a Fourier series for this signal.

(f) What is the Fourier transform of $\tilde{g}(t)$?

5.17. Derive the frequency domain sampling formula (4.17).

5.18. An infinite duration discrete time signal g_n is defined to be

$$g_n = \begin{cases} (-\frac{1}{2})^n & \text{if } n = 0, 1, 2, \ldots \\ 0 & \text{otherwise} \end{cases}.$$

An infinite duration continuous time signal $p = \{p(t); t \in \mathcal{R}\}$ has Fourier transform P.

(a) What is the Fourier transform $G(f)$ of g?

(b) Define the infinite duration continuous time signal $f(t)$ by

$$f(t) = \sum_{n=0}^{\infty} g_n p(t - nT).$$

Sketch the signal $f(t)$ for a simple p and find its Fourier transform $F(f)$ in terms of the given information. What happens if p is allowed to be a Dirac delta?

(c) For an ordinary signal p (not a Dirac delta), find the energy

$$\mathcal{E}_f = \int_{-\infty}^{\infty} f^2(t)\, dt?$$

5.19. Suppose that $g = \{g_n, n \in \mathcal{Z}\}$ is a discrete time signal. New signals y and w are defined for all integers n by

$$y_n = \begin{cases} g_n & \text{if } n \text{ is odd} \\ 0 & \text{if } n \text{ is even} \end{cases}$$

$$w_n = y(2n + 1).$$

Find the Fourier transform $W(f)$ of w in terms of $G(f)$.

Note: This is a challenging problem since you may *not* assume that G is bandlimited here. You can avoid the use of generalized functions if you see the trick, but straightforward analysis will lead you to an impulse train in the frequency domain.

5.20. Define a pulse train (pulses, not impulses!)

$$p(t) = \sum_{n=-\infty}^{\infty} \frac{1}{2\tau} \square_\tau(t - nT),$$

where T, $0 < \tau < T/2$ are parameters.

(a) Find a Fourier series for $p(t)$.

(b) Find the Fourier transform of $p(t)$.

(c) Suppose that $g = \{g(t); t \in \mathcal{R}\}$ is a band limited signal with Fourier transform G satisfying $G(f) = 0$ for $|f| \geq W$. Suppose that $T < \frac{1}{2W}$. Find the Fourier transform $R(f)$ of the signal $r(t) = (1 + mg(t))p(t)$, where m is a fixed constant. Sketch $R(f)$ for an example $G(f)$.

(d) Is the system mapping g into r linear? Time-invariant?

(e) A linear time invariant (LTI) filter $h_{BPF}(t)$ is defined by its Fourier transform

$$H_{BPF}(f) = \begin{cases} 1 & |f - f_0| \leq W \text{ or } |f + f_0| \leq W \\ 0 & \text{otherwise,} \end{cases}$$

where $f_0 > W$. (In practice usually $f_0 \gg W$.) A filter of this form is called an ideal *band pass filter* centered at f_0. Find the impulse response $h_{BPF}(t)$.

(f) Define $f_0 = M/T$ for some large integer M. Show that if $r(t)$ is the input to the ideal bandpass filter h_{BPF}, then the output $y(t)$ can be written as an amplitude modulation (AM) signal

$$y(t) = A[1 + mg(t)] \cos(2\pi f_0 t),$$

where you provide A in terms of the given parameters.

What happens in the limit as $\tau \to 0$?

Note: This problem is an old fashioned way of generating AM using a switch. Multiplication by $p(t)$ can be be accomplished by switching the waveform off (multiplying it by 0) and on (multiplying it by 1). This is called a *chopper* modulator.

5.21. List all the signals you know that are their own Fourier transform or generalized Fourier transform. What unusual properties do these signals have? (For example, what do the various properties derived for Fourier transforms of signals imply in this case.)

Chapter 6

Convolution and Correlation

We have thus far considered Fourier transforms of single signals and of linear combinations of signals. In this chapter we consider another means of combining signals: convolution integrals and sums. This leads naturally to the related topics of correlation and products of signals. As with the transforms themselves, the details of the various definitions may differ depending on the signal type, but the definitions and the Fourier transform properties will have the same basic form.

We begin with an introduction to the convolution operation in the context of perhaps its most well known and important application: linear time-invariant systems.

6.1 Linear Systems and Convolution

Recall from Chapter 1 that a system is a mapping $\mathcal{L}$ of an input signal $v = \{v(t); t \in \mathcal{T}_i\}$ into an output signal $w = \{w(t); t \in \mathcal{T}_o\} = \mathcal{L}(v)$. In this section we will consider the common special case where the input and output signals are of the same type; that is, $\mathcal{T}_i = \mathcal{T}_o$.

Recall also from Chapter 1 that a system is *linear* if given input signals $v^{(1)}$ and $v^{(2)}$ and complex numbers a and b, then $\mathcal{L}(av^{(1)} + bv^{(2)}) = a\mathcal{L}(v^{(1)}) + b\mathcal{L}(v^{(2)})$.

As with the linearity property of Fourier transforms, one can iterate the definition of linearity to argue that for any finite collection of input signals

$v^{(n)}; n = 1, \ldots, N$ and any complex constants $a_n; n = 1, \ldots, N$

$$\mathcal{L}\left(\sum_{n=1}^{N} a_n v^{(n)}\right) = \sum_{n=1}^{N} a_n \mathcal{L}(v^{(n)}). \tag{6.1}$$

This does not imply the corresponding result for infinite sums of signals and hence we often assume (without always explicitly stating it) that a linear system also has the *extended linearity* or *countable additivity* property: given an infinite collection of signals $v^{(n)}; n \in \mathcal{Z}$ and a complex-valued sequence $a_n; n \in \mathcal{Z}$, then

$$\mathcal{L}\left(\sum_{n=1}^{\infty} a_n v^{(n)}\right) = \sum_{n=1}^{\infty} a_n \mathcal{L}(v^{(n)}). \tag{6.2}$$

There is a further extension of extended linearity of which we shall also have need. It is best thought of as a limiting form of countable additivity: Suppose that we have a family of signals $v^{(r)}; r \in \mathcal{R}$, where as before $\mathcal{R}$ is the set of real numbers; that is, we now have an *uncountable infinity* or continuum of input signals, and a weighting function $a_r; r \in \mathcal{R}$. Then

$$\mathcal{L}\left(\int_{\mathcal{R}} a_r v^{(r)} \, dr\right) = \int_{\mathcal{R}} a_r \mathcal{L}(v^{(r)}) \, dr, \tag{6.3}$$

where the weighted linear combination of signals $\int_{\mathcal{R}} a_r v^{(r)} \, dr$ is written more completely as $\{\int_{\mathcal{R}} a_r v^{(r)}(t) \, dr; t \in \mathcal{R}\}$ in the infinite duration continuous time case. This complicated notation just means that one can pull integrals as well as sums from inside to the outside of a linear system operator.

As we have stated before, integrals and sums are linear operations. For this reason systems defined by integrals and sums of input functions are linear systems. For example, the systems with output u defined in terms of the input v by

$$u(t) = \int_{-\infty}^{\infty} v(\tau) h_t(\tau) \, d\tau$$

in the infinite duration continuous time case or the analogous

$$u_n = \sum_{k=-\infty}^{\infty} v_k h_{n,k}$$

in the discrete time case yield linear systems. In both cases $h_t(\tau)$ is a weighting which depends on the output time t and is summed or integrated over the input times τ. We shall see that these weighted integrals and sums are sufficiently general to describe all linear systems. A special case will yield the convolution operation that forms the focus of this chapter. First, however, some additional ideas are required.

The δ-Response

Suppose that we have a system $\mathcal{L}$ with input and output signals of the same type, e.g., they are both continuous time signals or both discrete time signals and the domains of definition are the same, say $\mathcal{T}_i = \mathcal{T}_o = \mathcal{T}$. Suppose that the input signal is a delta function at time τ; that is, if the system operates on discrete time signals, then the input signal $v = \{v_n; n \in \mathcal{Z}\}$ is a Kronecker delta delayed by τ, $v_n = \delta_{n-\tau}$, and if the system operates on continuous time signals, then the input signal $v = \{v(t); t \in \mathcal{R}\}$ is a Dirac delta delayed by τ, $v(t) = \delta(t - \tau)$. In both cases we can call the input signal $\delta^{(\tau)}$ to denote a delta delayed by τ. The output signal for this special case, $\{h(t, \tau); t \in \mathcal{T}\}$, is called the *delta response* or δ-*response*. For continuous time systems it is commonly called the *impulse response* and for discrete time systems it is often called a *unit sample response*. The name impulse response is also used for the discrete time case, but we avoid that use here as the word "impulse" or "unit impulse" is more commonly associated with the Dirac delta, a generalized function, than with the Kronecker delta, an ordinary function. While the two types of δ functions play analogous roles in discrete and continuous times, the Dirac delta or unit impulse is a far more complicated object mathematically than is the Kronecker delta or unit sample.

In discrete time

$$h(n, \tau) = \mathcal{L}_n(\{\delta_{k-\tau}; k \in \mathcal{T}\}) \tag{6.4}$$

and in continuous time

$$h(t, \tau) = \mathcal{L}_t(\{\delta(\rho - \tau); \rho \in \mathcal{T}\}). \tag{6.5}$$

Observe that if the system is time invariant, and if $\{w(t) = h(t, 0); t \in \mathcal{T}\}$ is the response to a δ at time $\tau = 0$, then shifting the δ must yield a response $\{w(t-\tau) = h(t, \tau); t \in \mathcal{T}\}$. Rewriting this as $h(t, \tau) = w(t-\tau)$ for all t and τ emphasizes the fact that the δ-response of a time invariant system depends on its arguments only through their difference. Alternatively, if a system is time invariant, then for all allowable t, τ and α

$$h(t - \alpha, \tau - \alpha) = h(t, \tau). \tag{6.6}$$

Both views imply that if a system is time invariant, then there is some function of a single dummy variable, say $\hat{h}(t)$, such that

$$h(t, \tau) = \hat{h}(t - \tau). \tag{6.7}$$

In fact, (6.6)–(6.7) imply that $\hat{h}(t - \tau) = h(t, \tau) = h(t - \tau, 0)$ and hence $\hat{h}(t) = h(t, 0)$. It is common practice, however, to use the same symbol for

h and $\hat{h}$ and to just write

$$h(t, \tau) = h(t - \tau) \tag{6.8}$$

if the system is time invariant. This is actually an abuse of notation as we are equating a function of two independent variables to a function of a single independent variable, but it is quite common. A function with the property of depending on two variables only through their difference is called a *Toeplitz function* [19].

Superposition

The δ-response plays a fundamental role in describing linear systems. To see why, consider the case of an infinite duration discrete time signal v as input to a linear system $\mathcal{L}$. Recall that

$$v_n = \sum_{k=-\infty}^{\infty} v_k \delta_{n-k}$$

and assume that the system satisfies the extended linearity property. Then the output of the system is given by

$$
\begin{aligned}
u_n &= \mathcal{L}_n(v) \\
&= \mathcal{L}_n(\{v_k; \, k \in \mathcal{Z}\}) \\
&= \mathcal{L}_n(\{ \sum_{l=-\infty}^{\infty} v_l \delta_{k-l}; \, k \in \mathcal{Z}\}) \\
&= \mathcal{L}_n(\sum_{l=-\infty}^{\infty} v_l \{\delta_{k-l}; \, k \in \mathcal{Z}\}) \\
&= \sum_{l=-\infty}^{\infty} v_l \mathcal{L}_n(\{\delta_{k-l}; \, k \in \mathcal{Z}\}) \\
&= \sum_{l=-\infty}^{\infty} v_l h_{n,l}.
\end{aligned}
\tag{6.9}
$$

The output at time n of an infinite duration discrete time linear system can be written as a weighted summation of the input values, where the weighting is given by the δ-response. Thus the output can always be found from the input using the δ-response and calculus. The sum is called the *superposition sum*.

A similar argument holds for the infinite duration continuous time case, where now the integral form of extended linearity is needed. In the continuous time case

$$v(t) = \int_{-\infty}^{\infty} v(\tau)\delta(t - \tau)\, d\tau$$

and the output of the system is given by

$$
\begin{aligned}
u(t) &= \mathcal{L}_t(v) \\
&= \mathcal{L}_t(\{v(r);\ r \in \mathcal{R}\}) \\
&= \mathcal{L}_t(\{\int_{-\infty}^{\infty} v(\tau)\delta(r - \tau)\, d\tau;\ r \in \mathcal{R}\}) \\
&= \mathcal{L}_t(\int_{-\infty}^{\infty} v(\tau)\{\delta(r - \tau);\ r \in \mathcal{R}\}\, d\tau) \\
&= \int_{-\infty}^{\infty} v(\tau)\mathcal{L}_t(\{\delta(r - \tau);\ r \in \mathcal{R}\})\, d\tau \\
&= \int_{-\infty}^{\infty} v(\tau)h(t, \tau)\, d\tau.
\end{aligned}
\tag{6.10}
$$

The integral above is called the *superposition integral*.

Suppose that the system has finite duration discrete time signals as input and output signals. The only difference from the discrete time case considered above is that now the shift is taken as a cyclic shift. For simplicity we omit the modular notation when writing delays, but all delays below are taken mod N, e.g., δ_{n-k} means $\delta_{(n-k)\bmod N}$ when considering signals of the form $\{v_n;\ n \in \mathcal{Z}_N\}$. As before

$$v_n = \sum_{k=0}^{N-1} v_k \delta_{n-k}.$$

The output of the system is given by

$$
\begin{aligned}
u_n &= \mathcal{L}_n(\{v_k;\ k \in \mathcal{Z}_N\}) \\
&= \mathcal{L}_n(\{\sum_{l=0}^{N-1} v_l \delta_{k-l};\ k \in \mathcal{Z}_N\}) \\
&= \mathcal{L}_n(\sum_{l=0}^{N-1} v_l\{\delta_{k-l};\ k \in \mathcal{Z}_N\}) \\
&= \sum_{l=0}^{N-1} v_l \mathcal{L}_n(\{\delta_{k-l};\ k \in \mathcal{Z}_N\})
\end{aligned}
$$

$$= \sum_{l=0}^{N-1} v_l h_{n,l}. \tag{6.11}$$

Note that here ordinary linearity suffices; that is, we need not assume extended linearity.

A similar form can be derived for the finite duration continuous time case.

We have seen that if a system is time invariant, then it must have a δ-response of the form $h(t, \tau) = h(t - \tau)$. Conversely, if a system has a δ-response of this form, then it follows from the superposition integral or sum that the system is also time invariant. Thus we can determine whether or not a system is time invariant by examination of its δ-response.

LTI Systems

Provided that the input and output signals to a linear system are of the same type, the system always satisfies either the superposition integral formula or the superposition sum formula expressing the output of the system as a weighted average (sum or integral) of the inputs. We now consider in more detail the simplifications that result when the system is also time invariant.

Suppose that a system $\mathcal{L}$ is both linear and time invariant, a special case which we refer to as an *LTI system* or *LTI filter*. Note that this is well-defined for all input and output signal types. Suppose further that the input and output time domains are the same so that the superposition integral or summation formula holds. Since in this case $h(t, \tau) = h(t - \tau)$, the superposition summation and integral reduce to simpler forms. For example, in the infinite duration discrete time case we have that

$$u_n = \sum_{l=-\infty}^{\infty} v_l h_{n,l} = \sum_{l=-\infty}^{\infty} v_l h_{n-l}. \tag{6.12}$$

This operation on the signals v and h to form the signal u is called the *convolution sum* and is denoted by

$$u = v * h. \tag{6.13}$$

Similarly, in the infinite duration continuous time case we have that

$$u(t) = \int_{-\infty}^{\infty} v(\tau) h(t, \tau) \, d\tau = \int_{-\infty}^{\infty} v(\tau) h(t - \tau) \, d\tau. \tag{6.14}$$

This operation combining a signal v with a signal h is called a *convolution integral*. It is the continuous time analog of the convolution sum and it is also denoted by

$$u = v * h. \qquad (6.15)$$

Similar forms hold for the finite duration examples with finite limits of summation or integration and a cyclic shift. (The details are given in the next section.)

We have now proved the following result.

Theorem 6.1 *If $\mathcal{L}$ is a linear time invariant (LTI) system for which the input and output signals have the same type and if h is the δ-response of the system, then*

$$\mathcal{L}(v) = v * h; \qquad (6.16)$$

that is, the output is given by the convolution of the input with the δ-response.

It should be emphasized that the result holds for both discrete and continuous time and for both finite and infinite duration signals. This result is the basis for the application of Fourier analysis to linear systems.

Having introduced the convolution operation, we next turn to developing its properties with respect to Fourier transforms.

6.2 Convolution

First suppose that $v = \{v(t); \ t \in \mathcal{R}\}$ and $h = \{h(t); \ t \in \mathcal{R}\}$ are two infinite duration continuous time signals. We formally define the *convolution* (or *convolution integral*) of these two signals by the signal $g = \{g(t); \ t \in \mathcal{R}\}$ given by

$$g(t) = \int_{-\infty}^{\infty} v(\zeta)h(t - \zeta)\,d\zeta; \ t \in \mathcal{R}, \qquad (6.17)$$

the integral of the product of one signal with the time reversed and shifted version of the other signal. We abbreviate this operation on signals by

$$g = v * h.$$

We also use the asterisk notation as $g(t) = v * h(t)$ when we wish to emphasize the value of the output signal at a specific time. The notation $g(t) = v(t) * h(t)$ is also common, but beware of the potential confusion of dummy variables: the convolution operation depends on the entire history of the two signals; that is, one is convolving $\{v(t); \ t \in \mathcal{T}\}$ with $\{h(t); \ t \in \mathcal{T}\}$, not just the specific output values $v(t)$ with $h(t)$.

We have already seen the primary example of convolution: if a signal v is put into a linear time-invariant system described by an *impulse response* $h = \{h(t); \ t \in \mathcal{R}\}$, then the output signal is the convolution $v * h$. Another common application of convolution is in probability theory: if v and h are two probability density functions describing two independent random variables X and Y, then the probability density function of the sum $X + Y$ is given by the convolution $v * h$. The same is true for probability mass functions. Convolution also cropped up in the development of inversion in (3.79), where a signal was convolved with a sinc function.

In the infinite duration discrete time case the definition of convolution is the same except that the integral becomes a sum: Given discrete time signals $v = \{v_n; \ n \in \mathcal{Z}\}$ and $h = \{h_n; n \in \mathcal{Z}\}$, then the *convolution* or *convolution sum* $g = \{g_n; \ n \in \mathcal{Z}\}$ of the two signals is defined by

$$g_n = \sum_{k=-\infty}^{\infty} v_k h_{n-k}; \ n \in \mathcal{Z}. \tag{6.18}$$

As in the continuous time case we use the shorthand $g = v * h$.

In the finite duration case we use the same trick that we used when extending the notion of the shift to finite duration signals. In fact, this trick is necessary because convolution is defined as a sum or integral of the product of a signal and a time reversed and shifted version of another signal. The basic idea is the same for both continuous and discrete time: Take the convolution of the periodic extensions of the finite duration signals over one period. In the discrete time case, given two signals $v = \{v_n; \ n = 0, 1, \ldots, N - 1\}$ and $h = \{h_n; \ n = 0, 1, \ldots, N - 1\}$, define the convolution $g = \{g_n; \ n = 0, \ldots, N - 1\} = v * h$ by

$$g_n = \sum_{k=0}^{N-1} v_k h_{(n-k) \bmod N} = \sum_{k=0}^{N-1} v_k \tilde{h}_{n-k} \tag{6.19}$$

where $\tilde{h}_n$ is the periodic extension of h_n. This form of convolution is called a *cyclic convolution* or *circular convolution*. Writing out the sums we have

$$
\begin{aligned}
g_0 &= v_0 h_0 + v_1 h_{N-1} + v_2 h_{N-2} + \cdots + v_{N-1} h_1 \\
g_1 &= v_0 h_1 + v_1 h_0 + v_2 h_{N-1} + \cdots + v_{N-1} h_2 \\
&\ \ \vdots \\
g_{N-1} &= v_0 h_{N-1} + v_1 h_{N-2} + v_2 h_{N-3} + \cdots + v_{N-1} h_0
\end{aligned}
$$

The h sequence is cyclically rotated to produce successive values of g_n.

The discrete time finite duration convolution is far more important in applications than the continuous time finite duration convolution because of its use in digital signal processing. For completeness, however, we observe that the same idea works for continuous time: given two finite duration continuous time signals $v = \{v(t); t \in [0, T)\}$ and $h = \{h(t); t \in [0, T)\}$, define the convolution

$$g = \{g(t); t \in [0, T)\} = v * h$$

by

$$g(t) = \int_0^T v(\zeta) h((t - \zeta) \bmod T) \, d\zeta = \int_0^T v(\zeta) \tilde{h}(t - \zeta) \, d\zeta \qquad (6.20)$$

where $\tilde{h}(t)$ is the periodic extension of $h(t)$.

⋆ Signal Algebra

Suppose that we now consider the space of all signals of the form $g = \{g(t); t \in \mathcal{R}\}$. While we will emphasize the infinite duration continuous time case in this section, the same results and conclusions hold in all cases for which we have defined a convolution operation. We have defined two operations on such signals: addition, denoted by $+$, and convolution, denoted by $*$. This resembles the constructions of arithmetic, algebra, and group theory where we have a collection of elements (such as numbers, polynomials, functions) and a pair of operations. A natural question is whether or not the operations currently under consideration have useful algebraic properties such as the commutative law, the distributive law, and the associative law. The following result answers this question affirmatively.

Theorem 6.2 *The convolution and addition operators on signals satisfy the following properties.*

1. *Commutative Law*
$$g * h = h * g. \qquad (6.21)$$

2. *Distributive Law*

$$f * (g + h) = f * g + f * h. \qquad (6.22)$$

3. *Associative Law*

$$f * (g * h) = (f * g) * h. \qquad (6.23)$$

Consider the first property in the case of infinite duration continuous time signals: the signal $f * h$ is defined by

$$\int_{-\infty}^{\infty} f(\zeta)h(t - \zeta)\, d\zeta.$$

Changing variables by defining $\eta = t - \zeta$ this becomes

$$\int_{\infty}^{-\infty} f(t - \eta)h(\eta)(-d\eta) = \int_{-\infty}^{\infty} f(t - \eta)h(\eta)\, d\eta,$$

which is just $h * f$, as claimed. The result follows for the other signal types similarly. The Distributive Law follows from the linearity of integration. The proof of the Associative Law is left as an exercise.

In order to have an algebra of signals with the convolution and sum operations, we also need an identity signal; that is, a signal such that if convolved with any other signal yields the other signal. (The signal that is identically 0 for all time is the additive identity.) This role is filled by the Kronecker delta function in discrete time and by the Dirac delta function in continuous time since if we define the signal δ by $\{\delta(t); t \in \mathcal{T}\}$ for continuous time or $\{\delta_n; n \in \mathcal{T}\}$ for discrete time, then $\delta * g = g$. For example, in the discrete time case

$$\sum_k \delta_{n-k} g_k = g_n.$$

To summarize the identity properties:

4. Identity

$$0 + g = g, \tag{6.24}$$

where 0 is the 0 signal (it is 0 for all t),

$$\delta * g = g. \tag{6.25}$$

A detail not yet treated which is needed for our demonstration that the space of signals (including generalized functions) is an algebra is the fact that we can convolve generalized functions with each other; that is, the convolution of two δ functions is well-defined. In fact, if δ is to play the role of the convolution identity, we should have that $\delta * \delta = \delta$. This is immediate for the Kronecker δ in discrete time. To verify it in the continuous time case suppose that $h_n(t)$ is a sequence of pulses yielding the Dirac delta in the sense of the limiting definition of a distribution:

$$\int_{-\infty}^{\infty} \delta(t)g(t)\, dt = \lim_{n \to \infty} \int_{-\infty}^{\infty} h_n(t)g(t)\, dt = g(0)$$

for $g(t)$ continuous at $t = 0$. We define $\delta * \delta$ by

$$
\int_{-\infty}^{\infty} (\delta * \delta)(t)g(t)\, dt \;=\; \lim_{n\to\infty} \int_{-\infty}^{\infty} h_n * h_n(t)g(t)\, dt
$$

$$
=\; \lim_{n\to\infty} \int_{-\infty}^{\infty} \left(\int_{-\infty}^{\infty} h_n(\zeta)h_n(t-\zeta)\, d\zeta \right) g(t)\, dt
$$

$$
=\; \lim_{n\to\infty} \int_{-\infty}^{\infty} d\zeta\, h_n(\zeta) \int_{-\infty}^{\infty} dt\, h_n(t-\zeta)g(t).
$$

The rightmost integral approaches $g(\zeta)$ in the limit and hence the overall integral approaches $g(0)$. Thus the convolution of two Dirac delta functions is another Dirac delta function.

The final requirement for demonstrating that our signal space indeed forms an algebra is the demonstration of an inverse for addition and for convolution. The additive inverse is obvious — the negative of a signal is its additive inverse $(g + (-g) = 0)$ — but the inverse with respect to convolution is not so obvious. What is needed is a means of finding for a given suitably well-behaved signal g another signal, say g^{-1}, with the property that $g * g^{-1} = \delta$. This is the signal space analog of the ordinary multiplicative inverse $a(1/a) = 1$. This property we postpone until we have proved the convolution theorem in a later section.

6.3 Examples of Convolution

Consider first the continuous time convolution of two signals $f(t)$ and $h(t)$ pictured in Figure 6.1 and defined by

$$
f(t) = \begin{cases} 0 & \text{if } t < 0; \\ 1 - t & \text{if } 0 \le t \le 1; \\ 0 & \text{if } t > 1 \end{cases}
$$

$$
h(t) = \sqcap(\frac{t-1}{2}).
$$

Note that in drawing the box function the end points can be determined easily as follows: The right edge of the box occurs when the argument $(t-1)/2 = 1/2$ and hence when $t = 2$. The left edge occurs when $(t-1)/2 = -1/2$ and hence when $t = 0$. To perform the convolution and find the signal

$$
g(t) = \int_{-\infty}^{\infty} f(r)h(t-r)\, dr
$$

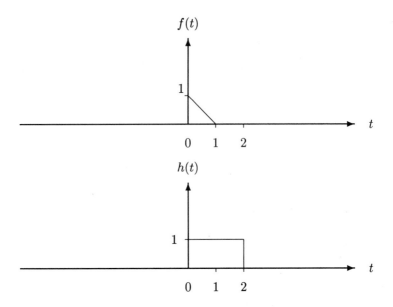

Figure 6.1: Example Waveforms

we need to integrate the product $f(r)h(t-r)$ over all r. Alternatively, we could integrate $f(t-r)h(r)$, but it is simpler to time-reverse the simplest of the signals, which in this case is the box.

The product of the signal $f(r)$ with the reversed and delayed signal $h(t-r)$ varies in shape depending on the value of t, which in turn affects the functional form of the convolution. We consider each possibility separately.

1. $t < 0$. Here the waveform $f(r)h(t-r) = 0$ for all r and hence its integral $g(t)$ is also 0 in this region. (See Figure 6.2.)

2. $0 \leq t < 1$. Here $f(r)h(t-r)$ is nonzero between 0 and t and its integral is the area under the product of the waveforms in as depicted in Fig. 6.3:

$$g(t) = \int_0^t (1-r)\, dr = (r - \frac{r^2}{2})|_0^t = t - \frac{t^2}{2}.$$

3. $1 \leq t < 2$. With reference to Figure 6.4, $g(t)$ is the area of the product of the waveforms in the region where both are nonzero, which is now

$$g(t) = \int_0^1 (1-r)\, dr = \frac{1}{2}.$$

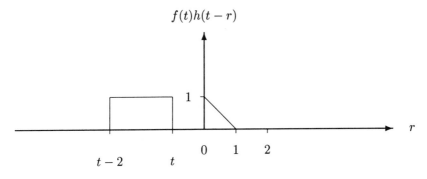

Figure 6.2: $t < 0$

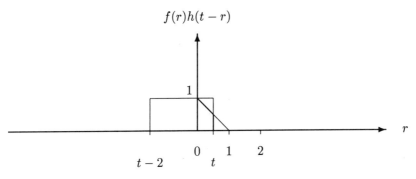

Figure 6.3: $0 \leq t < 1$

4. $2 \leq t < 3$. As illustrated in Figure 6.5

$$g(t) = \int_{t-2}^{1} (1-r)\, dr = \frac{1}{2} - (t-2) + \frac{1}{2}(t-2)^2.$$

5. $3 \leq t$. With reference to Figure 6.6, once again there is no overlap and $g(t) = 0$.

We can now summarize the convolution integral:

$$g(t) = \begin{cases} t - \frac{1}{2}t^2 & \text{if } 0 \leq t \leq 1; \\ \frac{1}{2} & \text{if } 1 \leq t \leq 2; \\ \frac{9}{2} - 3t + \frac{t^2}{2} & \text{if } 2 \leq t \leq 3; \\ 0 & \text{otherwise.} \end{cases}$$

$f(r)h(t-r)$

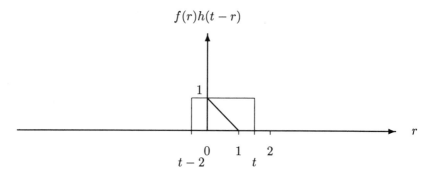

Figure 6.4: $1 \le t < 2$

$f(r)h(t-r)$

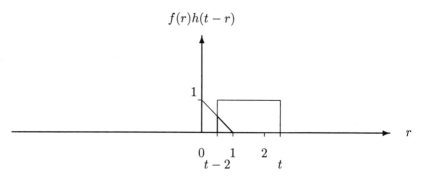

Figure 6.5: $2 \le t < 3$

$f(r)h(t-r)$

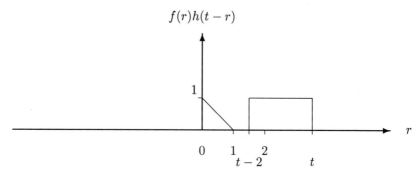

Figure 6.6: $3 \le t$

The final waveform is depicted in Figure 6.7. *Exercise:* Prove that the

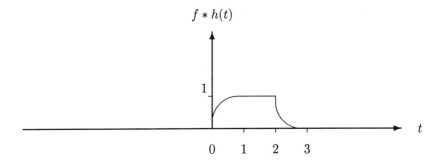

$$f * h(t)$$

Figure 6.7: $g(t) = f * h(t)$

convolution of $\sqcap(t)$ with itself is the triangle function

$$\wedge(t) = \begin{cases} 1 - |t| & \text{if } |t| < 1; \\ 0 & \text{otherwise .} \end{cases} \tag{6.26}$$

As a discrete time convolution example consider the signals

$$f_n = \begin{cases} \rho^n & \text{if } n = 0, 1, \ldots; \\ 0 & \text{otherwise,} \end{cases}$$

where $\rho \neq 1$, and

$$h_n = \begin{cases} 1 & n \in \mathcal{Z}_{N+1}; \\ 0 & \text{otherwise .} \end{cases}$$

To evaluate the convolution

$$g_n = \sum_k f_k h_{n-k}$$

observe that h_{n-k} is 1 if $n-k = 0, 1, \ldots, N$ and hence if $k = n, n-1, \ldots, n-N$. Thus the sum becomes

$$g_n = \sum_{k=n-N}^{n} f_k.$$

Again evaluation is eased by separately considering the possible cases:

1. $n < 0$

   In this case $g_n = 0$ since the summand is 0 for the indexes being summed over.

2. $0 \le n < N$

$$g_n = \sum_{k=0}^{n} \rho^k = \frac{1 - \rho^{n+1}}{1 - \rho}$$

Note that the above formula is not valid if $\rho = 1$.

3. $N \le n$

$$
\begin{aligned}
g_n &= \sum_{k=n-N}^{n} \rho^k \\
&= \rho^{n-N} \sum_{k=n-N}^{n} \rho^{k-(n-N)} \\
&= \rho^{n-N} \sum_{j=0}^{N} \rho^j \\
&= \rho^{n-N} \frac{1 - \rho^{N+1}}{1 - \rho}.
\end{aligned}
\tag{6.27}
$$

What happens when $\rho = 1$?

As a final example, we show that the continuous time infinite duration impulse train or III function can be combined with convolution to provide a convenient means of replicating a signal (or spectrum), e.g., of forming the periodic extension of a finite duration signal (or finite bandwidth spectrum). For simplicity we assume throughout that the signal is continuous. In this case recall that convolving an impulse with the signal simply produces the original signal:

$$\int_{-\infty}^{\infty} g(\tau)\delta(t - \tau)\, d\tau = g(t).$$

(This is just the sifting property.) Consider next convolving an infinite duration continuous time signal g with the ideal sampling function $\Psi_T(t) = \sum_{n=-\infty}^{\infty} \delta(t - nT)$ to form $g * \Psi_T$. This yields

$$
\begin{aligned}
g * \Psi_T(t) &= \int_{-\infty}^{\infty} g(\tau) \sum_{n=-\infty}^{\infty} \delta(t - nT - \tau)\, d\tau \\
&= \sum_{n=-\infty}^{\infty} \int_{-\infty}^{\infty} g(\tau)\delta(t - nT - \tau)\, dt
\end{aligned}
$$

$$= \sum_{n=-\infty}^{\infty} g(t - nT). \tag{6.28}$$

Thus the signal is the sum of an infinite number of replicas of g shifted by multiples of T. If $g(t) = 0$ for $t \notin [-T/2, T/2)$, then $g * \Psi_T$ will be an infinite sequence of periodic copies of the original signal. The same idea works in the frequency domain. The result can also be stated in terms of the III function by including the extra scaling factor of T.

6.4 The Convolution Theorem

One of the most important applications of the Fourier transform is for the computation of convolutions of all kinds. It might appear that the convolution integral or sum is not too complicated in the first place and hence a shortcut requiring the computation of Fourier transforms may in fact not be a shortcut. This may in fact be the case when convolving some simple signals, but when convolving more complicated signals whose transforms are known (perhaps in a handy table) and, more importantly, when doing multiple convolutions involving several signals, the Fourier transform can provide significantly simpler evaluations. Because of the importance of the result we state it formally as a theorem and discuss its proof for several signal types. The basic idea is simple to state before we get formal: the Fourier transform of the convolution of two signals is the product of the Fourier transforms of the two signals.

Theorem 6.3 *The Convolution Theorem*

Given two signals $g = \{g(t); t \in \mathcal{T}\}$ and $h = \{h(t); t \in \mathcal{T}\}$ with Fourier transforms $G = \{G(f); f \in \mathcal{S}\} = \mathcal{F}(g)$ and $H = \{H(f); f \in \mathcal{S}\} = \mathcal{F}(h)$, respectively, then

$$\mathcal{F}_f(g * h) = G(f)H(f). \tag{6.29}$$

Proof: First consider the case of infinite duration continuous time signals. In this case

$$
\begin{aligned}
\mathcal{F}_f(g * h) &= \int_{-\infty}^{\infty} e^{-i2\pi ft} \left(\int_{-\infty}^{\infty} g(\zeta) h(t - \zeta) \, d\zeta \right) dt \\
&= \int_{-\infty}^{\infty} g(\zeta) \left(\int_{-\infty}^{\infty} e^{-i2\pi ft} h(t - \zeta) \, dt \right) d\zeta.
\end{aligned}
$$

Changing variables with $t' = t - \zeta$ yields

$$
\begin{aligned}
\mathcal{F}_f(g * h) &= \int_{-\infty}^{\infty} g(\zeta) \left(\int_{-\infty}^{\infty} e^{-i2\pi f(t'+\zeta)} h(t') \, dt' \right) d\zeta \\
&= \int_{-\infty}^{\infty} g(\zeta) \left(e^{-i2\pi f\zeta} H(f) \right) d\zeta \\
&= H(f) \int_{-\infty}^{\infty} g(\zeta) e^{-i2\pi f\zeta} \, d\zeta = H(f)G(f),
\end{aligned}
$$

which proves the claim.

In the case of infinite duration discrete time signals the proof is the same except that the integrals over continuous variables are replaced by sums over discrete variables. For completeness we note this yields the sequence of equalities

$$
\begin{aligned}
\mathcal{F}_f(g * h) &= \sum_{n=-\infty}^{\infty} e^{-i2\pi fn} \left(\sum_{k=-\infty}^{\infty} g(k)h(n-k) \right) \\
&= \sum_{k=-\infty}^{\infty} g(k) \left(\sum_{n=-\infty}^{\infty} e^{-i2\pi fn} h(n-k) \right) \\
&= \sum_{k=-\infty}^{\infty} g(k) \left(\sum_{n'=-\infty}^{\infty} e^{-i2\pi f(n'+k)} h(n') \right) \\
&= \sum_{k=-\infty}^{\infty} g(k) \left(e^{-i2\pi fk} H(f) \right) \\
&= H(f) \sum_{k=-\infty}^{\infty} g(k)e^{-i2\pi fk} = H(f)G(f).
\end{aligned}
$$

In the case of the DFT the manipulations are similar with the sum over infinite limits replaced by the cyclic sum. As usual let $\tilde{h}$ be the periodic extension of h defined by $\tilde{h}(n) = h(n \bmod N)$. Then for any $l \in \mathcal{Z}_N$

$$
\begin{aligned}
\mathcal{F}_{\frac{l}{N}}(g * h) &= \sum_{n=0}^{N-1} e^{-i2\pi \frac{l}{N} n} \left(\sum_{k=0}^{N-1} g(k)\tilde{h}(n-k) \right) \\
&= \sum_{k=0}^{N-1} g(k) \left(\sum_{n=0}^{N-1} e^{-i2\pi \frac{l}{N} n} \tilde{h}(n-k) \right)
\end{aligned}
$$

$$= \sum_{k=0}^{N-1} g(k) \left(\sum_{n'=-k}^{N-1-k} e^{-i2\pi \frac{l}{N}(n'+k)} \tilde{h}(n') \right)$$

$$= \sum_{k=0}^{N-1} g(k) e^{-2\pi \frac{l}{N}k} \left(\sum_{n'=-k}^{N-1-k} e^{-i2\pi \frac{l}{N}n'} \tilde{h}(n') \right)$$

$$= \sum_{k=0}^{N-1} g(k) e^{-2\pi \frac{l}{N}k} \left(\sum_{n'=0}^{N-1} e^{-i2\pi \frac{l}{N}n'} \tilde{h}(n') \right)$$

$$= \sum_{k=0}^{N-1} g(k) e^{-2\pi \frac{l}{N}k} H(\frac{l}{N})$$

$$= H(\frac{l}{N}) G(\frac{l}{N}).$$

The above proofs make an important point: in all cases the proofs look almost the same; the only differences are minor. We used the functional notation $g(k)$ throughout instead of using g_k for the discrete time case to emphasize the similarity. We omit the proof for the case of finite duration continuous time signals since the modifications required to the above proofs should be clear.

We state without proof the dual result to the convolution theorem:

Theorem 6.4 *The Dual Convolution Theorem*
   Given two signals g and h with spectra G and H, then

$$\mathcal{F}(\{g(t)h(t); \ t \in \mathcal{T}\}) = cG * H; \tag{6.30}$$

where $c = 1$ for infinite duration signals, $1/N$ for the DFT of duration N, and $1/T$ for the CTFT of duration T. In words, multiplication in the time domain corresponds to convolution in the frequency domain.

The extra factor comes in from the Fourier inversion formula. For example, for the DFT case the Fourier transform of $\{g_n h_n; n = 0, ..., N - 1\}$ at frequency k/N is

$$\sum_{n=0}^{N-1} g_n h_n e^{-i2\pi n \frac{k}{N}} = \sum_{n=0}^{N-1} g_n [N^{-1} \sum_{m=0}^{N-1} H(\frac{m}{N}) e^{i2\pi m \frac{n}{N}}] e^{-i2\pi n \frac{k}{N}}$$

$$= N^{-1} \sum_{m=0}^{N-1} H(\frac{m}{N}) \sum_{n=0}^{N-1} g_n e^{-i2\pi n \frac{k-m}{N}}$$

$$= N^{-1} \sum_{m=0}^{N-1} H(\frac{m}{N}) G(\frac{k - m}{N} \bmod 1)$$

$$= N^{-1} H * G(\frac{k}{N}).$$

The convolution theorem provides a shortcut to computing Fourier transforms. For example, suppose that we wish to find the transform of $\wedge(t)$. A straightforward exercise shows that $\wedge(t)$ is the convolution of $\sqcap(t)$ with itself. Since the Fourier transform of $\sqcap(t)$ is $\operatorname{sinc}(f)$, we have easily that

$$\mathcal{F}_f(\{\wedge(t);\ t \in \mathcal{R}\}) = \operatorname{sinc}^2(f). \tag{6.31}$$

As a second and less obvious example, what is the Fourier transform of the convolution of $\operatorname{sinc}(t)$ with itself? Duality implies that the Fourier transform of $\operatorname{sinc}(t)$ is $\sqcap(f)$, and hence the convolution theorem implies that the transform of $\operatorname{sinc} * \operatorname{sinc}(t)$ is just the product of $\sqcap(f)$ with itself, which is $\sqcap(f)$ (except at the endpoints, which do not affect the Fourier transform). Thus

$$\mathcal{F}_f(\{\operatorname{sinc}(t); t \in \mathcal{R}\} * \{\operatorname{sinc}(t); t \in \mathcal{R}\}) = \sqcap(f), \tag{6.32}$$

and hence, after inverse transforming,

$$\operatorname{sinc} * \operatorname{sinc}(t) = \operatorname{sinc} t. \tag{6.33}$$

⋆ Convolution Inverses Revisited

The convolution theorem also permits us to complete the discussion of signal space algebra by showing how to define an inverse signal with respect to convolution. Suppose that we have a signal g. We wish to find a signal (if it exists) g^{-1} with the property that $g * g^{-1} = \delta$. (Note that even in ordinary algebra not every element of the space has an inverse, e.g., in the real numbers 0 has no inverse.) To accomplish this take the Fourier transform of both sides of the equality. The transform of a δ function (Kronecker in discrete time and Dirac in continuous time) is just a constant 1. If G^{-1} is the transform of the signal g^{-1}, then we have that $G(f)G^{-1}(f) = 1$ or

$$G^{-1}(f) = \frac{1}{G(f)};$$

that is, the inverse of g is just the inverse Fourier transform of one over the Fourier transform of g.

Clearly this inverse will not exist in general, e.g., if $G(f)$ has zeros then $1/G(f)$ is not well-behaved. For example, a band-limited function cannot have an inverse. It can have an inverse in an approximate sense, however. Suppose that $g(t)$ is an infinite duration continuous time band-limited signal with the property that $G(f) > 0$ if and only if $|f| < W$; that is, the spectrum is bandlimited and has no zeros inside of its bandwidth.

In this case $G(f) = G(f) \sqcap (f/2W)$ and we can define a "pseudo-inverse"
signal as the signal, say $h(t)$, having spectrum

$$H(f) = \frac{1}{G(f)} \sqcap (\frac{f}{2W}).$$

The signal $h(t)$ is now well-defined and now the convolution $g * h$ has spectrum $\sqcap(f/2W)$ and hence

$$g * h(t) = 2W \operatorname{sinc}(2Wt); \ t \in \mathcal{R}.$$

As $W \to \infty$, this convolution looks more and more like an impulse and the pseudo-inverse looks more like an inverse.

Note that the Dirac delta has an inverse: itself, since $\delta * \delta = \delta$. This is analogous to the number 1 being its own multiplicative inverse in the real number system.

As a simple example of an exact inverse, consider the two infinite duration discrete time signals

$$g_n = \begin{cases} \rho^n & n \geq 0 \\ 0 & \text{otherwise} \end{cases} ; \ f_n = \delta_n - \rho\delta_{n-1} = \begin{cases} 1 & n = 0 \\ -\rho & n = 1 \\ 0 & \text{otherwise} \end{cases}.$$

Then the convolution is

$$g * f(n) = \sum_{k=-\infty}^{\infty} g_k f_{n-k} = \begin{cases} 0 & n < 0 \\ 1 & n = 0 \\ 1 \times \rho^n - \rho \times \rho^{n-1} = 0 & \text{otherwise} \end{cases}$$

and hence

$$g * f(n) = \delta_n.$$

In this case

$$G(f) = \frac{1}{1 - \rho e^{-i2\pi f}}; \ F(f) = 1 - \rho e^{-i2\pi f}.$$

6.5 Fourier Analysis of Linear Systems

Since LTI systems are described by convolutions, the convolution theorem immediately provides a means of relating the input and output spectra of such a system.

Theorem 6.5 *Suppose that $\mathcal{L}$ is an LTI system with δ-response h. Let $H = \mathcal{F}(h)$ be the Fourier transform of the δ-response. If the input signal v has Fourier transform V, then the output signal w is the inverse Fourier transform of $W = \{W(f); f \in \mathcal{S}\}$, where*

$$W(f) = V(f)H(f). \tag{6.34}$$

The Fourier transform of the δ-response is called the *system function* or *transfer function* of the LTI system. Note that the transfer function is not in general defined if the system is not time invariant. It is well-defined if the system is time invariant but nonlinear, but in that case superposition does not hold and hence the transfer function is of little use. The power of the above result lies in the fact that it is much easier to multiply than convolve. This is particularly useful if a signal v is the input to a cascade of N linear systems with δ-responses $h_l(t)$; $l = 1, \ldots, N$. The output signal u is then the difficult N-fold convolution $h_1 * h_2 * \ldots * h_N * v$, but the overall transfer function is found easily by

$$H(f) = \prod_{l=1}^{N} H_l(f).$$

Eigenfunctions of LTI Systems

The second principal application of Fourier analysis to linear systems is a simple observation of the fact that delta functions (Kronecker in discrete time and Dirac in continuous time) are Fourier transform pairs with exponentials. This fact has an interesting implication to Fourier analysis itself in that it demonstrates why exponentials are so important. To state the result first requires a definition.

Given a system $\mathcal{L}$, an *eigenfunction* of $\mathcal{L}$ is an input signal e with the property that

$$\mathcal{L}(e) = \lambda e$$

for some complex constant λ; that is, an eigenfunction is a signal which is passed through a system without any change of shape, the sole effect of the system being to multiply the signal by a complex constant (not depending on time). For those old (or cultured) enough to remember the Firesign Theater, an eigenfunction can be viewed as an example of what they called "Fudd's third law": *What goes in must come out.* The complex constant λ is called an *eigenvalue* of $\mathcal{L}$.

Theorem 6.6 *Let $\mathcal{L}$ be an LTI system with common input and output signal types. Let $\mathcal{T}$ be the time domain of definition and $\mathcal{S}$ the frequency domain of definition. Then for any $f_0 \in \mathcal{S}$, the signal*

$$e = \{e^{i2\pi f_0 t}; t \in \mathcal{T}\}$$

is an eigenfunction of $\mathcal{L}$. The corresponding eigenvalue is $H(f_0)$, where H is the transfer function of the system.

Proof: Given an input signal v and δ-response h, the output signal is $w = v * h$. Transforming we have that $W(f) = V(f)H(f); f \in \mathcal{S}$. First suppose the system is an infinite duration discrete time system, in which case $\mathcal{S} = [-\frac{1}{2}, \frac{1}{2})$. The input signal $v_n = e^{i2\pi f_0 n}$ has Fourier transform $\delta(f - f_0)$, whence

$$W(f) = H(f)\delta(f - f_0) = H(f_0)\delta(f - f_0)$$

which implies that

$$w_n = H(f_0)e^{i2\pi f_0 n},$$

as claimed.

If the system is instead an infinite duration continuous time system, in which case $\mathcal{S}$ is the real line, then $v(t) = e^{i2\pi f_0 t}$ has Fourier transform $V(f) = \delta(f - f_0)$ and the same equality chain as above proves the result. If the system is finite duration and discrete time, then $\mathcal{S} = \{\frac{k}{N}; k \in \mathcal{Z}_N\}$ and $v_n = e^{i2\pi f_0 n}$ (with $f_0 = \frac{l}{N}$ for some $l \in \mathcal{Z}_N$) has Fourier transform $\delta_{f-\frac{l}{N}}$ for $f \in \mathcal{S}$. As previously (except that now only $f \in \{\frac{k}{N}; k \in \mathcal{Z}_N\}$ are possible)

$$W(f) = H(f)\delta_{f-f_0} = H(f_0)\delta_{f-f_0}$$

which implies that

$$w_n = H(f_0)e^{i2\pi f_0 n}, \; n \in \mathcal{Z}_N;$$

proving the result.

The finite duration continuous time case is left as an exercise.

The theorem indicates the reason why the complex exponentials are the most important functions in the study of LTI systems: they are the eigenfunctions of such systems. In general sines and cosines are *not* eigenfunctions. For this reason the decomposition of an arbitrary signal into a linear combination (by integration or summation) of complex exponentials is the most basic such decomposition; it breaks down an arbitrary signal into pieces that pass through a linear system without changing shape. Thus we can decompose an input signal into a linear combination of complex exponentials, infer the resulting output for each exponential (it is the same exponential times the transfer function at the same frequency), and then recombine these elementary outputs in a linear fashion (by summation or integration) to find the complete output to the original signal.

As a corollary to the theorem, observe that if an infinite duration continuous time LTI system has a purely real impulse response h, then an input of $\cos(2\pi f_0 t)$ produces an output of $|H(f_0)| \cos(2\pi f_0 t + \angle H(f_0))$, where $H(f_0) = |H(f_0)|e^{i\angle H(f_0)}$. (This follows by expanding the cosine as the sum of exponentials using Euler's relations, applying the eigenfunction property to each, and then using the fact that if h is real, then $H(-f) = H^*(f)$ to write the resulting sum as $\Re(H(f_0)e^{i2\pi f_0 t})$.) Thus if $H(f)$ also is purely real, then the cosine is in fact an eigenfunction. In general, however, transfer functions are not purely real and cosines are not eigenfunctions.

Another corollary to these results is that in an LTI system, only frequencies appearing at the input can appear at the output. This is not the case if the system is either nonlinear or time varying. For example, if the system output $w(t)$ for an input $v(t)$ is given by $w(t) = v^2(t)$ (a memoryless square law device), then if $v(t) = \cos(2\pi f_0 t)$ we have that

$$w(t) = \cos^2(2\pi f_0 t) = \frac{1}{2} + \frac{1}{2}\cos(4\pi f_0 t);$$

that is, new frequencies not appearing in the input do appear in the output. In this case the frequency has been doubled. We have seen that a DSB-SC modulation system is linear but time varying and that it too produces new frequencies at the output. In fact, any modulation system which performs frequency translation must either involve a nonlinear mapping or a time-varying mapping.

6.6 The Integral Theorem

The convolution theorem provides a form of converse result to the derivative theorem by relating the Fourier transform of the integral of a signal to the Fourier transform of the original signal.

Suppose that $g = \{g(t); t \in \mathcal{R}\}$ is an infinite duration continuous time signal and form the integral $\phi(t) = \int_{-\infty}^{t} g(\zeta) \, d\zeta$. Assuming that the integral is well-defined (the signal $g(t)$ is not too pathological), the Fourier transform of $\phi(t)$ can be easily found from the convolution theorem by using the fact that

$$\phi(t) = H * g(t) = \int_{-\infty}^{\infty} u_{-1}(t - \zeta)g(\zeta) \, d\zeta, \qquad (6.35)$$

where $u_{-1}(t)$ is the unit step function

$$u_{-1}(t) = \begin{cases} 1 & t \geq 0 \\ 0 & t < 0 \end{cases}.$$

Since the Fourier transform of the step function is $\frac{1}{2}\delta(f) - \frac{i}{2\pi f}(1 - \delta_f)$, the Fourier transform of $\phi(t)$ is given by the convolution theorem as

$$\Phi(f) = \frac{1}{2}G(0)\delta(f) + \frac{G(f)}{i2\pi f}(1 - \delta_f).$$

We have now proved the following result:

Theorem 6.7 *The Integral Theorem*
  Given an infinite duration continuous time integrable signal $g = \{g(t); t \in \mathcal{R}\}$, then

$$\mathcal{F}_f(\{\int_{-\infty}^{t} g(\zeta)\,d\zeta; \ t \in \mathcal{R}\}) = \frac{1}{2}G(0)\delta(f) + \frac{G(f)}{i2\pi f}(1 - \delta_f). \qquad (6.36)$$

The first term can be thought of as half the transform of the DC component of $g(t)$ represented by its area $\int_{-\infty}^{\infty} g(t)\,dt$. The second term shows that integration in the time domain corresponds to division by f in the frequency domain (except where $f = 0$).

The discrete time analog to the integration theorem is the Fourier transform of the sum (or discrete time integral) of a discrete time signal. Given $\{g_n; n \in \mathcal{Z}\}$, what is the Fourier transform of

$$\phi_n = \sum_{k=-\infty}^{n} g_k?$$

The answer is left as an exercise.

The integration and the summation properties have dual results formed by interchanging the roles of time and frequency. What are they?

6.7 Sampling Revisited

As another application of convolution and the convolution theorem we provide an alternative development of the sampling theorem. We do this in an apparently more general fashion in order to better consider what happens when the sampling frequency is not chosen in exactly the same fashion as in the previous derivation. Assume now that g is a continuous time infinite duration band-limited signal with W chosen so that $G(f) = 0$ for $|f| \geq W$ as before. Fix a sampling period T (which we do not assume to equal $1/2W$). Multiply the signal g by the scaled impulse train

$$\text{III}(\frac{t}{T}) = T \sum_{n=-\infty}^{\infty} \delta(t - nT)$$

to form the idealized sampled waveform

$$\hat{g}(t) = \text{III}(\frac{t}{T})g(t) = T \sum_{n=-\infty}^{\infty} g(nT)\delta(t - nT).$$

Using the convolution theorem, the fact that the III function is its own transform, and the scaling formula for delta functions yields

$$
\begin{aligned}
\hat{G}(f) &= \mathcal{F}_f(\hat{g}(t)) = \mathcal{F}_f(\text{III}(\tfrac{t}{T})g(t)) \\
&= T\text{III}(Tf) * G(f) \qquad\qquad (6.37) \\
&= \left(T\sum_{n=-\infty}^{\infty} \delta(Tf - n)\right) * G(f) \\
&= \left(\sum_{n=-\infty}^{\infty} \delta(f - \tfrac{n}{T})\right) * G(f) \\
&= \sum_{n=-\infty}^{\infty} G(f - \tfrac{n}{T}); \ f \in \mathcal{R}. \qquad (6.38)
\end{aligned}
$$

For example, given an input signal spectrum $G(f)$ having the shape depicted in Figure 6.8, then $\hat{G}$ is the sum of an infinite sequence of shifted

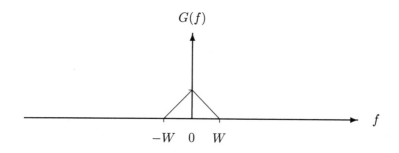

Figure 6.8: Input Spectrum

copies of the original spectrum G as indicated in Figure 6.9.

If we choose T so that $1/T > 2W$, the Nyquist frequency or Nyquist rate, then the copies of the spectra do not overlap and $\hat{G}$ is seen to consist of an infinite sequence of replicas of G centered at frequencies n/T for $n \in \mathcal{Z}$. Since these spectral "islands" are disjoint, we can recover the original signal g by passing the sampled signal $\hat{g}$ through an ideal low-pass filter, that is, an operation which multiplies the input spectrum by the transfer function

$$
H(f) = \sqcap(\tfrac{f}{2B}), \qquad\qquad (6.39)
$$

where B is chosen so that $W < B < \tfrac{1}{T} - W$, selecting only the central

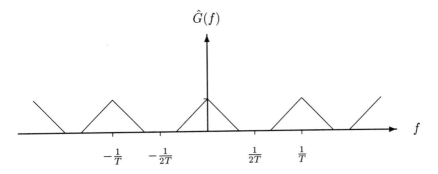

Figure 6.9: Replicated Spectra

island. Thus the output of this operation has spectrum

$$\hat{G}(f) \sqcap (\frac{f}{2B}) = G(f),$$

the spectrum of the original signal! This action is depicted in Figure 6.10 where the dashed box is the filter magnitude which selects only the central island.

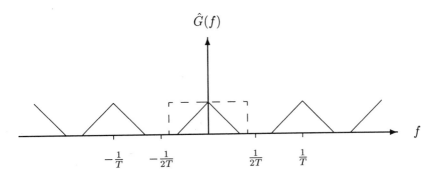

Figure 6.10: Signal Recovery

Again invoking the convolution theorem, taking inverse Fourier transforms then yields

$$g(t) \quad = \quad \hat{g}(t) * 2B \ \text{sinc}(2Bt)$$

$$= \left(T \sum_{n=-\infty}^{\infty} g(nT)\delta(t - nT) \right) * 2B \text{ sinc}(2Bt)$$

$$= 2BT \sum_{n=-\infty}^{\infty} g(nT) \text{ sinc}[2B(t - nT)],$$

which reduces to the previously derived sampling expansion (4.14) when $T = 1/2B$ and $B = W$.

The above derivation points out what happens if we sample too slowly in the sense that $1/T < 2W$. In this case the repeated copies of G in $\hat{G}$ overlap as shown in Figure 6.11, and there is no clear central island to

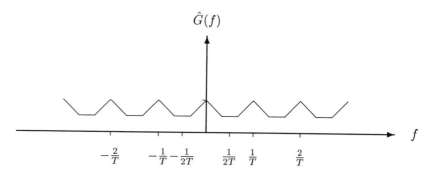

Figure 6.11: Overlapping Replicas

remove by low pass filtering. The figure shows the repeated copies and the final spectrum is the sum of these copies, indicated by the curve forming the "roof" of the overlapping islands. The low pass spectrum will be corrupted by portions of other islands and this will cause the resulting signal to differ from g. This distortion is called *aliasing* and some invariably occurs in any physical system since no physical signal can be perfectly band-limited.

The final comment above merits some elaboration. The basic argument is that all physical systems are time-limited assuming that the universe has a finite lifetime. A signal cannot be both time-limited and band-limited since, if it were, we could write for sufficiently large T and W that

$$g(t) = g(t) \sqcap (\frac{t}{2T}); \quad G(f) = G(f) \sqcap (\frac{f}{2W})$$

and hence, applying the convolution theorem to the relation on the left,

$$G(f) = \mathcal{F}_f(g(t)) = \mathcal{F}_f(g(t) \sqcap (\frac{t}{2T})) = G(f) * 2T \text{ sinc}(2Tf).$$

This yields a contradiction, however, since the convolution with a sinc function expands the bandwidth of $G(f)$, while multiplication by $\sqcap(f/2W)$ in general limits the extent of the spectrum.

As a final note, the sampling theorem states that a signal can be completely reconstructed from its samples provided $1/T > 2W$ and that it is not possible for $1/T < 2W$ because the resulting aliasing by overlapping spectral islands results in a distorted spectrum at low frequencies. In order to avoid such aliasing when sampling too slowly, the original signal can be first passed through a sharp low pass filter, that is, have its spectrum multiplied by $\sqcap(f/2W)$, and then this new signal will be recreated perfectly from its sampled version. The original low pass filtering introduces distortion, but it results in a signal that can be sampled without further distortion.

6.8 Correlation

Correlation is an operation on signals that strongly resembles convolution and which will be seen to have very similar properties. Its applications, however, are somewhat different. The principal use of correlation functions is in signal detection and estimation problems and in communications theory where they provide a measure of how similar a signal is to a delay of itself or to another signal. It also is crucial in defining *bandwidth* of signals and filters (as we shall see) and in describing the frequency domain behavior of the energy of a signal.

Suppose, as earlier, that we have two signals $g = \{g(t); \ t \in \mathcal{T}\}$ and $h = \{h(t); \ t \in \mathcal{T}\}$. The *crosscorrelation function* $\Gamma_{gh}(\tau)$ of g and h is defined for the various signal types as follows:

$$
\Gamma_{gh}(\tau) = \begin{cases}
\displaystyle\int_{-\infty}^{\infty} g^*(t-\tau)h(t)\,dt = \int_{-\infty}^{\infty} g^*(t)h(t+\tau)\,dt & \text{CTID}; \\[2ex]
\displaystyle\sum_{n=-\infty}^{\infty} g^*(n-\tau)h(n) = \sum_{n=-\infty}^{\infty} g^*(n)h(n+\tau) & \text{DTID}; \\[2ex]
\displaystyle\int_{0}^{T} \tilde{g}^*(t-\tau)h(t)\,dt = \int_{0}^{T} g^*(t)\tilde{h}(t+\tau)\,dt & \text{CTFD}; \\[2ex]
\displaystyle\sum_{n=0}^{N-1} \tilde{g}^*(n-\tau)h(n) = \sum_{n=0}^{N-1} g^*(n)\tilde{h}(n+\tau) & \text{DTFD},
\end{cases}
$$

where as usual $\tilde{g}$ denotes the periodic extension and where the finite duration correlation is a cyclic correlation (as was convolution). Analogous to the asterisk notation for convolution we abbreviate the correlation operation by a star: $\Gamma_{gh} = g \star h$. The argument of the correlation function (τ above) is often called the *lag*.

The *autocorrelation function* $\Gamma_g(\tau)$ of a signal g is the correlation of g with itself:

$$\Gamma_g(\tau) \equiv \Gamma_{gg}(\tau) = g \star g(\tau).$$

Note that the fact that one of the signals is conjugated is implied by, but not specifically stated in the shorthand notation. Both crosscorrelation and autocorrelation functions are referred to as simply *correlation functions*.

The convolution and correlation operations on signals are similar in that both take the product of two signals and then integrate (if continuous time) or sum (if discrete time). The operations differ in that correlation involves the complex conjugate of one signal (and convolution does not) and convolution reverses one signal in time (and correlation does not). In fact, we can express an crosscorrelation as a convolution as follows: given signals h and g, define g_- to be the signal g reversed in time; that is, if $g = \{g(t); t \in \mathcal{T}\}$, then $g_- = \{g(-t); t \in \mathcal{T}\}$. Then

$$g \star h = g_-^* * h. \tag{6.40}$$

A major difference between the convolution and correlation operations is that convolution is commutative, i.e., $g*h = h*g$, but in general $\Gamma_{gh} \neq \Gamma_{hg}$. In fact, in the continuous time infinite duration case

$$
\begin{aligned}
\Gamma_{hg}(\tau) &= \int_{-\infty}^{\infty} h^*(t) g(t + \tau)\, dt \\
&= \int_{-\infty}^{\infty} h^*(\zeta - \tau) g(\zeta)\, d\zeta \\
&= \left(\int_{-\infty}^{\infty} g^*(\zeta) h(\zeta - \tau)\, d\zeta \right)^* \\
&= \Gamma_{gh}^*(-\tau). \tag{6.41}
\end{aligned}
$$

The same result holds for discrete time and finite duration signals.

A function $\Gamma(\tau)$ with the property that $\Gamma(-\tau) = \Gamma^*(\tau)$ is said to be *Hermitian* and hence we have proved that autocorrelation functions are Hermitian. If g is a real function, this implies that the autocorrelation function is even.

All of the definitions for correlation functions were blithely written assuming that the various integrals or sums exist. As usual, this is trivial in the discrete time finite duration case. It can be shown that the other definitions all make sense (the integral or the limiting integral or sum exists) if the signals have finite energy. Recall that the energy of a signal g is defined by

$$\mathcal{E}_g = \begin{cases} \int\limits_{t \in \mathcal{T}} |g(t)|^2 \, dt & \text{continuous time;} \\ \sum\limits_{n \in \mathcal{T}} |g(n)|^2 & \text{discrete time.} \end{cases}$$

Note that the autocorrelation of a signal evaluated at 0 lag is exactly this energy; that is,

$$\mathcal{E}_g = \Gamma_g(0). \tag{6.42}$$

It is often convenient to normalize correlation functions by the signal energies. Towards this end we define the *correlation coefficient*

$$\gamma_{gh}(\tau) = \frac{\Gamma_{gh}(\tau)}{\sqrt{\mathcal{E}_g \mathcal{E}_h}} \tag{6.43}$$

and the corresponding normalized autocorrelation function

$$\gamma_g(\tau) = \gamma_{gg}(\tau) = \frac{\Gamma_g(\tau)}{\Gamma_g(0)}. \tag{6.44}$$

In order to derive one of the principal properties of correlation functions, an upper bound on the magnitude of the correlation in terms of the energies of the signal, we require one of the most useful inequalities of applied mathematics: the Cauchy-Schwarz inequality. Because of its importance we pause from the development to state and prove the inequality in its general form.

Theorem 6.8 *The Cauchy-Schwartz Inequality*
If $\{g(t); t \in \mathcal{T}\}$ and $\{h(t); t \in \mathcal{T}\}$ are two complex-valued continuous time functions on $\mathcal{T}$, then

$$\left| \int_{t \in \mathcal{T}} g(t)h(t) \, dt \right|^2 \leq \int_{t \in \mathcal{T}} |g(t)|^2 \, dt \int_{t \in \mathcal{T}} |h(t)|^2 \, dt \tag{6.45}$$

with equality if $g(t) = Kh^*(t)$ for some complex K; that is, g is a complex constant times the complex conjugate of h.
Similarly, if $\{g_n; n \in \mathcal{T}\}$ and $\{h_n; n \in \mathcal{T}\}$ are two complex-valued discrete time signals on $\mathcal{T}$, then

$$\left| \sum_{n \in \mathcal{T}} g_n h_n \right|^2 \leq \sum_{n \in \mathcal{T}} |g_n|^2 \sum_{n \in \mathcal{T}} |h_n|^2 \tag{6.46}$$

Proof: We prove the result only for the continuous time (integral) case. The corresponding result for discrete time follows in exactly the same manner by replacing the integrals by sums. Most proofs in the literature use a calculus of variations argument which is needlessly complicated. The proof below is much simpler. It is based on a simple trick and the fact that the answer is known and we need only prove it. (The calculus of variations is mainly useful when you do not know the answer first and need to find it.) Define as usual the energy of a signal by

$$\mathcal{E}_g = \int_{t \in T} |g(t)|^2 \, dt.$$

If either signal has infinite energy, then the inequality is trivially true. Hence we can assume that both signals have finite energy. Observe that obviously

$$\left(\frac{|g(t)|}{\sqrt{\mathcal{E}_g}} - \frac{|h(t)|}{\sqrt{\mathcal{E}_h}} \right)^2 \geq 0.$$

Expanding the square we have that

$$\frac{|g(t)|^2}{\mathcal{E}_g} + \frac{|h(t)|^2}{\mathcal{E}_h} - 2\frac{|g(t)||h(t)|}{\sqrt{\mathcal{E}_g \mathcal{E}_h}} \geq 0$$

or

$$\frac{|g(t)|^2}{\mathcal{E}_g} + \frac{|h(t)|^2}{\mathcal{E}_h} \geq 2\frac{|g(t)||h(t)|}{\sqrt{\mathcal{E}_g \mathcal{E}_h}}.$$

Integrating over t then yields

$$1 + 1 \geq \frac{2}{\sqrt{\mathcal{E}_g \mathcal{E}_h}} \int_{t \in T} |g(t)h(t)| \, dt.$$

The right hand side above can be bounded from below using the fact that for any complex valued function $x(t)$

$$\int_{t \in T} |x(t)| \, dt \geq \left| \int_{t \in T} x(t) \, dt \right| \qquad (6.47)$$

and hence setting $x(t) = g(t)h(t)$

$$\sqrt{\mathcal{E}_g \mathcal{E}_h} \geq \left| \int_{t \in T} g(t)h(t) \, dt \right|,$$

which proves the Cauchy-Schwartz inequality. If $g(t) = Kh^*(t)$, then both sides of the inequality equal $|K|\mathcal{E}_g^2$ and hence it is an equality. The integral inequality is due to Schwartz and the sum inequality is due to Cauchy. The name Cauchy-Schwartz inequality was originally given to the general inequality for inner products of which these are both special cases, but the name is often used for the special cases as well.

We now can apply the Cauchy-Schwarz inequality to derive the promised bound: we claim that

$$|\Gamma_{gh}(\tau)| \leq \sqrt{\mathcal{E}_g \mathcal{E}_h} \qquad (6.48)$$

and hence in particular that

$$|\Gamma_g(\tau)| \leq \mathcal{E}_g = \Gamma_g(0). \qquad (6.49)$$

Thus the autocorrelation function achieves its maximum at zero lag. We prove this for the continuous time case; the discrete time case follows similarly. Just replace $g(t)$ in the Cauchy-Schwartz inequality by $g^*(t)$ and $h(t)$ by $h(t + \tau)$ and we have that

$$|\Gamma_{gh}(\tau)| \leq \sqrt{\mathcal{E}_g \mathcal{E}_h}, \qquad (6.50)$$

since the energy of g^* is also the energy of $g(t)$ and the energy in $h(t + \tau)$ is also the energy in $h(t)$.

We summarize for latter use the principal properties of autocorrelation functions:

1. $\Gamma_g(0) = \mathcal{E}_g$, $\gamma_g(0) = 1$

2. $\Gamma_g(-\tau) = \Gamma_g^*(\tau)$, $\gamma_g(-\tau) = \gamma_g^*(\tau)$ (Hermitian)

3. $|\Gamma_g(\tau)| \leq |\Gamma_g(0)|$, $|\gamma_g(\tau)| \leq 1$.

4. The autocorrelation is shift invariant in the sense that if h is a shifted version of g ($h(t) = g(t - t_0)$ for some t_0), then $\Gamma_g(\tau) = \Gamma_h(\tau)$.

The notion of an autocorrelation function can be extended to signals with infinite energy but finite average power by a suitable normalization. For example, if

$$\int_{-\infty}^{\infty} |g(t)|^2 \, dt = \infty, \qquad (6.51)$$

but

$$\lim_{T \to \infty} \frac{1}{2T} \int_{-T}^{T} |g(t)|^2 dt < \infty, \qquad (6.52)$$

then we can define the generalized autocorrelation function

$$\Gamma_g(\tau) = \lim_{T \to \infty} \frac{1}{2T} \int_{-T}^{T} g^*(t - \tau) g(t) \, dt. \qquad (6.53)$$

As one might expect, the similarity between convolution and correlation will lead to similar properties when the Fourier transform is taken.

Theorem 6.9 *The Correlation Theorem*

$$\mathcal{F}_f(g \star h) = G^*(f)H(f); \tag{6.54}$$

that is, the Fourier transform of $\Gamma_{gh}(\tau)$ *is* $G^*(f)H(f)$. *Thus also*

$$\mathcal{F}_f(g \star g) = |G(f)|^2; \tag{6.55}$$

that is, the Fourier transform of the autocorrelation is the squared magnitude of the Fourier transform of the signal.

The proof parallels very closely the proof of the convolution theorem and hence is not presented. Alternatively, the result follows immediately from the representation of (6.40) of a correlation as a convolution combined with the fact that if g_- is the signal g reversed in time, then in the infinite duration continuous time case

$$
\begin{aligned}
\mathcal{F}_f(g_-^*) &= \int_{-\infty}^{\infty} g^*(-t)e^{-i2\pi ft}\, dt \\[2mm]
&= -\int_{\infty}^{-\infty} g^*(\zeta)e^{i2\pi f\zeta}\, d\zeta \\[2mm]
&= \int_{-\infty}^{\infty} g^*(\zeta)e^{i2\pi f\zeta}\, d\zeta \\[2mm]
&= \left(\int_{-\infty}^{\infty} g(\zeta)e^{-i2\pi f\zeta}\, d\zeta\right)^* = G^*(f).
\end{aligned}
$$

Thus

$$
\begin{aligned}
\mathcal{F}_f(g \star h) &= \mathcal{F}_f(g_-^* * h) \\
&= G^*(f)H(f).
\end{aligned}
$$

An implication of the correlation theorem is that the Fourier transform of the autocorrelation is real and nonnegative.

Note that all phase information in the spectrum $G(f)$ is lost in $|G(f)|^2$. This implies that many functions (differing from one another only in phase) have the same autocorrelation function. Thus the mapping of $g(t)$ into $\Gamma_g(\tau)$ is many to one. In general, without further a priori information or restrictions, a unique $g(t)$ cannot be found from $\Gamma_g(\tau)$.

6.9   Parseval's Theorem Revisited

The correlation theorem stated that given two signals g and h with transforms $G(f)$ and $H(f)$, then the Fourier transform of their crosscorrelation $\Gamma_{gh} = g \star h$ is given by $G^*(f)H(f)$. For example, in the infinite duration continuous time case this is

$$\Gamma_{gh}(\tau) = \int_{-\infty}^{\infty} g^*(t)h(t+\tau)\, dt = \int_{-\infty}^{\infty} G^*(f)H(f)e^{i2\pi f\tau}\, df.$$

Application of the inversion formula then implies that $\Gamma_{gh}(\tau)$ must be the inverse transform of $G^*(f)H(f)$. Applying this result to the special case where $\tau = 0$ immediately yields the general form of Parseval's theorem of Theorem 4.5. The general form is

$$\Gamma_{gh}(0) = \mathcal{F}_0^{-1}(\{G^*(f)H(f);\ f \in \mathcal{S}\}). \tag{6.56}$$

6.10   ⋆ Bandwidth and Pulsewidth Revisited

The properties of autocorrelation functions can be used to provide additional definitions of bandwidth and pulsewidth of signals and to relate the two quantities.

Autocorrelation Width

Because of the shortcomings of the definition of equivalent width, it is desirable to find a better definition not having these problems. Toward this end we introduce the *autocorrelation width* of a signal defined (easily) as the equivalent width of the autocorrelation of the signal. Since the autocorrelation function has its maximum at the origin, the autocorrelation width of g is defined by

$$W_{\Gamma_g} = \frac{\int_{-\infty}^{\infty} \Gamma_g(t)\, dt}{\Gamma_g(0)}.$$

From the correlation theorem, the Fourier transform of $\Gamma(t)$ is $|G(f)|^2$ and hence from the zeroth moment property

$$W_{\Gamma_g} = \frac{|G(0)|^2}{\Gamma_g(0)} = \frac{|G(0)|^2}{\int_{-\infty}^{\infty} |G(f)|^2\, df},$$

where the area property was used to express the denominator in terms of the spectrum. The right hand side is just one over the equivalent width of the magnitude squared spectrum, $|G(f)|^2$, and hence as with equivalent width,

$$W_{\Gamma_g} W_{|G|^2} = 1,$$

the corresponding widths in the time and frequency domain are inversely proportional.

The autocorrelation width of a signal does not give a nonsensical result if the signal has zero area, but it can still give nonsensical results. Consider, for example, the autocorrelation width of the signal $\text{sinc}(t) \cos(10\pi t)$, which can be thought of as an amplitude modulated sinc function. The spectrum of this signal can be found from the modulation theorem to be

$$G(f) = \frac{1}{2} \sqcap (f - 5) + \frac{1}{2} \sqcap (f + 5).$$

Thus $|G(0)|^2 = 0$ and hence the autocorrelation width of this signal is 0.

The autocorrelation width does provide a meaningful result for the special case of a Dirac delta. The autocorrelation function of a delta function is another delta function (as with the convolution of two delta functions) and hence $W_{\Gamma_g} = 1/\infty = 0$, which is what one would expect for a Dirac delta.

Mean-Squared Width

Yet another definition of width is the standard deviation of the instantaneous power or the *mean squared width*, which we denote Δt_g. It is defined as the square root of the variance

$$
\begin{aligned}
(\Delta t_g)^2 &= \sigma_{|g|^2}^2 \\
&= <t^2>_{|g|^2} - <t>_{|g|^2}^2 \\
&= \frac{\int_{-\infty}^{\infty} t^2 |g(t)|^2 \, dt}{\int_{-\infty}^{\infty} |g(t)|^2 \, dt} - \left(\frac{\int_{-\infty}^{\infty} t |g(t)|^2 \, dt}{\int_{-\infty}^{\infty} |g(t)|^2 \, dt} \right)^2 .
\end{aligned}
$$

Unlike the previous moments we have considered, the magnitude square of the signal is always nonnegative and hence the weighting of the integrand is nonnegative. This width is independent of the origin (not affected by translation) and it is strictly positive for all physical functions. Its primary difficulty is that it can be infinite for a signal with a physically finite area. Consider, for example, the signal $g(t) = \text{sinc}(t)$. We have that

$$\int_{-\infty}^{\infty} t^2 \text{sinc}^2(t) \, dt = \frac{1}{\pi^2} \int_{-\infty}^{\infty} \sin^2(\pi t) \, dt = \infty.$$

The corresponding frequency domain width is the standard deviation of the energy spectrum, the square root of the variance

$$(\Delta f_G)^2 = \sigma_{|G|^2}^2 = \frac{\int_{-\infty}^{\infty} f^2 |G(f)|^2 \, df}{\int_{-\infty}^{\infty} |G(f)|^2 \, df} - \left(\frac{\int_{-\infty}^{\infty} g |G(f)|^2 \, df}{\int_{-\infty}^{\infty} |G(f)|^2 \, df} \right)^2.$$

These two quantities have a famous relation to each other called *the uncertainty relation* which is given in the following theorem.

Theorem 6.10 *The Uncertainty Relation*
  For any continuous time infinite duration signal $g(t)$ with spectrum $G(f)$, the timewidth-bandwidth product $\Delta t_g \Delta f_G$ (called the uncertainty product) satisfies the following inequality:

$$\Delta t_g \Delta f_G \geq \frac{1}{4\pi}. \tag{6.57}$$

The bound is achieved with equality by the Gaussian signal.

The uncertainty relation can be interpreted as saying that one cannot have too narrow a width in both domains simultaneously. This is not as simple a relation as the corresponding time-bandwidth products of the other width definitions, but at least it provides a lower bound to the width in one domain in terms of the inverse of the width in the other.

Proof of the Uncertainty Relation: For convenience we assume that

$$\int_{-\infty}^{\infty} |g(t)|^2 \, dt = \int_{-\infty}^{\infty} |G(f)|^2 \, df = 1$$

$$\int_{-\infty}^{\infty} t |g(t)|^2 \, dt = \int_{-\infty}^{\infty} f |G(f)|^2 \, df = 0.$$

Define

$$(\Delta t)^2 = \int_{-\infty}^{\infty} t^2 |g(t)|^2 \, dt$$

$$(\Delta f)^2 = \int_{-\infty}^{\infty} f^2 |G(f)|^2 \, df.$$

The goal is then to prove that

$$(\Delta t)(\Delta f) \geq \frac{1}{4\pi}.$$

1. If $\Delta t = \infty$, the relationship is obviously true, so we assume that $\Delta t < \infty$. If this is true, then $\lim_{t \to \infty} t |g(t)|^2 = 0$ since otherwise the integral would blow up.

2. Applying the Cauchy-Schwarz inequality yields

$$\left| \int_{-\infty}^{\infty} (tg(t))(\frac{d}{dt}g^*(t))\, dt \right|^2 \leq (\int_{-\infty}^{\infty} t^2 |g(t)|^2 dt)(\int_{-\infty}^{\infty} \left| \frac{dg(t)}{dt} \right|^2 dt)$$

with equality if $tg(t) = K\frac{dg(t)}{dt}$.

3. To evaluate the right-hand side of the previous equation, first note that $\int_{-\infty}^{\infty} t^2 |g(t)|^2 dt = (\Delta t)^2$. Next observe that since $\frac{dg(t)}{dt} \to i2\pi f G(f)$, Rayleigh's theorem implies that

$$\int_{-\infty}^{\infty} \left| \frac{dg(t)}{dt} \right|^2 dt = \int_{-\infty}^{\infty} |i2\pi f G(f)|^2 ds = 4\pi^2 (\Delta f)^2.$$

Thus

$$| \int_{-\infty}^{\infty} (tg(t))(\frac{d}{dt}g^*(t))\, dt|^2 \leq 4\pi^2 (\Delta t)^2 (\Delta f)^2.$$

4. Now consider the left-hand side. For any complex number z we have that

$$|z| = \frac{1}{2}(|z| + |z^*|) \geq \frac{1}{2}|z + z^*| = |\Re(z)|$$

whence

$$| \int_{-\infty}^{\infty} (tg(t))(\frac{d}{dt}g^*(t))\, dt| \geq \frac{1}{2} \left| \int_{-\infty}^{\infty} (tg^*(t)\frac{d}{dt}g(t) + tg(t)\frac{d}{dt}g^*(t))\, dt \right|.$$

Furthermore, we have that

$$g^* \frac{dg}{dt} + g \frac{dg^*}{dt} = \frac{d}{dt}gg^* = \frac{d}{dt}|g|^2.$$

Thus we have

$$\frac{1}{4}| \int_{-\infty}^{\infty} t\frac{d}{dt}|g(t)|^2 dt|^2 \leq 4\pi^2 (\Delta t)^2 (\Delta f)^2.$$

5. Integrating the expression inside the magnitude of the left-hand side of the previous equation by parts yields

$$\int_{-\infty}^{\infty} t\frac{d}{dt}|g(t)|^2 dt = \int_{-\infty}^{\infty} td|g|^2 = t|g|^2|_{-\infty}^{\infty} - \int_{-\infty}^{\infty} |g|^2 dt.$$

But $\int_{-\infty}^{\infty} |g|^2 dt = 1$ and from point 1, $t|g|^2 = 0$ at $-\infty$ and ∞.

6. Finally, we have

$$\frac{1}{4}(-1)^2 \leq 4\pi^2(\Delta t)^2(\Delta f)^2$$

or

$$\Delta t \Delta f \geq \frac{1}{4\pi}.$$

When does $\Delta t \Delta f$ actually equal $1/4\pi$? In the first inequality (step (2)), equality is achieved if and only if $tg(t) = k\frac{d}{dt}g(t)$ for some complex constant k. In the second inequality (step (4)) equality is achieved if and only if $Z = \int_{-\infty}^{\infty} tg(t)\frac{d}{dt}g(t)^* \, dt$ is real valued. Thus if $tg(t) = k\frac{d}{dt}g(t)$ for some *real* constant k, then both conditions are satisfied since then $Z = \int_{-\infty}^{\infty} t^2|g(t)|^2 \, dt/k$. Thus the lower bound will hold if

$$k\frac{d}{dt}g(t) - tg(t) = 0$$

for some real-valued k. The solution to this differential equation is

$$g(t) = ce^{\frac{t^2}{2k}}.$$

In addition we require that $\lim_{|t|\to\infty} t|g(t)| = 0$ or $2k = -1/\alpha$ (for $\alpha > 0$) and $\int_{-\infty}^{\infty} |g(t)|^2 \, dt = 1$ or $c = \sqrt[4]{2\alpha/\pi}$. Thus the lower bound is achieved by a signal having the form

$$g(t) = \sqrt[4]{\frac{2\alpha}{\pi}}e^{-\alpha t^2}$$

for $\alpha > 0$. It can be shown that a signal achieving the lower bound necessarily has this form.

6.11 ⋆ The Central Limit Theorem

The convolution of $\sqcap(t)$ with itself yields $\wedge(t)$. Let us denote this self convolution by $\sqcap(t)^{*2}$. If in turn we convolve $\wedge(t)$ with $\sqcap(t)$, we get a new signal that is more spread out and smoother. As this is the convolution of three copies of the signal we denote the result by $\sqcap(t)^{*3}$. We can continue convolving further copies of $\sqcap(t)$ with itself in this way. The question now is, if we convolve $\sqcap(t)$ with itself n times to produce $\sqcap(t)^{*n}$, does the result tend to a limit as n gets large? More generally, what happens when we convolve an arbitrary signal with itself n times? Does it converge to a limit and is the limit different for different signals? The answer is called the central limit theorem and states that if the signal is suitably normalized in time scale and magnitude and satisfies certain constraints to be discussed, then the result of convolving it with itself many times will be a Gaussian signal, regardless of the original shape. The moment properties play a key role in proving this result. The key applications of approximating the result of a multiple convolution of a signal with itself are in systems theory and probability theory where the convolutions correspond respectively to cascading many identical linear filters and summing many identically distributed independent random variables.

Theorem 6.11 *The Central Limit Theorem*
   Given an infinite duration continuous time signal $g(t)$ with Fourier transform $G(f)$, suppose that $G(f)$ has the property that for small f

$$G(f) \approx a - cf^2 \qquad (6.58)$$

for positive a and c.
   Then

$$\lim_{n \to \infty} \frac{(\sqrt{n}g(\sqrt{n}t))^{*n}}{a^n} = \sqrt{\frac{\pi a}{c}}e^{-\pi^2 \frac{a}{c} t^2}. \qquad (6.59)$$

   The proof of the theorem will take the remainder of the subsection. First, however, some comments are in order.
   The small f approximation may seem somewhat arbitrary, but it can hold under fairly general conditions. For example, suppose that $G(f)$ has a Taylor series at $f = 0$:

$$G(f) = \sum_{k=0}^{\infty} b_k f^k,$$

where

$$b_k = \frac{G^{(k)}(0)}{k!}$$

(the derivatives of all orders exist and are finite). Suppose further that

   1. $b_1 = G'(0) = 0$ (This is assumed for convenience.)

   2. $b_0 = G(0) > 0$. Define $a = b_0$.

   3. $b_2 = G''(0)/2 < 0$. Define $c = -b_2$.

   These assumptions imply (6.58), an equation which is commonly written as

$$G(f) = a - cf^2 + o(f^2),$$

where $o(f^2)$ means a term that goes to zero faster than f^2. In fact this is all we need and we could have used this as the assumption.

An additional implication of assuming that $G(f)$ has a Taylor series is that the entire spectrum can be computed from the moments by using the moment properties; that is, since

$$G^{(k)}(0) = (\frac{2\pi}{i})^k < t^k >_g,$$

then

$$G(f) = \sum_{k=0}^{\infty} \frac{(-2\pi i)^k}{k!} < t^k >_g f^k = \sum_{k=0}^{\infty} \frac{(-2\pi i f)^k}{k!} < t^k >_g .$$

This formula provides one of the reasons that moments are of interest–they can be used to compute (or approximate) the original signal or spectrum if it is nice enough to have a Taylor series.

The condition that $b_1 = G'(0) = 0$ is satisfied by any real even $G(f)$. From the first moment property this implies also that $< t >_g = 0$.

Proof of the Central Limit Theorem: From the stretch theorem, the signal $\sqrt{n}g(\sqrt{n}t)$ has Fourier transform $G(f/\sqrt{n})$. From the convolution theorem, $(\sqrt{n}g(\sqrt{n}t))^{*n}$ has Fourier transform $G^n(f/\sqrt{n})$. Thus

$$\mathcal{F}_f(\frac{(\sqrt{n}g(\sqrt{n}t))^{*n}}{a^n}) = \frac{G^n(f/\sqrt{n})}{a^n}$$

$$= \frac{1}{a^n}\left(a - c(\frac{f}{\sqrt{n}})^2 + o(\frac{f^2}{n})\right)^n$$

$$= \left(1 - \frac{c}{a}\frac{f^2}{n} + \frac{1}{a}o(\frac{f^2}{n})\right)^n,$$

where the $o(f^2/n)$ notation means a term that goes to zero with increasing n faster than f^2/n. From elementary real analysis, as $n \to \infty$ the rightmost term goes to $e^{-cf^2/a}$. This result is equivalent to the fact that for small ϵ,

$$\ln(1 - \epsilon) \approx -\epsilon.$$

Thus for large n

$$\frac{G^n(f/\sqrt{n})}{a^n} \approx (1 - \frac{c}{a}\frac{f^2}{n})^n \approx e^{-\frac{cf^2}{a}}.$$

The inverse Fourier transform of $e^{-\frac{cf^2}{a}} = e^{-\pi(\sqrt{\frac{c}{a\pi}}f)^2}$ is $\sqrt{\frac{\pi a}{c}}e^{-\pi(\sqrt{\frac{\pi a}{c}}t)^2}$. We now assume that we can interchange limits and transforms to infer that the inverse transform of a limit of the transforms is the limit of the inverse transforms. Applying this to $\dfrac{\left(\sqrt{n}g(\sqrt{n}t)\right)^{*n}}{a^n}$ should give us the limit of the $\dfrac{\left(\sqrt{n}g(\sqrt{n}t)\right)^{*n}}{a^n}$, which yields

$$\lim_{n\to\infty} \frac{\left(\sqrt{n}g(\sqrt{n}t)\right)^{*n}}{a^n} = \sqrt{\frac{\pi a}{c}}e^{-\pi(\sqrt{\frac{\pi a}{c}}t)^2},$$

which completes the proof of the theorem.

The central limit theorem can be used to provide an approximation for an n-fold self convolution $g(t)^{*n}$ by suitable scaling. We have seen that

$$G^n\left(\frac{f}{\sqrt{n}}\right) \approx a^n e^{-\frac{c}{a}f^2}$$

and hence

$$G^n(f) \approx a^n e^{-\frac{c}{a}nf^2}.$$

Inverting this result we have that

$$g(t)^{*n} \approx \frac{a^{n+\frac{1}{2}}}{\sqrt{n}}\sqrt{\frac{\pi}{c}}e^{-\frac{\pi^2 a}{cn}t^2}.$$

As an example consider the signal

$$g(t) = \frac{\sqcap\left(\frac{t}{2}\right)}{\pi\sqrt{1-t^2}}.$$

Does $g(t)^{*n}$ converge to a Gaussian signal? If so, what is its approximate form? If not, why not? From the Fourier transform tables, the transform of $g(t)$ is $J_0(2\pi f)$. Using the series representation Eq. 1.13 for Bessel functions, it can be verified that $G(f)$ indeed has finite derivatives of all orders and that

$$G(0) = a = 1 > 0,$$

$$G'(0) = 0,$$

and

$$c = -\frac{G''(0)}{2} = \pi^2 > 0.$$

From Eq. 1.13, for small f $J_0(2\pi f) \approx 1 - \pi^2 f^2$ and hence $a = 1$ and $c = \pi^2$.

Since all of the conditions are met, the CLT holds and we have for large n that

$$g(t)^{*n} \approx \frac{1}{\sqrt{\pi n}}e^{-\frac{t^2}{n}}.$$

As another example, consider the same questions asked about $g(t) = J_0(2\pi t)$. Here the transform is

$$G(f) = \frac{\sqcap(\frac{f}{2})}{\pi\sqrt{1 - f^2}}.$$

Here $G(0) > 0$ as required, but $G''(0) > 0$ violates the sufficient condition and hence the CLT cannot be applied. What is happening here is that for small f

$$G(f) = \frac{1}{\pi}(1 - f^2)^{-\frac{1}{2}} \approx \frac{1}{\pi} + \frac{1}{2\pi}f^2.$$

Thus $a = 1/\pi$ and $c = -1/\pi$. A positive c is required, however, for the derivation to hold.

6.12 Problems

6.1. A certain discrete time system takes an input signal $x = \{x_n; n \in \mathbb{Z}\}$ and forms an output signal $y = \{y_n; n \in \mathbb{Z}\}$ in such a way that the input and output are related by the formula

$$y_n = x_n - ay_{n-1}, \ n \in \mathbb{Z};$$

where $|a| < 1$.

(a) If the signals x and y have Fourier transforms X and Y, respectively, find $Y(f)/X(f)$.

(b) Find the δ-response of the system, that is, the response to a Kronecker delta $\delta = \{\delta_n; n \in \mathbb{Z}\}$.

(c) Express the output y as a convolution of the input x and a signal h (which you must describe).

(d) Is the system linear? Time invariant?

6.2. The formula (3.79) is a convolution. Use the convolution theorem to relate the Fourier transform G_a of g_a to G, the Fourier transform of g. Describe what happens to both G_a and g_a as $a \to \infty$.

6.3. Suppose that an infinite duration, continuous time linear time invariant system has transfer function $H(f)$ given by

$$H(f) = -i\text{sgn}(f) = \begin{cases} e^{\frac{-i\pi}{2}} & \text{if } f > 0 \\ e^{\frac{i\pi}{2}} & \text{if } f < 0 \end{cases}.$$

This filter causes a $90°$ phase shift for negative frequencies and a $-90°$ phase shift for positive frequencies. Let $h(t)$ denote the corresponding impulse response. Let $v(t)$ be a real valued signal with spectrum $V(f)$ which is bandlimited to $[-W, W]$. Define the signal $\hat{v} = v * h$, the convolution of the input signal and the impulse response.

(a) Show that

$$\hat{v}(t) = \frac{1}{\pi} \int_{\lambda \in \mathcal{R}, \lambda \neq t} \frac{v(\lambda)}{t - \lambda} d\lambda,$$

where the integral is interpreted in the Cauchy sense. The negative of $\hat{v}(t)$ is called the *Hilbert transform* of the signal $v(t)$.

(b) Suppose that the spectrum of $v(t)$ has the shape given in Figure 6.8. For $f_c \gg W$ sketch the spectrum of the waveform

$$v_c(t) = v(t) \cos(2\pi f_c t) + \hat{v}(t) \sin(2\pi f_c t).$$

This is called *single-sideband amplitude modulation* or SSB for short. How does the bandwidth of the above signal compare with that of ordinary double sideband-suppressed carrier (DSB) modulation $y(t) = v(t) \cos(2\pi f_c t)$?

6.4. Using naked brute force, prove that convolution satisfies the associativity property:
$$f * (g * h) = (f * g) * h$$

6.5. Prove the dual convolution theorem (6.30).

6.6. Evaluate the convolution
$$e^{-at^2} * e^{-bt^2}.$$

6.7. Evaluate the following convolutions using Fourier transforms.

(a) $J_0(2\pi t) * \text{sinc}(2t)$
(b) $\sin(\pi t) * \Lambda(t)$.
(c) $\text{sinc}^2(2t) * \text{sinc}(2t)$
(d) $\cos(\pi t) * e^{-\pi t^2}$

6.8. The error function $\text{erf}(t)$ is defined by

$$\text{erf}(t) = \frac{2}{\sqrt{\pi}} \int_0^t e^{-\zeta^2} d\zeta.$$

Find the Fourier transform of $\text{erf}(t)$.

6.9. Find and sketch the following convolutions and cross-correlations. All signals are infinite duration continuous time signals.

(a) $t^2 u_{-1}(t) * e^t u_{-1}(t)$

(b) $\sqcap(t + 1/2) * \sqcap(3t + 1)$

(c) $\sqcap(t - 1) \star \sqcap(3t + 3)$

(d) $\sqcap(t) * \sqcap(t) * \sqcap(t)$

6.10. Find and sketch the following convolutions:

(a) $\{tu_{-1}(t); t \in \mathcal{R}\} * \{e^t u_{-1}(t); t \in \mathcal{R}\}$

(b) $\{\sqcap(t - 1); t \in \mathcal{R}\} * \{\sqcap(2t + 4); t \in \mathcal{R}\}$

(c) $\{a^n u_{-1}(n); n \in \mathcal{Z}\} * \{u_{-1}(n); n \in \mathcal{Z}\}$

6.11. Find and sketch the following convolutions and cross-correlations:

(a) $\sqcap(t - 1) * \sqcap(2t + 3)$

(b) $\sqcap(t - 1) \star \sqcap(2t + 3)$

(c) $\sqcap(t + 1) \star \sqcap(2t - 3)$

6.12. Find the convolution $\wedge(t) * \wedge(t)$.

6.13. Given a signal $g = \{g(t); t \in \mathcal{R}\}$, denote by g_τ the shifted signal defined by $g_\tau(t) = g(t - \tau)$. Show that

$$\Gamma_g(t) = \Gamma_{g_\tau}(t);$$

that is, shifting does not affect autocorrelation.

6.14. Define the discrete time signals ($\mathcal{T} = \mathcal{Z}$)

$$g_n = \begin{cases} 1, & n=0,1,2,3,4 \ ; \\ 0, & \text{otherwise} \end{cases}$$

and

$$h_n = \begin{cases} (\frac{1}{2})^n, & n=0,1,\ldots ; \\ 0, & \text{otherwise.} \end{cases}$$

(a) Find the autocorrelation function $\Gamma_g(n)$ of g and provide a labeled sketch.

(b) Sketch the signals $x_n = g_{n+2}$ and $y_n = h_{2n}$. Find the Fourier transforms of these two signals.

6.15. Find the autocorrelation function and the Fourier transform of the signal $\{e^{i2\pi k/8}; \ k = 0, 1, \ldots, 7\}$. Find the circular convolution of this signal with itself and the Fourier transform of the resulting signal.

6.16. Find and sketch the autocorrelations and the crosscorrelation of the discrete time signals

$$g_n = \begin{cases} r^n, & n = 0, 1, \ldots; \\ 0, & \text{otherwise} \end{cases}$$

(where $|r| < 1$) and

$$h_n = \begin{cases} e^{-in}, & n = -N, -N + 1, \ldots, -1, 0, 1, \ldots, N; \\ 0, & \text{otherwise.} \end{cases}$$

Verify in this example that

$$|\Gamma_{gh}(t)|^2 \leq \Gamma_g(0)\Gamma_h(0).$$

6.17. Evaluate the continuous time convolution $\delta(t - 1) * \delta(t - 2) * \sqcap(t - 1)$.

6.18. Find a continuous time signal $g(t)$ whose autocorrelation function is

$$\Gamma_g(t) = \wedge(2(t - 4)) + 2 \wedge (2t) + \wedge(2(t + 4)).$$

Is the answer unique?

6.19. Suppose that $h = f * g$, where the signals are infinite duration continuous time. Prove that the area under h equals the product of the areas under f and g; that is,

$$\int_{-\infty}^{\infty} h(t) \, dt = \left(\int_{-\infty}^{\infty} f(t) \, dt \right) \left(\int_{-\infty}^{\infty} g(t) \, dt \right).$$

6.20. Evaluate the continuous time convolutions $\delta(2t + 1) * \sqcap(t/3)$ and $\delta(t - 0.5) * \delta(t + 1) * \text{sinc}(t)$.

6.21. Two finite duration discrete time complex-valued signals $g = \{g_n; \ n = 0, 1, \ldots, N - 1\}$ and $h = \{h_n; \ n = 0, 1, \ldots, N - 1\}$ are to be convolved (using cyclic or circular convolution) to find a new signal $y = g * h$. This can be done in two ways:

- by brute force evaluation of the convolution sum, or
- by first taking the DFT of each signal, then forming the product, and then taking the IDFT to obtain $g * h$.

Assuming that an FFT algorithm is used to perform both the DFT and IDFT, estimate the number of real multiplies required for each approach. Do *not* count the multiplication by $1/N$ in the inverse DFT (the IFFT). (In practice, N is a power of 2 and this can be accomplished by bit shifting.)

Specialize your answer to the case $N = 1024$.

6.22. The *ambiguity function* of a signal $\{g(t); \; t \in \mathcal{R}\}$ is defined by

$$\chi(\tau, f) = \int_{-\infty}^{\infty} g^*(\zeta) g(\zeta + \tau) e^{-i2\pi f \zeta} \, d\zeta.$$

(This function is important in radar signal analysis.)

(a) What is the maximum value of $|\chi(\tau, f)|$ over all τ and f. When is it achieved?

(b) Evaluate $\chi(\tau, f)$ for $g(t) = \sqcap(t)$.

6.23. Define two discrete time, infinite duration signals $x = \{x_n; \; n \in \mathcal{Z}\}$ and $y = \{y_n; \; n \in \mathcal{Z}\}$ by

$$x_n = \rho^n u_{-1}(n),$$

where $u_{-1}(n)$ is the unit step function (1 if $n \geq 0$ and 0 otherwise) and $|\rho| < 1$,

$$y_n = \delta_n - \delta_{n-1},$$

where δ_n is the Kronecker delta (1 if $n = 0$ and 0 otherwise).

Find the Fourier transforms of the following signals:

(a) x and y

(b) $\{x_{n-5}; \; n \in \mathcal{Z}\}$

(c) $\{x_{5-n}; \; n \in \mathcal{Z}\}$

(d) $\{x_{|n|}; \; n \in \mathcal{Z}\}$

(e) $\{x_n \cos(2\pi n/9); \; n \in \mathcal{Z}\}$

(f) $\{x_n y_n; \; n \in \mathcal{Z}\}$

(g) $x * y$ (the convolution of x and y)

(h) $x + y$

(i) $\{x_n^2; \; n \in \mathcal{Z}\}$

6.24. Suppose that you are told two continuous time, infinite duration signals $x = \{x(t); \ t \in \mathcal{R}\}$ and $h = \{h(t); \ t \in \mathcal{R}\}$ are convolved to form a new signal $y = x * h = \{y(t); \ t \in \mathcal{R}\}$. You are also told that $x(t) = e^{-t}u_{-1}(t)$ and

$$y(t) = \square_{1/2}(t - \frac{1}{2}) - e^{-t}u_{-1}(t) + e^{-(t-1)}u_{-1}(t-1).$$

(a) Find the Fourier transform $Y(f)$ of y.

(b) Find and sketch the signal h. (This is an example of *deconvolution*.)

6.25. Find the circular convolution of the sequence $\{1,1,0,1,0,0,0,0\}$ with itself.

6.26. Let $g(t)$ be a bandlimited signal with 0 spectrum $G(f)$ for $|f| \geq W$. We wish to sample $g(t)$ at the slowest possible rate that will allow recovery of

$$\int_{-\infty}^{\infty} g(t)\, dt.$$

We do not wish to reconstruct $g(t)$, only to know its area. What is the minimum sampling rate that will allow recovery of the above area? How do we express the above integral in terms of the samples $g(t)$?

6.27. Evaluate the following integrals using Fourier transforms:

(a)

$$\int_{-\infty}^{\infty} \left(\frac{2}{1+(2\pi t)^2}\right)^2 dt.$$

(b)

$$\int_{-\infty}^{\infty} J_0(2\pi t)\frac{J_1(2\pi t)}{2t}\, dt$$

(c)

$$\int_{-\infty}^{\infty} (\text{sinc}(t+\frac{1}{2}) + \text{sinc}(t-\frac{1}{2}))^2\, dt.$$

6.28. Evaluate the following integral using Fourier transforms (there is an easy way).

$$\int_{-\infty}^{\infty} t^2 \text{sinc}^4(t)\, dt.$$

6.29. Suppose that an infinite duration continuous time signal $g(t)$ has a zero spectrum outside the interval $(-\frac{1}{2}, \frac{1}{2})$. It is asserted that the magnitude of such a function can never exceed the square root of its energy. For what nontrivial signal $g(t)$ is equality achieved? That is, for what function is the magnitude of $g(t)$ actually equal to $\sqrt{\mathcal{E}_g}$ for some t? Prove your answer. (*Hint:* $G(f) = \sqcap(f)G(f)$.)

6.30. A signal $g(t)$ is known to have a spectrum that is non-zero only in the range $4B < |f| < 6B$. This signal could, of course, be recovered from samples taken at a rate at least as large as $12B$ samples per second. It is claimed that $g(t)$ can be recovered from samples taken at a rate much slower than $12B$ samples per second using a single linear time-invariant interpolation filter. What is the slowest rate for which this is true? Describe the interpolation filter.

6.31. A bandlimited signal of maximum frequency W is to be sampled using a sampling signal $\text{III}(t/T)$ where $1/T < 2W$. Specify the bandwidth of the widest rectangular low-pass filter to be used prior to sampling which will eliminate aliasing. If no filter is used prior to sampling, specify the widest rectangular low-pass filter that can be used after sampling to provide an output free from aliasing.

6.32. A continuous time signal $g(t)$ is bandlimited and has 0 spectrum $G(f)$ for all f with $|f| \geq W$.

   (a) Instead of ideal sampling with impulses, the signal is sampled using a train of very narrow pulses, each with pulse shape $p(t)$ and spectrum $P(f)$. The resulting sampled pulse train is given by

$$\hat{g}(t) = g(t)T \sum_{n=-\infty}^{\infty} p(t - nT).$$

   This is sometimes called *natural sampling*. What is the effect of the nonideal pulse shape on the procedure for recovering $g(t)$ from $\hat{g}(t)$? State any assumptions you make relating T, W, and $p(t)$.

   (b) Next suppose that the signal is first multiplied by an ideal impulse train to produce a sampled waveform

$$y(t) = T \sum_{n=-\infty}^{\infty} g(nT)\delta(t - nT),$$

but $y(t)$ then is passed through a filter whose output spectrum $U(f)$ is given by $Y(f)P(f)$ so that the final sampled signal is

$$u(t) = T \sum_{n=-\infty}^{\infty} g(nT)p(t - nT).$$

How can the original signal be recovered from this signal?

6.33. What is the convolution of the infinite duration continuous time signals

$$\frac{1}{1 + [2\pi(t - 1)]^2}$$

and

$$\frac{1}{1 + [2\pi(t + 1)]^2}?$$

6.34. Name two different nontrivial signals or generalized signals g that have the property that $g * g = g$, i.e., the signal convolved with itself is itself. (A trivial signal is zero everywhere.) What can you say about the spectrum of such a signal?

6.35. Suppose that g is an arbitrary continuous time infinite duration signal and a new signal y is formed by convolving g with a box function $\square_{1/2}$. What can be said about $Y(k)$ for integer k?

6.36. You are given a linear, time-invariant system with transfer function $H(f) = e^{-\pi f^2}$, $f \in \mathcal{R}$.

  (a) Suppose the input to the system is $x(t) = e^{-\pi t^2}$, $t \in \mathcal{R}$. Write the output $v(t)$ as a convolution integral of the input with another function $h(t)$, which you must specify.

  (b) Find the output $v(t)$ with the input as given in part (a).

  (c) A signal $g(t) = 3 \cos 3\pi t$, $t \in \mathcal{R}$, is input to the system. Find the output.

  (d) The output to a particular input signal $y(t)$ is $e^{-\pi t^2}$, $t \in \mathcal{R}$. What is the input $y(t)$?

  (e) Now suppose that the signal $x(t)$ of part (a) is instead put into a filter with impulse response $\square_{1/2}(t)$ to form an output $w(t)$. Evaluate the integrals

$$\int_{-\infty}^{\infty} w(t)\, dt$$

and

$$\int_{-\infty}^{\infty} tw(t)\, dt.$$

6.37. Suppose that we approximate the impulse train interpretation of the sampling theorem by substituting narrow tall square pulses for the Dirac deltas as follows. Define the pulse $p(t)$ to be $1/\Delta$ for $|t| \leq \Delta/2$ and then define the pulse train

$$s(t) = \sum_{n=-\infty}^{\infty} p(t - nT).$$

Assume that $g = \{g(t); t \in \mathcal{R}\}$ is a bandlimited signal with a Fourier transform G such that $G(f)$ is nonzero only for $|f| \leq W$. Define the signal $y(t) = g(t)s(t)$. (If the $p(t)$ were indeed impulses, then this would be the idealized impulsive PAM signal.)

Find the Fourier transform $Y(f)$ of y and show that if T is chosen large enough (and specify what that means), then g can be recovered from y. Show how to accomplish this recovery.

6.38. An applications problem (courtesy of Rick Wesel):

Modem Training Tones. You may have noticed that when a modem first begins transmission it makes a tonal sound before the white-noise "shhhhh" sound of normal data transmission. In this problem you will use a simplified model of modem training to determine (quite accurately) the frequencies and relative power of these "training tones" for the V.29 modem.

The V.29 modem sends a training signal that alternates between two phase-amplitude points of a 1700 Hz sinusoid (the carrier). The switching between the two points occurs at the symbol rate of 2400 Hz.

For our calculation, one of these points will have zero amplitude, and the other point will be a unit amplitude cosine. Make the (obviously untrue) assumption that the training signal is bi-infinite to simplify your calculations. When the cosine is switched back on, it picks up where it would have been if it had been left on, *not* where it was when it was switched off. Note that the training signal can then be modeled as $s(t) = \cos(2\pi 1700t)p(t)$, where $p(t)$ is a square wave oscillating between 0 and 1 and holding each value for the duration of a single (ASCII) symbol.

(a) How many tones are there in this signal?

(b) What are the frequencies of the three most powerful tones?

(c) What are the frequencies of the next two most powerful tones?

(d) What is the power difference in dB between the most powerful of the three terms and the next two most powerful?

6.39. Let $p = \{p(t); \ t \in [0, 2)\}$ be the signal defined by

$$p(t) = \begin{cases} +1 & 0 \le t < 1 \\ -1 & 1 \le t < 2 \end{cases}.$$

(a) Find the Fourier transform P of p.

(b) Find a Fourier series representation for p.

(c) Find a Fourier series representation for $\tilde{p}$, the periodic extension of p with period 2. Provide a labeled sketch of $\tilde{p}$.

(d) Find the Fourier transform $\tilde{P}$ of $\tilde{p}$.

(e) Suppose that $\tilde{p}$ is put into a filter with an impulse response h whose Fourier transform H is defined by

$$H(f) = \begin{cases} 1 & |f \pm \frac{5}{2}| \le \frac{1}{8} \\ 0 & \text{otherwise} \end{cases}$$

to produce an output $y = \{y(t); \ t \in \mathcal{R}\}$. (This is called an ideal bandpass filter.) Find a simple expression for $y(t)$.

6.40. Suppose that a discrete time system with input $x = \{x_n; \ n \in \mathcal{Z}\}$ and output $y = \{y_n; \ n \in \mathcal{Z}\}$ is defined by the difference equations

$$y_n = \alpha x_n - \sum_{k=1}^{M} a_k y_{n-k}; \ n \in \mathcal{Z},$$

where $\alpha > 0$ and the a_k are all real. This system is linear and time invariant and hence has a Kronecker delta response h with DTFT H. Assume that h is a causal filter, that is, that $h_n = 0$ for $n < 0$. (Physically, the difference equations cannot produce an output before time 0 if the input is 0 for all negative time.) Assume also that the a_k are such that h is absolutely summable.

(a) Find an expression for $H(f)$ in terms of α and

$$A(f) = 1 + \sum_{k=1}^{M} a_k e^{-i2\pi fk}$$

(A filter of this form is called an *autoregressive filter*.)

Observe that if a_k is extended to all integer k by defining $a_0 = 1$ and $a_k = 0$ if k is not in $\{0, 1, \ldots, M\}$, then

$$A(f) = \sum_{k=-\infty}^{\infty} a_k e^{-i2\pi f k},$$

and $A(f)$ can be considered as the DTFT of the time limited signal $a = \{a_n; \ n \in \mathcal{Z}\}$.

(b) Is h real?

(c) Define the signal $w = h * a$, i.e.,

$$w_n = \sum_{k=0}^{M} h_{n-k} a_k, \tag{6.60}$$

where $a_0 = 1$. Find a *simple* expression for w_n.

(d) What is h_0 in terms of α and the a_k?

(e) Suppose that a Kronecker delta is the input to h. Show that the autocorrelation function of the output signal satisfies

$$\Gamma_h(l - k) = \sum_{n=-\infty}^{\infty} h_{n-k} h_{n-l}; \ l, k \in \mathcal{Z}. \tag{6.61}$$

(Recall that the autocorrelation function of a signal is given by $\Gamma_h(k) = \sum_n h_n^* h_{n+k}$.)

(f) Multiply both sides of (6.60) by h_{n-l} and sum over n to obtain a *simple* expression for

$$\beta_l = \sum_{k=0}^{M} \Gamma_h(k - l) a_k.$$

Specialize this expression to the two cases $l = 0$ and $l > 0$.

(g) Now suppose that we observe a discrete time signal s_n from an unknown source. We measure its autocorrelation $\Gamma_s(k)$ for $M+1$ values $k = 0, 1, \ldots, M$. We can form a *model* of this signal called the *correlation matching autoregressive model* by asking for what autoregressive filter h will it be true that

$$\Gamma_h(k) = \Gamma_s(k); \ k = 0, 1, \ldots, M \ ? \tag{6.62}$$

This filter h will have the property that when a Kronecker delta is input, the autocorrelation of the output will equal ("match") the measured autocorrelation for the first $M + 1$ values.

Use the previous parts of this problem to provide a set of equations in terms of α, the a_k, and Γ_s that will satisfy (6.62).

Note: What you have done above is to derive the Yule-Walker or Normal equations that are at the basis of LPC (linear predictive coding) of speech [20, 1, 22]. Intuitively, forcing the autocorrelation of the model to match the measured short-term autocorrelation of the genuine speech will result in the model producing a sound similar to the original speech.

These equations can be rapidly solved numerically using an algorithm called the Levinson-Durbin algorithm. These methods are also useful in a variety of other signal processing problems not related to speech.

6.41. Consider the signals $g = \{g(t) = \lambda e^{-\lambda t} u_{-1}(t); t \in \mathcal{R}\}$ and $h(t) = \{\Box_{1/2}(t); t \in \mathcal{R}\}$. λ is a positive real parameter.

Evaluate the following moments:

(a) $\int_{-\infty}^{\infty} t^k g(t)\, dt$ for $k = 0, 1$.

(b) $\int_{-\infty}^{\infty} t^k h(t)\, dt$ for $k = 0, 1$.

(c) $\int_{-\infty}^{\infty} f^2 H(f)\, df$, where H is the Fourier transform of h.

(d) Define y as $g * h$. Find an expression for $y(t)$ for all t.

(e) Evaluate

$$\int_{-\infty}^{\infty} y(t)\, dt,$$

where y is as in the previous part.

6.42. Suppose that $g = \{g_n; n \in \mathcal{Z}\}$ is a discrete time signal with Fourier transform $G = \{G(f); f \in [-1/2, 1/2)\}$ defined by

$$G(f) = \begin{cases} 1 & |f| \le 1/6 \\ 0 & \text{otherwise} \end{cases}.$$

(a) Suppose that a system with input g has output y defined by $y_n = g_n \cos(2\pi \frac{1}{3} n)$; $n \in \mathcal{Z}$. Find $Y(f)$ and provide a labeled sketch.

(b) Is the system described in the previous part linear? Time-invariant?

(c) Suppose that a system with input g has output w defined by $w_n = g_{3n}$; $n \in \mathcal{Z}$. Find $W(f)$ and provide a labeled sketch.

(d) Is the system described in the previous part linear? Time-invariant?

(e) Suppose that a system with input g has output z defined by $z_n = g_n^2$; $n \in \mathcal{Z}$. Find $Z(f)$ and provide a labeled sketch.

(f) Is the system described in the previous part linear? Time-invariant?

(g) What is the Kronecker delta response to each of the above three systems?

6.43. A continuous time infinite duration signal g is bandlimited to $[-.5, .5]$, i.e., its Fourier transform G satisfies $G(f) = 0$ for $f \geq .5$. The signal is passed through a filter with impulse response h, which is also bandlimited to $[-.5, .5]$. The output of the system, y, is sampled at a rate of 1 sample per second.

(a) Using the sampling theorem, express $g(t)$ and $h(t)$ in terms of their respective samples if they are both sampled at 1 sample per second.

(b) Express the sampled values of the filter output $y(t)$ in terms of the sampled values of g and h. (Be as rigorous as possible.) Your solution should look familiar.

(c) Let $\{\hat{G}(f); f \in [-1/2, 1/2)\}$ denote the Fourier transform of the discrete time signal $\{g(n); n \in \mathcal{Z}\}$. Express G in terms of $\hat{G}$.

6.44. Suppose that a continuous time signal $g(t)$ is sampled to form a discrete time signal $\gamma_n = g(nT_s)$. Consider also the idealized sampled waveform $\hat{g}(t) = g(t)\text{III}(t/T_s)$, that is, the continuous time waveform with the samples imbedded on impulses. Let $\tilde{\Gamma}(f)$ denote the periodic extension of the DTFT of γ and let $\hat{G}$ denote the CTFT of $\hat{g}$. Is it true that

$$\hat{G}(f) = T_s\tilde{\Gamma}(T_s f)? \tag{6.63}$$

6.45. Given a finite duration signal, show that the generalized autocorrelation function of its periodic extension is just the periodic extension of the autocorrelation function of the finite duration signal.

6.46. Calculate the following integrals.

(a)

$$\int_{-\frac{1}{2}}^{\frac{1}{2}} \left[\frac{\sin(2\pi t(N + 1/2))}{\sin(\pi t)}\right]^2 dt$$

(b)
$$\int_{-\infty}^{\infty} \frac{2}{1 + (2\pi t)^2} \operatorname{sinc}(t)\, dt$$

(c)
$$\int_{-\infty}^{\infty} \left[\operatorname{sinc}(t) + e^{-|t|}\operatorname{sgn}(t)\right]^2 dt$$

(d)
$$\int_{-\infty}^{\infty} 8i\, \frac{\sin^2 \pi f/2}{\pi f} e^{i2\pi ft}\, df,$$

for real t.

(e)
$$\int_{-\frac{1}{2}}^{\frac{1}{2}} \frac{\frac{1}{a}}{1 + \frac{e^{-i2\pi f}}{a}} e^{i2\pi 4f}\, df,$$

where $|a| > 1$.

6.47. Does the signal $\operatorname{sinc}^3(t)$ satisfy the conditions for the central limit theorem?

6.48. For the function $g(t) = \wedge(t)\cos(\pi t)$; $t \in \mathcal{R}$, find

(a) The Fourier transform of g.
(b) The equivalent width of g.
(c) The autocorrelation width of g.

6.49. For the function $g(t) = \operatorname{sinc}^2(t)\cos(\pi t)$; $t \in \mathcal{R}$, find

(a) The Fourier transform of g.
(b) The equivalent width of g.
(c) The autocorrelation width of g.

6.50. Find the equivalent width and the autocorrelation width of $g(t) = \sqcap(t) + \delta(t - 2)$.

6.51. Prove (4.61)

6.52. For the function $g(t) = J_0(2\pi t)$,

(a) Find the equivalent width and the autocorrelation width.
(b) Does $g(t)$ obey the central limit theorem? If yes, what is an approximate expression for $g(t)^{*n}$ for large n? If no, why not?

6.53. Find the Fourier transform of the $g(t) = J_1(2\pi t)/(2t)$.

   (a) Find the equivalent width and the autocorrelation width.

   (b) Does $g(t)$ obey the central limit theorem? If yes, what is an approximate expression for $g(t)^{*n}$ for large n? If no, why not?

6.54. Given the signal $g(t) = 2e^{-2|t|}$; $t \in \mathcal{R}$,

   (a) Find its area.

   (b) Find its second moment.

   (c) Find its equivalent width.

   (d) Find its autocorrelation width.

   (e) Does the central limit theorem apply to this signal? If so, find an approximation for $g(t)^{*20}$, the convolution of $g(t)$ with itself 20 times.

6.55. Given a band-limited signal $g(t)$ with spectrum $G(f) = G(f) \sqcap (\frac{f}{2W})$, define the semi-inverse signal $\hat{g}(t)$ as the inverse Fourier transform of $(1/G(f)) \sqcap (\frac{f}{2W})$. Evaluate σ_g^2, $\sigma_{\hat{g}}^2$, and $\sigma_{g*\hat{g}}^2$ in terms of $G(f)$ and its derivatives.

Chapter 7

Two Dimensional Fourier Analysis

The basic definitions for two-dimensional (2D) Fourier transforms were introduced in Chapter 2 as Fourier transforms of signals with two arguments.

As in the one-dimensional case, a variety of two-dimensional signal types are possible. In this chapter we focus on a particular signal type, and demonstrate the natural extensions of many of the one-dimensional results to two-dimensional Fourier transforms. We here consider briefly several extensions of previously described results to the two-dimensional case. More extensive treatments of two-dimensional Fourier analysis to images may be found in Goodman [18] and Bracewell [8].

For this chapter we consider two dimensional signals having continuous "time" and infinite duration. The word "time" is in quotes because in most two-dimensional applications the parameters correspond to space rather than time. Thus a signal will have the form

$$g = \{g(x,y); \ x \in \mathcal{R}, y \in \mathcal{R}\}.$$

As we only consider this case in this chapter, we can safely abbreviate the full notation to just $g(x,y)$ when appropriate. Repeating the definition for the two dimensional Fourier transform using this notation, we have

$$G(f_X, f_Y) = \int_{-\infty}^{\infty} \int_{-\infty}^{\infty} g(x,y) e^{-i2\pi(f_X x + f_Y y)} \, dx \, dy \qquad (7.1)$$

7.1   Properties of 2-D Fourier Transforms

In general the properties of 2-D Fourier transforms are similar to corresponding properties of the 1-D transform. In this section, several of these properties are collected. They are stated without proof as the proofs are straightforward extensions of the corresponding 1-D results.

1. *Linearity*

   If $g_k(x, y) \supset G_k(f_X, f_Y)$ for $k = 1 \ldots K$, then

   $$\sum_{k=1}^{K} \alpha_k g_k(x, y) \supset \sum_{k=1}^{K} \alpha_k G_k(f_X, f_Y). \tag{7.2}$$

   Thus, just as in the 1-D case, if we can decompose a complicated function into a sum of simpler functions whose Fourier transforms we know, we can easily find the Fourier transform of the more complicated function by adding known transforms of the simpler functions.

2. *Symmetry*

   If $g(-x, -y) = g^*(x, y)$, then $G(f_X, f_Y)$ is entirely real-valued. If $g(-x, -y) = -g^*(x, y)$, then $G(f_X, f_Y)$ is entirely imaginary.

3. *Inversion (Fourier Integral Theorem)*

   If $g(x, y) \supset G(f_X, f_Y)$, then

   $$g(x, y) = \int_{-\infty}^{\infty} \int_{-\infty}^{\infty} G(f_X, f_Y) e^{i2\pi(f_X x + f_Y y)} \, df_X \, df_Y \tag{7.3}$$

   at points of continuity of $g(x, y)$ if

   (a)
   $$\int_{-\infty}^{\infty} \int_{-\infty}^{\infty} |g(x, y)| \, dx \, dy < \infty$$

   and

   (b) $g(x, y)$ is smooth, e.g., it is piecewise continuous or it has no infinite discontinuities and only a finite number of maxima and minima in any finite area.

At a point of discontinuity, we recover the *angular average* of $g(x, y)$ at that point as indicated in Fig. 7.1. This is best expressed in polar coordinates, where the angular average can be written $\lim_{r \to 0} \int_0^{2\pi} g(r, \theta) \, d\theta$ and r represents radius measured from the point of discontinuity.

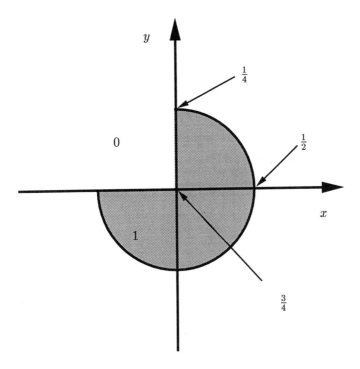

Figure 7.1: Values recovered at two dimensional discontinuities. The function is one in the shaded area and zero outside that area.

4. *Stretch (Similarity)*

   If $g(x,y) \supset G(f_X, f_Y)$, then

   $$g(ax, by) \supset \frac{1}{|ab|} G(\frac{f_X}{a}, \frac{f_Y}{b}). \tag{7.4}$$

   In the 2-D case, a stretch or a contraction of a function can be considered to be a *magnification* or a *demagnification* of that function.

5. *Shift*

   If $g(x,y) \supset G(f_X, f_Y)$, then

   $$g(x - a, y - b) \supset G(f_X, f_Y)e^{-i2\pi(f_X a + f_Y b)}. \tag{7.5}$$

   Thus a shift in the (x, y) domain results in the introduction of a linear phase function in the (f_X, f_Y) domain.

6. *Parseval's Theorem*

$$\int_{-\infty}^{\infty}\int_{-\infty}^{\infty} g_1(x,y)g_2^*(x,y)\,dx\,dy =$$
$$\int_{-\infty}^{\infty}\int_{-\infty}^{\infty} G_1(f_X,f_Y)G_2^*(f_X,f_Y)\,df_X\,df_Y. \qquad (7.6)$$

When the functions g_1 and g_2 are both equal to the same function g, Parseval's theorem reduces to a statement that the energy of g can be calculated as the volume lying under either $|g|^2$ or $|G|^2$.

7. *Convolution*

$$\mathcal{F}_{f_X,f_Y}\left\{\int_{-\infty}^{\infty}\int_{-\infty}^{\infty} g(\zeta,\eta)h(x-\zeta,y-\eta)\,d\zeta\,d\eta\right\} = G(f_X,f_Y)H(f_X,f_Y).$$
$$(7.7)$$

Two-dimensional convolution serves the same basic purpose as one-dimensional convolution: it represents a general linear invariant filtering of the input signal to produce an output signal. The operation is depicted in Fig. 7.2.

7.2   Two Dimensional Linear Systems

Consider a 2-D system that maps an input $\{v(x,y); x,y \in \mathcal{R}\}$ into an output $\{w(x,y); x,y \in \mathcal{R}\}$. If the system is linear, we can characterize it by an impulse response $h(x,y;\zeta,\eta)$, the response at output coordinates (x,y) to an input δ-function at coordinates (ζ,η).

As an aside, the unit impulse in two dimensions is represented by the symbol $\delta(x,y)$. By analogy with the 1-D case, its critical property is the sifting property,

$$\int_{-\infty}^{\infty}\int_{-\infty}^{\infty} \delta(x-\zeta,y-\eta)p(\zeta,\eta)\,d\zeta\,d\eta = p(x,y)$$

valid at all points of continuity of $p(x,y)$. In some cases it may be convenient to consider the 2-D unit impulse to be separable in the (x,y) coordinate system, leading to the representation

$$\delta(x,y) = \delta(x)\delta(y).$$

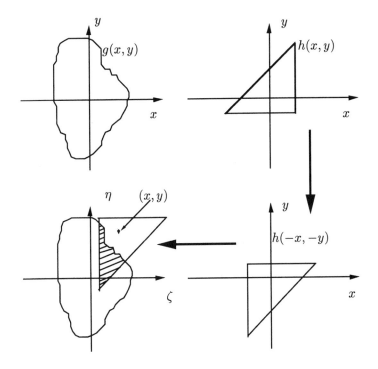

Figure 7.2: 2-D convolution

However, other forms of the 2-D unit impulse are also possible. For example, a circularly symmetric form could be defined through a limiting sequence of circ functions, i.e.

$$\delta(x,y) = \lim_{n\to\infty} \frac{n}{\pi} \text{circ}(\sqrt{(nx)^2 + (ny)^2})$$

where the factor $1/\pi$ assures unit volume. Of course the above equation should not be taken literally, for the limit does not exist at $x = y = 0$, but equality of the left and right will hold if the limit is applied to integrals of the right hand side.

In two dimensions, the unit impulse response of a linear system is often called the *point-spread function*. By direct analogy with the 1-D case:

$$w(x,y) = \int_{-\infty}^{\infty} \int_{-\infty}^{\infty} h(x,y;\zeta,\eta)v(\zeta,\eta)\, d\zeta\, d\eta, \qquad (7.8)$$

a 2-D superposition integral. If the impulse response simply shifts with movements of the impulse and does not change its shape, that is, if it is *space invariant,* then

$$h(x, y; \zeta, \eta) = h(x - \zeta, y - \eta) \tag{7.9}$$

and

$$w(x, y) = \int_{-\infty}^{\infty} \int_{-\infty}^{\infty} h(x - \zeta, y - \eta) v(\zeta, \eta) \, d\zeta \, d\eta, \tag{7.10}$$

a 2-D convolution integral. A space invariant 2-D filter is a shift invariant system where the shift is in space instead of in time.

As in the 1-D case, convolution is greatly simplified if we represent it as its frequency domain equivalent. By the convolution theorem,

$$W(f_X, f_Y) = H(f_X, f_Y) V(f_X, f_Y), \tag{7.11}$$

where

$$H(f_X, f_Y) = \mathcal{F}_{f_X, f_Y}(h)$$

is the 2-D transfer function.

Example

As an example of a linear 2-D system, consider the imaging system of Fig. 7.3. The system maps an object into an out-of-focus image (the distances z of the object plane and the image plane from the imaging lens do not satisfy the lens law). As a consequence, the image of the object is blurred. Our goal is to represent this blur operation as a linear, invariant system.

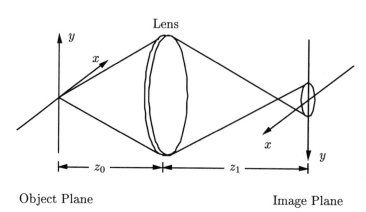

Figure 7.3: Defocused optical system

According to the laws of geometrical optics, for a point source object and a perfect imaging lens, the optical rays diverging from the object will be focused by the lens to a perfect geometrical focus in the correct image plane, i.e. in the plane where the lens law

$$\frac{1}{z_o} + \frac{1}{z_i} = \frac{1}{f}$$

is satisfied, where z_o and z_i are the distances of the object and the image planes from the lens, respectively, and f is the focal length of the lens.

In the case of current interest, the image plane is slightly too far from the lens, with the result that the rays pass through perfect focus in front of the image plane, and diverge from that focus to form an illuminated region that has the same shape as the aperture of the lens used for imaging. For a circular lens, the system has the impulse response pictured in Fig. 7.4. The impulse response is uniformly bright (or of uniform intensity) over a

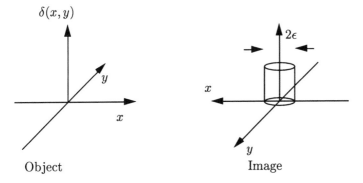

Figure 7.4: System impulse response

circle of radius ϵ, where the value of ϵ depends on geometrical factors and, in particular, the degree of defocus (a larger defocus implies a larger ϵ). We represent the imaging system acting on the point source to produce a circular defocused image through the equation

$$\mathcal{L}\{\delta(x,y)\} = k\text{circ}(\frac{r}{\epsilon}) = h(r).$$

Image inversion can be shown to destroy space invariance according to the strict definition (you prove it). The same is true for image magnification or demagnification. However, the system can be made space invariant by redefining the image coordinate axes, inverting their directions and normalizing their scale to assure a magnification of unity. In this current case, the object and image planes are assumed to be equal distances from the lens, so the magnification of the system happens to be unity. (The magnification can be shown to be the ratio of the distance of the image plane from the lens to the distance of the object plane from the lens.) We assume that proper normalizations of the object coordinate system have been carried out to make the system space invariant.

To find the transfer function $H(\rho) = \mathcal{F}_\rho\{h(r)\}$ recall that

$$\text{circ}(r) \supset \frac{J_1(2\pi\rho)}{\rho}. \tag{7.12}$$

The scaling theorem for the zero-order Hankel transform can be stated

$$g(ar) \supset \frac{1}{a^2}G(\frac{\rho}{a}), \tag{7.13}$$

and therefore

$$\text{circ}(\frac{r}{\epsilon}) \supset \epsilon^2 \frac{J_1(2\pi\epsilon\rho)}{\epsilon\rho}. \tag{7.14}$$

Thus

$$H(\rho) = k'\left(2\frac{J_1(2\pi\epsilon\rho)}{2\pi\epsilon\rho}\right), \tag{7.15}$$

where k' is a constant.

It is convenient to normalize the transfer function to have value unity at the origin:

$$\mathcal{H}(\rho) = \frac{H(\rho)}{H(0)} = 2\frac{J_1(2\pi\epsilon\rho)}{2\pi\epsilon\rho}. \tag{7.16}$$

Note the following properties of this transfer function:

1. The first zero occurs at $\rho_0 = 0.610/\epsilon$. The larger ϵ (i.e. the larger defocus), the smaller ρ_0 and hence the smaller the bandwidth of the system.

2. A negative transfer function at a particular frequency can be thought of as a 180° phase reversal of sines or cosines at that frequency, or equivalently a contrast reversal of a spatial frequency component.

3. If $\mathcal{H}(\rho) = |\mathcal{H}(\rho)|e^{i\phi(\rho)}$, then an input of the form $a + b\cos(2\pi(f_X x + f_Y y))$ yields an output of $a+b|\mathcal{H}(\rho)|\cos(2\pi(f_X x + f_Y y)+\phi(\rho))$, where $\phi(\rho)$ is the phase of the transfer function at radial frequency ρ. The physical significance of this result is two-fold. First the contrast of the cosinusoidal component of image intensity (defined as the difference between the maximum and minimum intensities normalized by the sum of the maximum and minimum intensities) is reduced. Equivalently, the contrast can be expressed as the ratio of the peak intensity

to the average intensity. Thus

$$C = \frac{\text{peak intensity}}{\text{average intensity}} = \frac{b}{a} \to \frac{b}{a}|\mathcal{H}(\rho)|.$$

Thus the magnitude of the normalized transfer function represents the amount by which the contrast of the sinusoidal component of intensity is reduced. Note that at any zero of the magnitude of the transfer function, the sinusoidal part of the intensity distribution completely disappears, leaving only the constant background or average intensity.

Second, the phase of the cosinusoidal component of image intensity is changed in proportion to the phase of the transfer function, which means that the cosinusoidal intensity is shifted by some fraction $\frac{\phi(\rho)}{2\pi}$ of its two-dimensional period.

7.3 Reconstruction from Projections

The field of radiology has undergone dramatic changes since the invention and commercialization of computerized tomography as a measurement and diagnostic technique. Using such techniques it is now possible to obtain images of portions of the body that were not accessible by previously used techniques, and to differentiate different types and constituents of tissue with sensitivity and discrimination not previously possible. In this section the principles of tomographic measurement and image reconstruction will be discussed, not only because the subject is intrinsically important, but also because the problem provides instructive connections between 1-D and 2-D Fourier analysis.

Figure 7.5 depicts a simplified model of data collection techniques used in computerized tomography. For the purpose of detailed analysis of the technique, consider an (x, y) coordinate system attached to the patient to be scanned. The (x', y') coordinate system is attached to the scanning geometry and rotates with the rotation of the measurement apparatus. For future use observe that the relations between the two coordinate systems are

$$\begin{align}
x' &= x\cos\theta + y\sin\theta \tag{7.17}\\
y' &= -x\sin\theta + y\cos\theta \tag{7.18}\\
x &= x'\cos\theta - y'\sin\theta \tag{7.19}\\
y &= x'\sin\theta + y'\cos\theta \tag{7.20}
\end{align}$$

The data is gathered with a combination of linear scanning and rotation. To begin, the angle θ is fixed, and the X-ray source and detector are scanned

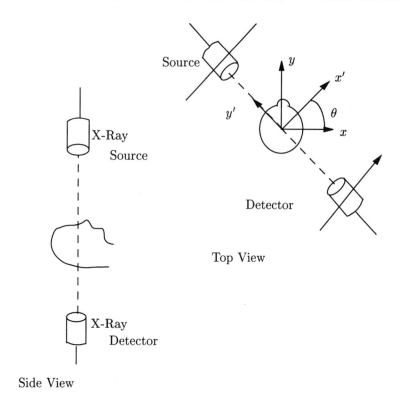

Figure 7.5: Computerized tomography data collection

linearly in the chosen direction in synchronism. The result of such a scan is a measurement of the total attenuation observed by the X-ray source-detector pair at each position along the linear scan. After each such scan, the angle θ is incremented and the scanning operation is repeated, until the angle has been incremented through π radians or 180 degrees.

Let I_0 represent the constant X-ray intensity emitted by the source. Let $I(x')$ represent the X-ray intensity falling on the detector when it is at location x' along the scan. Let $g(x, y)$ be the X-ray attenuation coefficient of the head or body in a transverse slice (constant z). Pathology will be indicated by unusual structure of $g(x, y)$.

The measured X-ray intensity at scanner coordinates x' and θ will be

$$I_\theta(x') = I_0 e^{-\int_{-\infty}^{\infty} g(x' \cos\theta - y' \sin\theta, x' \sin\theta + y' \cos\theta)\, dy'}. \qquad (7.21)$$

The integral is required because absorption occurs along the entire y' path of the X-ray beam.

After detection, the ratio

$$- \ln \frac{I_\theta(x')}{I_0} \qquad (7.22)$$

is formed, yielding the measured *projection* of the X-ray attenuation coefficient

$$p_\theta(x') = \int_{-\infty}^{\infty} g(x' \cos\theta - y' \sin\theta, x' \sin\theta + y' \cos\theta) \, dy'. \qquad (7.23)$$

To illustrate the concept of a projection, consider the projection of the function circ(r) of Fig. 7.6 onto the x'-axis. Since circ(r) is circularly symmetric, projection onto the x'-axis is independent of θ. The shaded plane

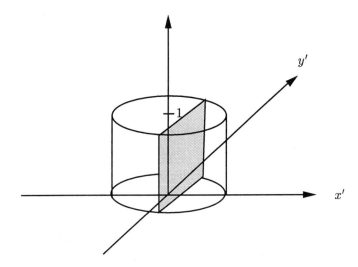

Figure 7.6: Projection of circ(r)

cutting through the cylinder has area $p_\theta(x')$ for the specific x'. To find this area we need the height of the rectangle, which is unity, and its width. To find the width refer to Fig. 7.7. The result is:

$$p_\theta(x') = \text{height} \times \text{width} = 1 \times (2\sqrt{1 - (x')^2}).$$

While the tomographic apparatus measures a series of projections of the X-ray attenuation coefficient onto a series of rotated axes, it is not those projections that are of ultimate interest, but rather the two-dimensional

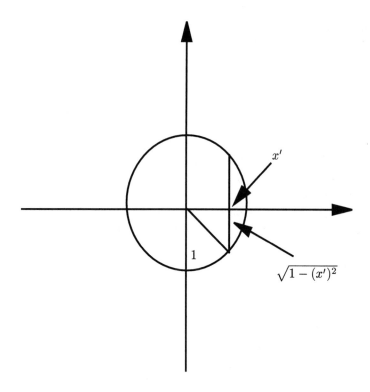

Figure 7.7: Width of rectangle

distribution of X-ray absorption that gave rise to those projections. Thus the mathematical challenge is to find a way of calculating the "image" $g(x, y)$ of the attenuation coefficient from the projection data that has been collected. This problem is known as the *inversion problem*, and a solution is of course critical to utility of the tomographic method.

7.4 The Inversion Problem

Given a set of measured projections $p_\theta(x')$ (for all $\theta \in [0, \pi)$) through $g(x, y)$, how do we recover $g(x, y)$? Alternatively, how do we reconstruct the two dimensional signal $g(x, y)$ from its projections? The answer is given by the *Projection-Slice theorem* (also known as the Central Slice theorem).

Theorem 7.1 *The 1-D Fourier transform $P_\theta(f)$ of a projection $p_\theta(x')$ is identical to the 2-D Fourier transform $G(f_X, f_Y)$ of $g(x,y)$ evaluated along a slice through the origin at angle $+\theta$ to the f_X axis:*

$$P_\theta(f) = G(s\cos\theta, s\sin\theta). \qquad (7.24)$$

Proof:

$$
\begin{aligned}
P_\theta(f) &= \int_{-\infty}^{\infty} p_\theta(x')e^{-i2\pi x'f}\,dx' \\
&= \int_{-\infty}^{\infty}\left(\int_{-\infty}^{\infty} g(x'\cos\theta - y'\sin\theta, x'\sin\theta + y'\cos\theta)\,dy'\right)e^{-i2\pi f x'}\,dx'.
\end{aligned}
$$

This integral is to be carried out over the *entire* (x', y') plane. Equivalently, we can integrate over the entire (x, y) plane

$$P_\theta(f) = \int_{-\infty}^{\infty}\int_{-\infty}^{\infty} g(x,y)e^{-i2\pi(x\cos\theta + y\sin\theta)s}\,dx\,dy.$$

Thus $P_\theta(f) = G(s\cos\theta, s\sin\theta)$. $\square$

Figure 7.8 illustrates the projection-slice theorem. The projection through an attenuation function onto the axis at angle θ to the x axis is shown, as is also the slice in the frequency domain at angle θ to the f_X axis.

The projection slice theorem also works in reverse. A slice of $g(x,y)$ at angle θ to the x axis in the space domain has a 1-D transform that is the projection of $G(f_X, f_Y)$ onto an axis at angle θ to the f_X axis in the frequency domain.

In conclusion, the above analysis shows that if we can determine the projections $p_\theta(x')$ for all θ and x', then we can determine the spectrum $G(f_X, f_Y)$ for all (f_X, f_Y). With an inverse 2-D transform we can then recover the image $g(x,y)$ of interest.

7.5   Examples of the Projection-Slice Theorem

Example 1. As a first example, consider a circularly symmetric function $g(x,y) = g_R(r)$ with known radial profile $g_R(r)$. We wish to determine $g_R(r)$ from a single projection $p_\theta(x')$ known to be

$$p_\theta(x') = 2\operatorname{sinc}(2x').$$

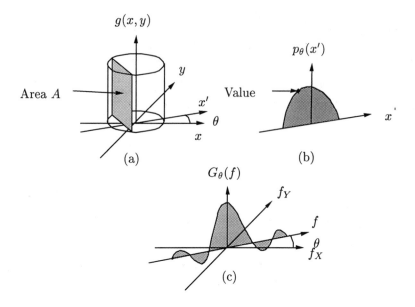

Figure 7.8: Illustration of the projection-slice theorem. (a) Calculation of the projection at one point for the attenuation function $g(x,y)$; (b) The complete projection through $g(x,y)$ at angle θ; (c) the Fourier transform of the proection, which is a central slice through the 2-D transform of $g(x,y)$.

Find the radial profile $g_R(r)$.

    Solution: We outline the various steps in the solution as follows:

1. $P_\theta(f) = \mathcal{F}_f\{2\ \text{sinc}(2x')\} = \sqcap(f/2)$.

2. From the projection slice theorem, $P_\theta(f)$ is a slice through a circularly symmetric 2-D spectrum. Clearly, then,

$$G(\rho) = \text{circ}\rho.$$

3. The inverse zero-order Hankel transform can then be taken to yield the original function,

$$g_R(r) = \frac{J_1(2\pi r)}{r}.$$

    Note that a single projection at any angle suffices for recovery of any circularly symmetric function.

Example 2. Show that any function $g(x,y) = g_X(x)g_Y(y)$ separable in rectangular coordinates can be reconstructed from *two* projections.

Solution:

1. Any function that is separable in the (x, y) domain has a Fourier transform that is separable in the (f_X, f_Y) domain. Therefore

$$g(x, y) = g_X(x)g_Y(y) \supset G_X(f_X)G_Y(f_Y).$$

2. First find the projection through $g(x, y)$ for $\theta = 0$:

$$p_0(x) = \int_{-\infty}^{\infty} g(x, y) \, dy = g_X(x) \int_{-\infty}^{\infty} g_Y(y) \, dy.$$

In addition, Fourier transform the resulting projection using frequency variable f_X

$$p_0(x) \supset G_X(f_X)G_Y(0).$$

Thus the 1-D transform of this first projection yields information regarding the f_X dependence of the spectrum, up to an as yet unknown multiplier $G_Y(0)$.

3. Next find the projection through $g(x, y)$ at angle $\theta = \pi/2$. We obtain

$$p_{\frac{\pi}{2}}(x') = \int_{-\infty}^{\infty} g(-y', x') dy' = \int_{-\infty}^{\infty} g_X(-y') \, dy' g_Y(x').$$

A 1-D transform of this projection using frequency variable f_Y yields

$$p_{\frac{\pi}{2}}(x') \supset G_X(0)G_Y(f_Y).$$

Thus the 1-D transform of this projection with respect to variable x' yields information about the f_Y dependence of the spectrum, up to an unknown multiplier $G_X(0)$.

4. Now the undefined multipliers must be found. Note that the value of the 1-D spectrum of either projection at zero frequency yields the product of these two constants,

$$P_0(0) = P_{\frac{\pi}{2}}(0) = G_X(0)G_Y(0).$$

5. Divide out the unknown multipliers by taking the ratio

$$\frac{P_0(f_X)P_{\frac{\pi}{2}}(f_Y)}{P_0(0)} = \frac{G_X(f_X)G_Y(0)G_X(0)G_Y(f_Y)}{G_X(0)G_Y(0)} = G_X(f_X)G_Y(f_Y).$$

6. Finally, inverse transform to obtain

$$G_X(f_X)G_Y(f_Y) \subset g_X(x)g_Y(y) = g(x, y).$$

Example 3. Find an expression for any projection through the function

$$g_R(r) = \pi a J_0(2\pi a r).$$

Solution:
It can be shown (see, e.g., Bracewell, p. 249)

$$\pi a J_0(2\pi a r) \supset \frac{1}{2}\delta(\rho - a) = G(\rho).$$

Therefore

$$P_\theta(f) = G(f) = \frac{1}{2}\delta(s - a) + \frac{1}{2}\delta(s + a)$$

for any θ and hence

$$p_\theta(x') = \cos 2\pi a x'.$$

7.6   Reconstruction by Convolution and Back-projection

We can summarize the straightforward approach to reconstructing a signal based on the projection-slice theorem as follows:

1. Collect projections at a discrete set of closely spaced angles.

2. Fourier transform the projections, obtaining samples of $G(f_X, f_Y)$ on the grid shown in Fig. 7.9.

3. Interpolate these samples to a rectangular grid.

4. Apply the inverse 2-D Fourier transform.

The straightforward approach has two serious problems:

1. Interpolation is computationally expensive.

2. We cannot begin the final inverse FFT until all data has been gathered and transformed.

This leads to an alternative approach, developed by Bracewell in the 1950s, called *convolution and backprojection*.   The goal is to reconstruct $g(x, y)$ from the projections $p_\theta(x')$ taken at all θ between 0 and π. We will neglect the fact that θ is sampled discretely.
The starting point is the definition of the inverse 2-D Fourier transform,

$$g(x, y) = \int_{-\infty}^{\infty} \int_{-\infty}^{\infty} G(f_X, f_Y) e^{i2\pi(f_X x + f_Y y)} \, df_X \, df_Y.$$

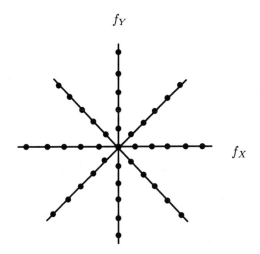

Figure 7.9: Sampling grid for $G(f_X, f_Y)$ in Reconstruction

Convert from (f_X, f_Y) coordinates to (s, θ), where s scans $\mathcal{R}$ and θ runs from 0 to π:

$$
\begin{aligned}
f_X &= s \cos \theta \\
f_Y &= s \sin \theta \\
df_X \, df_Y &= |s| \, ds \, d\theta.
\end{aligned}
$$

Then by straightforward substitution,

$$
g(x, y) = \int_0^{2\pi} d\theta \int_{-\infty}^{\infty} ds |s| G(s \cos \theta, s \sin \theta) e^{i 2\pi s(x \cos \theta + y \sin \theta)};
$$

hence

$$
g(x, y) = \int_0^{2\pi} d\theta \int_{-\infty}^{\infty} ds |s| P_\theta(s) e^{i 2\pi s(x \cos \theta + y \sin \theta)}.
$$

This result can be rewritten in the following form:

$$
g(x, y) = \int_0^{2\pi} d\theta \, f_\theta(x \cos \theta + y \sin \theta) \tag{7.25}
$$

where

$$
f_\theta(x') = \mathcal{F}_{x'}^{-1}\{|s| P_\theta(s)\}. \tag{7.26}
$$

The function $f_\theta(x')$ is called the *filtered backprojection* of the projection $p_\theta(x')$. To explain the name and interpret the result, suppose that we obtain a projection $p_\theta(x')$ at angle θ; then the backprojection of $p_\theta(x')$ is the 2-D function $p_\theta(x\cos\theta + y\sin\theta)$. The backprojection can be obtained by taking the 1-D function $p_\theta(x')$ and "spreading" it uniformly across the 2-D plane, with the spreading operation being titled at angle θ with respect to the x' axis, as shown in Figure 7.10 for the specific case of a backprojection angle that is 22.5 degrees counterclockwise from the y axis..

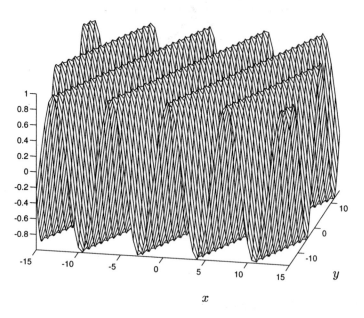

Figure 7.10: Illustration of a backprojection. The function being backprojected is a 2D cosine oriented at angle 22.5^o to the x axis.

We can not reconstruct $g(x,y)$ simply by adding up backprojections; that is,

$$g(x,y) \neq \int_0^{2\pi} d\theta\, p_\theta(x\cos\theta + y\sin\theta).$$

However we *can* reconstruct $g(x,y)$ if we filter each projection first, before backprojecting. The filter required is one dimensional and has a transfer function $|s|$. This transfer function compensates for the fact that simple (unfiltered) backprojection gives too heavy a weighting near the origin in the (f_X, f_Y) plane.

Because the 2-D object spectrum is approximately bandlimited, so too are the Fourier transforms of the projections. Therefore the transfer function $H(f) = |s|$ need not rise indefinitely. A transfer function

$$\tilde{H}(f) = |s| \sqcap \left(\frac{s}{2f_c} \right)$$

will do, where f_c is chosen just large enough so that all significant Fourier components of the spectrum will be included. Recognizing that the resulting transfer function is the difference between a rectangle function of height and width f_c and a triangle function of height and width f_c, the impulse response is found to be

$$\tilde{h}(x) = 2f_c^2 \operatorname{sinc}(2f_c x) - f_c^2 \operatorname{sinc}^2(f_c x).$$

The procedure used in practice would be

1. Collect a projection at angle θ.

2. While collecting it, convolve with $\tilde{h}(x)$ in real time.

3. As soon as convolution is finished, backproject. Accumulate.

4. Repeat for all $0 \leq \theta \leq \pi$.

A procedure based on backprojection and convolution was used in early computerized tomography X-ray scanners. However, modern X-ray tomography machines typically have more complex data gathering apparatus than described here, often gathering several different measurements simultaneously to increase the scanning speed. More complex reconstruction algorithms have been developed for use with such machines.

7.7 ⋆ Two-Dimensional Sampling Theory

Just as a bandlimited 1-D function of a single independent variable can be reconstructed from samples taken at a uniform rate that exceeds a critical lower bound, so too a 2-D function that is bandlimited in the (f_X, f_Y) plane can be perfectly reconstructed from samples that are uniformly dense provided that density exceeds a certain lower bound. However, the theory of 2-D sampling is richer than the theory of 1-D sampling, due to the extra degree of freedom afforded by the second dimension. Two dimensional sampling theory provides interesting examples of the similarities and differences between 1-D theory and 2-D theory, and therefore is briefly discussed here.

To embark on a discussion of 2-D sampling, we first need a clear definition of a *bandlimited* function in two dimensions. A function $g(x,y)$ is called bandlimited if its 2-D Fourier spectrum $G(f_X, f_Y)$ is identically zero for all radial frequencies ρ greater than some finite limit ρ_c. There is considerable flexibility as to the exact shape of the region on which the spectrum is non-zero, as indicated in Figure 7.11. The only requirement is that it vanish eventually at large enough radius in the frequency domain.

■ = Frequency-domain area occupied by spectrum

◯ = Minimum circular region within which the spectrum is non-zero

Figure 7.11: Various types of bandlimiting in two dimensions.

The geometrical pattern in which the samples are taken (i.e. the sampling grid) is also quite flexible. Most common is the use of a *rectangular* sampling function defined by

$$s(x,y) = \text{III}\left(\frac{x}{X}\right)\text{III}\left(\frac{y}{Y}\right) = XY \sum_{n=-\infty}^{\infty}\sum_{m=-\infty}^{\infty}\delta(x-nX)\,\delta(y-mY) \quad (7.27)$$

and illustrated in Figure 7.12. However, other sampling grids are also possible. For example, the sampling function

$$s(x,y) = \text{III}\left(\frac{x+y}{\sqrt{2}L_1}\right)\text{III}\left(\frac{x-y}{\sqrt{2}L_2}\right) \quad (7.28)$$

can be thought of as resulting from a rotation of the (x,y) axes with respect to the delta functions by 45 degrees, thus preserving the volume under the delta functions but changing their locations. The arguments of the two III functions are simultaneously zero at the new locations

$$x_{n,m} = \frac{nL_1 + mL_2}{\sqrt{2}}, \quad y_{n,m} = \frac{nL_1 - mL_2}{\sqrt{2}} \quad (7.29)$$

where n and m run over the integers. The above locations are the new locations of the delta functions in this modified sampling grid.

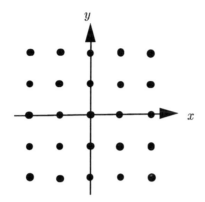

Figure 7.12: Rectangular sampling grid. δ functions are located at each dot.

When a sampling function is multiplied by a signal $g(x, y)$, the product operation in the time domain results in a convolution operation in the frequency domain. Thus the Fourier transform of the sampled function becomes

$$\hat{G}(f_X, f_Y) = S(f_X, f_Y) * G(f_X, f_Y).$$

For the sampling function defined on the original rectangular grid, the spectrum of the sampled data becomes

$$\hat{G}(f_X, f_Y) = \sum_{n=-\infty}^{\infty} \sum_{m=-\infty}^{\infty} \delta(f_X - \frac{n}{X}) \delta(f_Y - \frac{m}{Y}) * G(f_X, f_Y)$$

$$= \sum_{n=-\infty}^{\infty} \sum_{m=-\infty}^{\infty} G\left(f_X - \frac{n}{X}, f_Y - \frac{m}{Y}\right). \qquad (7.30)$$

As in the 1-D case, the effect of the sampling operation is to replicate the signal spectrum an infinite number of times in the frequency domain, as shown in Figure 7.13.

In order for the zero-order ($n = 0, m = 0$) term in the frequency domain to be recoverable, it is necessary that there be no overlap of the various spectral terms. This will be assured when the sampling intervals X and Y are chosen sufficiently small. For the original rectangular sampling grid, if all frequency components of $g(x, y)$ lie within a rectangle of dimensions $2B_X \times 2B_Y$ in the f_X and f_Y directions, respectively, centered on the origin in the frequency domain, then sampling intervals $X = \frac{1}{2B_X}$ and $Y = \frac{1}{2B_Y}$ or

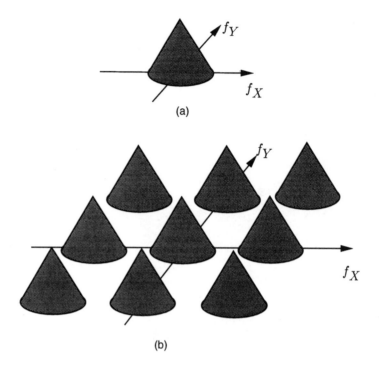

Figure 7.13: Replication in the frequency domain caused by sampling in the space domain. (a) Original spectrum, (b) Spectrum after sampling

smaller will suffice. Under such conditions, the spectral islands will separate and the $n = 0, m = 0$ island will be recoverable by proper filtering.

To recover $g(x, y)$, the sampled data $\hat{g}(x, y)$ is passed through a 2-D linear invariant filter which passes the $(n = 0, m = 0)$ spectral island without change, and completely eliminates all other islands. For this sampling grid, a suitable linear filter is one with transfer function

$$H(f_X, f_Y) = \sqcap \left(\frac{f_X}{2B_X} \right) \sqcap \left(\frac{f_Y}{2B_Y} \right).$$

Thus the identities

$$G(f_X, f_Y) = \sqcap \left(\frac{f_X}{2B_X} \right) \sqcap \left(\frac{f_Y}{2B_Y} \right) \sum_{n=-\infty}^{\infty} \sum_{m=-\infty}^{\infty} G(f_X - nB_X, f_Y - mB_Y)$$

and the space domain equivalent

$$g(x, y) \quad = \quad \hat{g}(x, y) * 4B_X B_Y \, \text{sinc}(2B_X x) \, \text{sinc}(2B_Y y) \tag{7.31}$$

$$= \sum_{n=-\infty}^{\infty} \sum_{m=-\infty}^{\infty} g\left(\frac{n}{2B_X}, \frac{m}{2B_Y}\right) \text{sinc}\left(x - \frac{n}{2B_X}\right) \text{sinc}\left(y - \frac{m}{2B_Y}\right)$$

hold provided $X \leq \frac{1}{2B_X}, Y \leq \frac{1}{2B_Y}$. This equation is the 2-D equivalent of (4.15) which was valid in one dimension. Note that at least $4B_X B_Y$ samples per unit area are required for reconstruction of the original function.

This discussion would not be complete without some mention of the wide variety of other possible sampling theorems that can be derived in the 2-D case. For example, suppose that the reconstruction or interpolation filter were chosen to be circularly symmetric, rather than separable in the (x, y) coordinate system. A filter with transfer function

$$H(f_X, f_Y) = H(\rho) = \text{circ}(\rho/\rho_c)$$

could be applied to the sampled data, with the cutoff frequency ρ_c chosen to equal the diagonal width of the rectangle used above, i.e. $\rho_c = \sqrt{B_X^2 + B_Y^2}$. This filter, like the one considered earlier, will completely isolate the zero-order spectral island. Noting that

$$\frac{J_1(2\pi r)}{r} \supset \text{circ}(\rho),$$

and assuming that the intervals between samples satisfy $X = Y = \frac{1}{2\rho_c}$, it is a simple matter to show that the original function $g(x, y)$ can be recovered with another interpolation function, i.e.

$$g(x, y) = \sum_{n=-\infty}^{\infty} \sum_{m=-\infty}^{\infty} g\left(\frac{n}{\rho_c}, \frac{m}{\rho_c}\right) \frac{\pi}{4}\left[2\frac{J_1\left(2\pi\rho_c\sqrt{(x - \frac{n}{\rho_c})^2 + (y - \frac{m}{\rho_c})^2}\right)}{2\pi\rho_c\sqrt{(x - \frac{n}{\rho_c})^2 + (y - \frac{m}{\rho_c})^2}}\right],$$

$$(7.32)$$

where the factor $\pi/4$ is the ratio of the unit rectangular cell in the 2-D frequency domain replication grid $(4\rho_c^2)$ to the frequency domain area covered by the transfer function of the interpolation filter $(\pi\rho_c^2)$. Thus it is clear that even for a fixed choice of sampling grid there are many possible forms for the sampling theorem.

If the sampling grid is allowed to change, then additional forms of the sampling theorem can be found. For example, an *hexagonal* sampling grid plays an important role in the sampling of functions with circularly band-limited spectra. Such a grid provides the densest possible packing of circular regions in the frequency domain.

We will not pursue the subject of sampling further here. The purpose of the discussion has been to point out the extra richness of the theory that occurs when the dimensionality of the signals is raised from 1 to 2.

7.8   Problems

7.1. Find the following convolutions.

(a) $g(r) = \frac{J_1(2\pi r)}{r} * \frac{1}{2}\frac{J_1(\pi r)}{r}$.

(b) $g(x, y) = \sqcap(x) \sqcap (y) * \sqcap(x) \sqcap (y)$.

7.2. Evaluate the integral

$$\int_{-\infty}^{\infty} \int_{-\infty}^{\infty} \frac{J_1(2\pi\sqrt{x^2 + y^2})}{\sqrt{x^2 + y^2}} e^{-i\pi(x+y)} \, \mathrm{sinc}\, x \, \mathrm{sinc}\, y \, dx \, dy.$$

7.3. Find the 2-D Fourier transforms of

(a) $\delta(r - r_0)$ (r_0 a constant).

(b) $e^{-\pi r^2}$ (be clever).

(c) $\sqcap(r - 1)$.

(d) $\cos(2\pi ax)\cos(2\pi by)$.

(e) $\cos(2\pi(ax + by))$.

7.4. (a) Find an expression for any projection through the 2-D function

$$g(r) = \pi a J_0(2\pi ar).$$

(b) A projection through a 2-D circularly symmetric function is found to be

$$p(x) = e^{-\pi ax^2}.$$

Specify the two dimensional function through which the projection was taken.

7.5. The projection through a certain circularly symmetric function is known to be $p(x') = \cos 2\pi x' \, \mathrm{sinc}\,(x')$. Specify the original circularly symmetric function.

7.6. Find an expression for any projection through the circularly symmetric function $g_R(r) = (1 + r^2)^{-\frac{3}{2}}$.

Chapter 8

Memoryless Nonlinearities

We have seen that Fourier analysis is a powerful tool for describing and analyzing linear systems. A particularly important application has been that of sampling continuous time signals to produce discrete time signals and the quantifying of the conditions under which no information is lost by sampling. The purpose of this chapter is twofold. First, we demonstrate that Fourier analysis can also be a useful tool for analyzing simple nonlinear systems. The techniques used in this chapter are a relatively minor variation of techniques already seen, but this particular application of Fourier theory is often overlooked in the engineering literature. Although a standard component of courses is devoted to nonlinear systems, the relative scarcity of such courses and the lack of examples in engineering transform texts has led to a common belief of near mythological nature that Fourier methods are useful only in linear systems. Using an idea originally due to Rice [28] and popularized by Davenport and Root [15] as the "transform method," we show how the behavior of memoryless nonlinear systems can be studied by applying the Fourier transform to the nonlinearity rather than to the signals themselves.

The second goal of the chapter is to consider the second step in the conversion of continuous signals to digital signals: quantization. A uniform quantizer provides an excellent and important example of a memoryless nonlinearity and it plays a fundamental role at the interface of analog and digital signal processing. Just as sampling "discretizes" time, quantization converts a continuous amplitude signal into a discrete amplitude signal. The combination of sampling and quantization produces a signal that is discrete in both time and amplitude, that is, a digital signal. A benefit of focusing on this example is that we can use the tools of Fourier analysis to consider the accuracy of popular models of and approximations for the behavior of quantization error, the distortion that occurs in a signal when it is mapped from an analog amplitude into a discrete approximation.

8.1   Memoryless Nonlinearities

A memoryless system is one which maps an input signal $v = \{v(t); t \in \mathcal{T}\}$ into an output signal $w = \{w(t); t \in \mathcal{T}\}$ via a mapping of the form

$$w(t) = \alpha_t(v(t)); t \in \mathcal{T}$$

so that the output at time t depends only on the current input and not on any past or future inputs (or outputs). We will emphasize real-valued nonlinearities, i.e., $\alpha_t(v) \in \mathcal{R}$ for all $v \in \mathcal{R}$. When α_t does not depend on t, the memoryless nonlinearity is said to be *time invariant* and we drop the subscript.

Let $\mathcal{A} \subset \mathcal{R}$ denote the range of possible values for the input signal $v(t)$, that is, $v(t) \in \mathcal{A}$ for all $t \in \mathcal{R}$. Depending on the system, $\mathcal{A}$ could be the entire real line $\mathcal{R}$ or only some finite length interval of the real line, e.g., $[-V, V]$. We assume that $\mathcal{A}$ is a continuous set since our principal interest will be an α_t that quantizes the input, i.e., that maps a continuous input into a discrete output.

The function α_t maps $\mathcal{A}$ into the real line; that is, α can be thought of as being a signal $\alpha_t = \{\alpha(v); v \in \mathcal{A}\}$. This simple observation is the fundamental idea needed. Since α_t is a signal, it has a Fourier transform

$$A_t(f) = \int_{\mathcal{A}} \alpha_t(x) e^{-i2\pi x f} \, dx.$$

Keep in mind that here t can be considered as a fixed parameter. If the nonlinearity is time-invariant, the t disappears.

Assuming that the usual technical conditions are met, the Fourier transform can be inverted to recover α, at least at its points of continuity. First suppose that $\mathcal{A} = \mathcal{R}$, the case where the range has "infinite duration" (infinite length is better terminology here). In this case we have that

$$\alpha_t(x) = \int_{-\infty}^{\infty} A_t(f) e^{i2\pi x f} \, df; \ x \in \mathcal{R}. \tag{8.1}$$

If, on the other hand, $\mathcal{A}$ has finite length, say L (we don't use T since the parameter no longer corresponds to time), then the inversion is a Fourier series instead of a Fourier integral:

$$\alpha_t(x) = \sum_{n=-\infty}^{\infty} \frac{A_t(\frac{n}{L})}{L} e^{i2\pi \frac{n}{L} x}; \ x \in \mathcal{A}. \tag{8.2}$$

This representation is also useful if $A = R$, but the nonlinearity α_t is a periodic function of its argument with period L.

Thus we can represent the memoryless nonlinearity as a weighted linear combination of exponentials. Why is this of interest? Observe that we can now write the output of a memoryless nonlinearity as follows: Since $w(t) = \alpha_t(v(t))$,

$$w(t) = \begin{cases} \int_{-\infty}^{\infty} A_t(f) e^{i2\pi v(t)f} \, df & \text{if } A = R \\ \sum_{n=-\infty}^{\infty} \frac{A_t(\frac{n}{L})}{L} e^{i2\pi \frac{n}{L} v(t)} & \text{if } A \text{ has finite width } L \text{ or } \alpha_t(u) \\ & \text{is periodic in } u \text{ with period } L. \end{cases}$$

(8.3)

Thus we have what resembles a Fourier integral or Fourier series representation of the output in terms of the input, even though the system is nonlinear! It is, of course, not an ordinary Fourier integral or series because of the appearance of the input signal in the exponents. In fact, one can view the complex exponential terms containing the input as a form of *phase modulation* or PM of the input. Nonetheless, this expansion of the output can be useful in analyzing the behavior of the system, as shall be seen. This method of analyzing is sometimes called the *transform method* because of its reliance on the Fourier transform.

This is about as far as we can go in general without narrowing the development down to a particular memoryless nonlinearity or input signal. First, however, we observe a basic fact about combining sampling and memoryless operations. Suppose that our system first takes a continuous time signal $u(t)$, samples it to form a discrete time signal $u_n = u(nT_s)$, and then passes the sampled signal through a discrete time memoryless time-invariant nonlinearity α to form the final signal $w_n = \alpha(u(nT_s))$. Now suppose that instead the order of the two operations is reversed and we first pass the continuous time signal $u(t)$ through a continuous time nonlinearity α (the same functional form as before) to form $\alpha(u(t))$ and then we sample to form the final output $w_n = \alpha(u(nT_s))$. Regardless of the order, the outputs are the same. For this reason we often have our choice of considering continuous or discrete time when analyzing nonlinearities.

8.2 Sinusoidal Inputs

The transform method does not provide simple general results for Fourier representations of output signals of memoryless nonlinear systems in terms of the Fourier representation of the input. Such generality and simplicity is usually only possible for linear systems. A general result can be obtained for memoryless nonlinearities operating on sinusoidal inputs. This result is not as important as the corresponding result for linear systems because sinusoids are not the fundamental building blocks for nonlinear systems that they are for linear systems. Nonetheless it is an important class of "test signals" that are commonly used to describe the behavior of nonlinear systems.

Suppose now that $u(t) = a \sin(2\pi f_0 t)$ for $t \in \mathcal{T}$ and suppose that the Fourier series representation of the memoryless nonlinearity is given. Then

$$w(t) = \sum_{n=-\infty}^{\infty} \frac{A_t(\frac{n}{L})}{L} e^{i2\pi \frac{n}{L} a \sin(2\pi f_0 t)}. \tag{8.4}$$

The exponential term is itself now periodic in t and can be further expanded in a Fourier series, which is exactly the Jacobi-Anger formula of (3.91):

$$e^{i2\pi \frac{n}{L} a \sin(2\pi f_0 t)} = \sum_{k=-\infty}^{\infty} J_k(2\pi \frac{n}{L} a) e^{i2\pi f_0 t k}, \tag{8.5}$$

where J_k is the kth order Bessel function of (1.12). Incorporating (8.5) into (8.4) yields

$$
\begin{aligned}
w(t) &= \sum_{n=-\infty}^{\infty} \frac{A_t(\frac{n}{L})}{L} \sum_{k=-\infty}^{\infty} J_k(2\pi \frac{n}{L} a) e^{i2\pi f_0 t k} \\
&= \sum_{k=-\infty}^{\infty} e^{i2\pi f_0 t k} \sum_{n=-\infty}^{\infty} \frac{A_t(\frac{n}{L})}{L} J_k(2\pi \frac{n}{L} a).
\end{aligned}
\tag{8.6}
$$

If the nonlinearity is time-invariant and hence A_t does not depend on t, then this result has the general form of a Fourier series

$$w(t) = \sum_{k=-\infty}^{\infty} b_k e^{i2\pi f_0 t k} \tag{8.7}$$

with coefficients

$$b_k = \sum_{n=-\infty}^{\infty} \frac{A(\frac{n}{L})}{L} J_k(2\pi \frac{n}{L} a). \tag{8.8}$$

An interesting technicality arises in (8.7). If $\mathcal{T} = \mathcal{R}$, then (8.7) is indeed a Fourier series for a continuous time periodic function having period $1/f_0$. In the discrete time case of $\mathcal{T} = \mathcal{Z}$, however, we can only say that (8.7) *has the form* of a Fourier series because the sum is not actually a Fourier series because $w(n)$ is not a periodic function in general. It will be periodic only if $2\pi f_0$ is a rational number. If f_0 is irrational e_n is not periodic since, for example, $\sin(2\pi f_0 n)$ is only 0 if $n = 0$. Another difference with the ordinary Fourier series is that the frequencies do not have the form n/N for an integer period N (since there is no period). A series of this form behaves in many ways like a Fourier series and is an example of a generalized Fourier series or a Bohr-Fourier series. The signal e_n turns out to be an example of an *almost periodic function*. See, e.g., Bohr for a thorough treatment [4].

In both the continuous and discrete time case, (8.7) immediately gives the Fourier transform of the output signal as

$$W(f) = \sum_{k=-\infty}^{\infty} b_k \delta(f - kf_0). \qquad (8.9)$$

Thus the memoryless nonlinearity produces an infinite collection of harmonics of the original input signal, that is, an infinite sum of sinusoids with frequencies equal to integer multiples of the fundamental frequency f_0 of the original signal. Recall that a linear time invariant system could not have any output frequencies that were not also present at the input, but that a linear time varying system (such as amplitude modulation) could (and does) produce such frequencies.

We have not been rigorous in the above derivation in that we have not proved that the exchange of order of summation in (8.6) is valid. This would have to be done for a specific nonlinearity in order to prove the result.

8.3   Phase Modulation

The simplest example of the previous derivation is that of phase modulation. In Chapter 4 we considered amplitude modulation of an input signal where the output of a system was formed by multiplying the input signal by a sinusoid. In particular, an input signal $g = \{g(t); \ t \in \mathcal{R}\}$ resulted in an output signal $g_e(t) = ag(t)e^{i2\pi f_0 t}$. This complex signal provided a useful means of handling the transform manipulations. In real life the modulated waveform would be formed by taking the real part to form the signal $g_c(t) = \Re g_e(t) = ag(t)\cos(2\pi f_0 t)$. In contrast to this linear system, phase modulation is formed by modulating the carrier phase rather than its amplitude:

$$g_p(t) = a\cos(2\pi f_c t + \Delta g(t)) = \Re(ae^{i(2\pi f_c t + \Delta g(t))}).$$

This system is nonlinear, but memoryless. If the real part is ignored for convenience, then the nonlinearity is complex. It is also a time varying nonlinearity. To apply the results of the previous section, observe that $g_p(t)$ is the real part of the signal $y(t) = ae^{i(2\pi f_c t + \Delta g(t))} = ae^{i2\pi f_c t}e^{i\Delta g(t)}$. This is a memoryless nonlinearity with

$$\alpha_t(u) = ae^{i2\pi f_c t}e^{i\Delta u}$$

which has the form of (8.2) with $\Delta = 2\pi/L$

$$\frac{A_t(\frac{n}{L})}{L} = \begin{cases} e^{i2\pi f_c t} & n = 1 \\ 0 & \text{otherwise,} \end{cases}$$

so that only one term in the Fourier series is needed. If the input signal is taken as a sinusoid $g(t) = a\sin(2\pi f_0 t)$ we then have from (8.6) that

$$\begin{aligned} w(t) &= e^{i2\pi f_c t} \sum_{k=-\infty}^{\infty} J_k(\Delta a) e^{i2\pi f_0 t k} \\ &= \sum_{k=-\infty}^{\infty} J_k(\Delta a) e^{i2\pi(f_c + k f_0)t} \end{aligned}$$

and hence the spectrum is given by

$$W(f) = \sum_{k=-\infty}^{\infty} J_k(\Delta a)\delta(f - (f_c + k f_0)), \qquad (8.10)$$

a weighted sum of Dirac delta functions including the carrier frequency and all sidebands formed by adding or subtracting integral multiples of the input signal frequency to the carrier. This is a complicated spectrum, even though the input signal is simple! A single frequency component at the input results in an infinite collection of components at different frequencies at the output. This is typical behavior for a nonlinear system.

8.4   Uniform Quantization

A uniform quantizer can be thought of as a form of rounding off. A specified input range, say $[-b, b]$ is divided up into M bins of equal size Δ so that

$$\Delta = \frac{2b}{M}.$$

The number $R = \log_2 M$ is called the *rate* or *bit rate* of the quantizer and is measured in bits per sample. We usually assume for convenience that M is a power of 2 and hence R is an integer. If the input falls within a bin, then the corresponding quantizer output is the midpoint (or Euclidean centroid) of the bin. If the input is outside of the bins, its representative value is the midpoint of the closest bin. Thus a uniform quantizer is represented as a "staircase" function. Fig. 8.1 provides an example of a uniform quantizer $M = 8$ and hence $R = 3$ and $b = 4\Delta$. The operation of the quantizer can also be summarized as in Table 8.1.

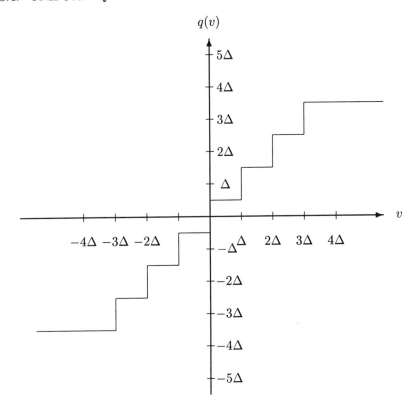

Figure 8.1: Uniform Quantizer

Since quantization introduces error, it is of interest to study the behavior of the error. Note that unlike the sampling case, information has genuinely been lost by quantization and there is in general no way to recover it. Hence one must include the distortion generated by quantization into any noise analysis of the system and any practical system must have enough bits (a large enough M) to ensure acceptable quality. Typical digitized images and pulse coded modulated speech have eight bits per sample. Computerized tomography images have 12 bits per sample. High fidelity digital audio typically has 14 bits.

Given an input v and its quantized value $q(v)$, define the *quantizer error*

$$\epsilon(v) = q(v) - v. \tag{8.11}$$

If the quantizer is applied to a discrete time signal $\{v_n\}$, then the corre-

If input is	then output is
$[3\Delta, \infty)$	$7\frac{\Delta}{2}$
$[2\Delta, 3\Delta)$	$5\frac{\Delta}{2}$
$[\Delta, 2\Delta)$	$3\frac{\Delta}{2}$
$[0, \Delta)$	$\frac{\Delta}{2}$
$[-\Delta, 0)$	$-\frac{\Delta}{2}$
$(-2\Delta, -\Delta)$	$-3\frac{\Delta}{2}$
$(-3\Delta, -2\Delta)$	$-5\frac{\Delta}{2}$
$(-\infty, -3\Delta)$	$-7\frac{\Delta}{2}$

Table 8.1: Uniform Quantizer, $M = 8$

sponding error sequence is

$$\epsilon_n = \epsilon(v_n) = q(v_n) - v_n. \tag{8.12}$$

We write it in this form so that the quantizer output can be written as its input plus an error term, viz.

$$q(v_n) = v_n + \epsilon_n. \tag{8.13}$$

This yields the so-called "additive noise model" of quantization error as depicted in Fig. 8.2.

This is not really a "model" at all since it is simply an alternative way of writing the equation defining the quantizer error. The modeling comes in when one makes statistical assumptions about the error behaving like random noise. Here, however, we will not consider such assumptions but will focus on exactly derived properties. The quantizer error signal is the output of primary interest here as it quantifies how well the quantizer approximates the original signal. Thus in this section the memoryless non-linearity of primary interest is not the quantizer itself, but the quantizer error function.

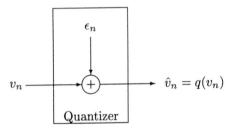

Figure 8.2: Additive noise model of a quantizer

We further simplify matters by focusing on normalized signals. In particular, define the normalized error

$$e(v) = \frac{\epsilon(v)}{\Delta} = \frac{q(v)}{\Delta} - \frac{v}{\Delta}.$$

The error function $e(v)$ is plotted as a function of v in Fig. 8.3 for the case of $M = 8$.

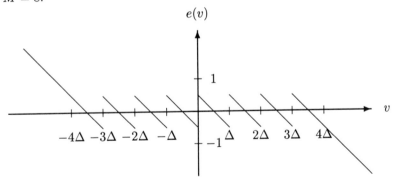

Figure 8.3: Normalized Quantizer Error Function

Several important facts can be inferred from the picture. First, the error function satisfies

$$|e(v)| \leq \frac{1}{2} \text{ if } |v| \leq 4\Delta.$$

In other words, the normalized error cannot get too big provided the input does not lie outside of the M quantizer bins. For general M the condition becomes

$$|e(v)| \leq \frac{1}{2} \text{ if } |v| \leq \frac{M}{2}\Delta = b,$$

that is, provided the input does not exceed the nominal range b. When an input falls outside this range $[-b, b]$ we say that the quantizer is *overloaded* or *saturated*. When a quantizer is not overloaded, the normalized quantization error magnitude cannot exceed $1/2$. In this case the quantization error is often called *granular noise*. If the quantizer is overloaded, the error is called *overload noise*. We here consider only granular noise and assume that the input range is indeed $[-b, b]$ and hence $e(v) \in [-1/2, 1/2]$. If this is not true in a real system, it is often forced to be true by clipping or limiting the input signal to lie within the range of the quantizer. Sometimes it is useful to model the quantizer as having an infinite number of levels, in which case the no-overload region is the entire real line.

Next observe that $e(u)$ is a periodic function of u for $u \in [-b, b]$ and that its period is Δ. In other words, the error is periodic within the no-overload region. In fact it can be shown by direct substitution and some algebra that

$$e(u) = \frac{1}{2} - \frac{u}{\Delta} \bmod 1; \ u \in [-b, b]. \tag{8.14}$$

This periodic signal can then be expanded in a Fourier series as

$$e(u) = \sum_{\substack{l=-\infty \\ l \neq 0}}^{\infty} \frac{1}{i2\pi l} e^{\frac{i2\pi l u}{\Delta}} = \sum_{l=1}^{\infty} \sin\left(2\pi l \frac{u}{\Delta}\right). \tag{8.15}$$

This series gives the error function except at the points of discontinuity. This is the Fourier series of the time-invariant memoryless nonlinearity $\alpha(u) = e(u)$ of interest in this section.

Now suppose that the input to the quantizer is a sampled signal $u_n = u(nT_s)$ as previously. We now have that the resulting normalized error sequence $e_n = e(u_n)$ is given by

$$e_n = \sum_{\substack{l=-\infty \\ l \neq 0}}^{\infty} \frac{1}{i2\pi l} e^{\frac{i2\pi l u_n}{\Delta}}. \tag{8.16}$$

This application of Fourier series to the error function in a uniform quantizer was first developed by Clavier, Panter and Grieg in 1947 [11, 12] and it was also suggested by Rice (as reported by Bennett in 1948 [2]).

Suppose next that the input is a sinusoid $u_n = a\sin(2\pi f_0 n)$, where $a \leq b$ is chosen so as not to overload the quantizer. Making the identifications $L = \Delta$ and

$$\frac{A(\frac{n}{L})}{L} = \begin{cases} \frac{1}{i2\pi n} & n \neq 0 \\ 0 & n = 0 \end{cases}$$

we immediately have from (8.7–8.8) that the normalized quantizer error signal is given by

$$e_n = \sum_{k=-\infty}^{\infty} b_k e^{i2\pi f_0 n k} \tag{8.17}$$

and that its Fourier transform is therefore

$$E(f) = \sum_{k=-\infty}^{\infty} b_k \delta(f - kf_0), \tag{8.18}$$

where here the frequency shift is modulo the frequency domain of definition, and where the coefficients are

$$b_k = \sum_{\substack{l=-\infty \\ l \neq 0}}^{\infty} \frac{1}{i2\pi l} J_k\left(2\pi l \frac{a}{\Delta}\right). \tag{8.19}$$

The formula for b_k can be simplified since

$$J_k(z) = (-1)^k J_k(-z)$$

and hence

$$b_k = \begin{cases} \sum_{\substack{l=-\infty \\ l \neq 0}}^{\infty} \frac{1}{i2\pi l} J_k\left(2\pi l \frac{a}{\Delta}\right) & k \text{ odd} \\ 0 & k \text{ even} \end{cases} \tag{8.20}$$

which yields

$$e_n = \sum_{m=-\infty}^{\infty} e^{in2\pi f_0(2m-1)} \left(\sum_{l \neq 0} \frac{1}{i2\pi l} J_{2m-1}\left(2\pi l \frac{a}{\Delta}\right) \right). \tag{8.21}$$

This formula has the form

$$e_n = \sum_{m=-\infty}^{\infty} e^{i\lambda_m n} c_m, \tag{8.22}$$

where

$$\lambda_m = f_0(2m - 1)$$

(taken mod 1 so that it is in $[0, 1)$) and

$$c_m = \sum_{l \neq 0} \frac{1}{i2\pi l} J_{2m-1}\left(2\pi l \frac{A}{\Delta}\right).$$

As previously discussed, this is actually a Bohr-Fourier or generalized Fourier series because e_n is not periodic.

We thus have a (generalized) Fourier series for the error signal when a sinusoid is input to a sampler and a uniform quantizer. The Fourier transform can be defined in the same way as it was for periodic signals: it is a sum of impulses at frequencies λ_m having area c_m. Note that although a single frequency sinusoid is put into the system, an infinity of harmonics is produced–a behavior not possible with a time invariant linear system.

As one might guess, this technique is not useful for all memoryless nonlinearities. It only works if the input/output mapping in fact has a Fourier transform or a Fourier series. This can fail in quite ordinary cases, for example a square-law device $\alpha(x) = x^2$ acting on an unbounded input cannot be handled using ordinary Fourier transforms. Laplace transforms can play a useful role for such cases. See, for example, Davenport and Root [15].

8.5 Problems

1. Frequency modulation (FM) of an input signal $g = \{g(t); \ t \in \mathcal{R}\}$ is defined by the signal

$$y_{\mathrm{FM}}(t) = \cos\left(2\pi f_c t + \Delta_{\mathrm{FM}} \int_0^t g(\tau)\,d\tau\right); \ t > 0.$$

Suppose that the input signal g is a sinusoid. Find the Fourier transform of y_{FM}.

2. Prove Equation (8.14).

3. A *hard limiter* is a memoryless nonlinear device with input/output relation

$$w(t) = a\,\mathrm{sgn}(u(t)) = \begin{cases} +a & u(t) > 0 \\ 0 & u(t) = 0 \\ -a & u(t) < 0 \end{cases}.$$

Find an exponential series for the output of a hard limiter when the input is a sampled sinusoid.

4. Suppose that a uniform quantizer has an odd number of levels with the middle level at the origin. Derive the series representation of the output when the input is a sampled sinusoid.

5. An alternative form of a hard limiter which arises in imaging takes the form

$$w(t) = \begin{cases} a & u(t) \in [0, E] \\ 0 & \text{otherwise} \end{cases}.$$

Find an exponential series for the output when the input is a sampled sinusoid.

6. Suppose that an image g is transformed using a unitary transform T into $G = TG$, which is quantized to form $\hat{G}$, which is inverse transformed to form $\hat{g}$. Prove that the mean squared error in the two domains is the same, that is, that

$$\sum_n \sum_k |g_{n,k} - \hat{g}_{n,k}|^2 = \sum_n \sum_k |G_{n,k} - \hat{G}_{n,k}|^2.$$

Appendix A

Fourier Transform Tables

We here collect several of the Fourier transform pairs developed in the book, including both ordinary and generalized forms. This provides a handy summary and reference and makes explicit several results implicit in the book. We also use the elementary properties of Fourier transforms to extend some of the results.

We begin in Tables A.1 and A.2 with several of the basic transforms derived for the continuous time infinite duration case. Note that both the Dirac delta $\delta(x)$ and the Kronecker delta δ_x appear in the tables. The Kronecker delta is useful if the argument x is continuous or discrete for representations of the form $h(x) = f(x)\delta_x + g(x)(1 - \delta_x)$ which means that $h(0) = f(0)$ and $h(x) = g(x)$ when $x \neq 0$.

The transforms in Table A.2 are all obtained from transforms in Table A.1 by the duality property, that is, by reversing the roles of time and frequency.

Several of the previous signals are time-limited (i.e., are infinite duration signals which are nonzero only in a finite interval) and hence have corresponding finite duration signals. The Fourier transforms are the same for any fixed real frequency f, but we have seen that the appropriate frequency domain S is no longer the real line but only a discrete subset. Table A.3 provides some examples.

Table A.4 collects several discrete time infinite duration transforms. Remember that for these results a difference or sum in the frequency domain is interpreted modulo that domain.

Table A.5 collects some of the more common closed form DFTs.

Table A.6 collects several two-dimensional transforms.

g	$\mathcal{F}(g)$
$\{\sqcap(t); t \in \mathcal{R}\}$	$\{\mathrm{sinc}(f); f \in \mathcal{R}\}$
$\{\square_T(t); t \in \mathcal{R}\}$	$\{2T\,\mathrm{sinc}(2Tf); f \in \mathcal{R}\}$
$\{e^{-t}u_{-1}(t); t \in \mathcal{R}\}$	$\{\frac{1}{1+2\pi i f}; f \in \mathcal{R}\}$
$\{e^{-\|t\|}; t \in \mathcal{R}\}$	$\{\frac{2}{1+(2\pi f)^2}; f \in \mathcal{R}\}$
$\{t \sqcap (t - \frac{1}{2}); t \in \mathcal{R}\}$	$\{\frac{1}{2}\delta_f + \left(\frac{ie^{-i2\pi f}}{2\pi f} + \frac{e^{-i2\pi f}-1}{(2\pi f)^2}\right)(1 - \delta_f); f \in \mathcal{R}\}$
$\{\wedge(t); t \in \mathcal{R}\}$	$\{\mathrm{sinc}^2(f); f \in \mathcal{R}\}$
$\{e^{-\pi t^2}; t \in \mathcal{R}\}$	$\{e^{-\pi f^2}; f \in \mathcal{R}\}$
$\{e^{+i\pi t^2}; t \in \mathcal{R}\}$	$\{\frac{1}{\sqrt{-i}}e^{-i2\pi f^2}; f \in \mathcal{R}\}$
$\{\mathrm{sgn}(t); t \in \mathcal{R}\}$	$\{\frac{-i}{\pi f}; f \in \mathcal{R}\}$
$\{u_{-1}(t); t \in \mathcal{R}\}$	$\{\frac{1}{2}\delta(f) - \frac{i}{2\pi f}(1 - \delta_f); f \in \mathcal{R}\}$
$\{\delta(t - t_0); t \in \mathcal{R}\}$	$\{e^{-i2\pi f t_0}; f \in \mathcal{R}\}$
$\{\delta(at + b); t \in \mathcal{R}\}$	$\{\frac{1}{\|a\|}e^{-i2\pi f \frac{b}{a}}; f \in \mathcal{R}\}$
$\{\delta(t); t \in \mathcal{R}\}$	$\{1; f \in \mathcal{R}\}$
$\{\delta'(t); t \in \mathcal{R}\}$	$\{i2\pi f; f \in \mathcal{R}\}$
$\{\Psi_T(t) = \sum_{n=-\infty}^{\infty} \delta(t - nT); t \in \mathcal{R}\}$	$\{\frac{1}{T}\Psi_{1/T}(f); f \in \mathcal{R}\}$
$\{\mathrm{III}(t); t \in \mathcal{R}\}$	$\{\mathrm{III}(f); f \in \mathcal{R}\}$
$\{\frac{1}{2}(\delta(t + \frac{1}{2}) + \delta(t - \frac{1}{2})); t \in \mathcal{R}\}$	$\{\cos(\pi f); f \in \mathcal{R}\}$
$\{\frac{1}{2}(\delta(t + \frac{1}{2}) - \delta(t - \frac{1}{2})); t \in \mathcal{R}\}$	$\{i\sin(\pi f); f \in \mathcal{R}\}$
$\{\mathrm{sech}(\pi t); t \in \mathcal{R}\}$	$\{\mathrm{sech}(\pi f); f \in \mathcal{R}\}$
$\{J_0(2\pi t); t \in \mathcal{R}\}$	$\begin{cases} \frac{1}{\pi\sqrt{1-f^2}} & f \in \mathcal{R}, \|f\| < 1 \\ 0 & \text{otherwise} \end{cases}$

Table A.1: Continuous Time, Infinite Duration

g	$\mathcal{F}(g)$
$\{\text{sinc}(t); t \in \mathcal{R}\}$	$\{\sqcap(f); f \in \mathcal{R}\}$
$\{\text{sinc}^2(t); t \in \mathcal{R}\}$	$\{\wedge(f); f \in \mathcal{R}\}$
$\{\frac{1}{\pi t}; t \in \mathcal{R}\}$	$\{-i\text{sgn}(f); f \in \mathcal{R}\}$
$\{e^{-i2\pi f_0 t}; t \in \mathcal{R}\}$	$\{\delta(f + f_0); f \in \mathcal{R}\}$
$\{1; t \in \mathcal{R}\}$	$\{\delta(f); f \in \mathcal{R}\}$
$\{\cos(\pi t); t \in \mathcal{R}\}$	$\{II(f); f \in \mathcal{R}\}$
$\{\sin(\pi t); t \in \mathcal{R}\}$	$\{iI_I(f); f \in \mathcal{R}\}$
$\{\frac{1}{2}\delta(t) + \frac{i}{2\pi t}(1 - \delta_t); t \in \mathcal{R}\}$	$\{u_{-1}(f); f \in \mathcal{R}\}$

Table A.2: Continuous Time, Infinite Duration (Duals)

g	$\mathcal{F}(g)$
$\{\sqcap(t); t \in [-\frac{T}{2}, \frac{T}{2})\}, T \geq 1$	$\{\operatorname{sinc}(f); f \in \{\frac{k}{T}; k \in \mathcal{Z}\}\}$
$\{\wedge(t); t \in [-\frac{T}{2}, \frac{T}{2})\}, T \geq 2$	$\{\operatorname{sinc}^2(f); f \in \{\frac{k}{T}; k \in \mathcal{Z}\}\}$
$\{\delta(t); t \in [-\frac{T}{2}, \frac{T}{2})\}$	$\{1; f \in \{\frac{k}{T}; k \in \mathcal{Z}\}\}$
$\{II(t); t \in [-\frac{T}{2}, \frac{T}{2})\}, T > 1$	$\{\cos(\pi f); f \in \{\frac{k}{T}; k \in \mathcal{Z}\}\}$
$\{I_I(t); t \in [-\frac{T}{2}, \frac{T}{2})\}, T > 1$	$\{i \sin(\pi f); f \in \{\frac{k}{T}; k \in \mathcal{Z}\}\}$
$\{t; t \in [0,1)\}$	$\{\frac{1}{2}\delta_k + \frac{i}{2\pi k}(1 - \delta_k); k \in \mathcal{Z}\}$

Table A.3: Continuous Time, Finite Duration

g	$\mathcal{F}(g)$
$\{r^n u_{-1}(n);\ n \in \mathcal{Z}\}\ (|r| < 1)$	$\{\frac{1}{1-re^{-i2\pi f}};\ f \in [-\frac{1}{2}, \frac{1}{2})\}$
$\{\square_N(n);\ n \in \mathcal{Z}\}$	$\{\frac{\sin(2\pi f(N+\frac{1}{2}))}{\sin(\pi f)};\ f \in [-\frac{1}{2}, \frac{1}{2})\}$
$\{\mathrm{sgn}(n);\ n \in \mathcal{Z}\}$	$\{\frac{1-e^{i2\pi f}}{1-\cos 2\pi f};\ f \in [-\frac{1}{2}, \frac{1}{2})\}$
$\{\delta_{n-n_0};\ n \in \mathcal{Z}\}$	$\{e^{-i2\pi f n_0};\ f \in [-\frac{1}{2}, \frac{1}{2})\}$
$\{\delta_n;\ n \in \mathcal{Z}\}$	$\{1;\ f \in [-\frac{1}{2}, \frac{1}{2})\}$
$\{e^{-i2\pi \frac{k}{N} n};\ n \in \mathcal{Z}\}$	$\{\delta(f + \frac{k}{N});\ f \in [-\frac{1}{2}, \frac{1}{2})\}$

Table A.4: Discrete Time, Infinite Duration

g	$\mathcal{F}(g)$
$\{r^n;\ n \in \mathcal{Z}_N\}$	$\{\frac{1-r^N}{1-re^{-i2\pi k/N}};\ k \in \mathcal{Z}_N\}$
$\{1;\ n \in \mathcal{Z}_N\}$	$\{N\delta_k;\ k \in \mathcal{Z}_N\}$
$\{\delta_n;\ n \in \mathcal{Z}_N\}$	$\{1;\ k \in \mathcal{Z}_N\}$
$\{\delta_{n-k};\ n \in \mathcal{Z}_N\}$	$\{e^{-i2\pi \frac{k}{N}};\ k \in \mathcal{Z}_N\}$

Table A.5: Discrete Time, Finite Duration (DFT)

Function	Transform
$\{\exp[-\pi(x^2 + y^2)];\ x, y \in \mathcal{R}\}$	$\{\exp\left[-\pi\left(f_X^2 + f_Y^2\right)\right];\ f_X, f_Y \in \mathcal{R}\}$
$\{\sqcap(x) \sqcap (y);\ x, y \in \mathcal{R}\}$	$\mathrm{sinc}(f_X)\ \mathrm{sinc}(f_Y);\ f_X, f_Y \in \mathcal{R}\}$
$\{\Lambda(x)\,\Lambda(y);\ x, y \in \mathcal{R}\}$	$\{\mathrm{sinc}^2(f_X)\ \mathrm{sinc}^2(f_Y);\ f_X, f_Y \in \mathcal{R}\}$
$\{\delta(x, y);\ x, y \in \mathcal{R}\}$	$\{1;\ f_X, f_Y \in \mathcal{R}\}$
$\{\exp[i\pi(x + y)];\ x, y \in \mathcal{R}\}$	$\{\delta(f_X - 1/2, f_Y - 1/2);\ f_X, f_Y \in \mathcal{R}\}$
$\{\mathrm{sgn}(x)\,\mathrm{sgn}(y);\ x, y \in \mathcal{R}\}$	$\{\frac{1}{i\pi f_X}\,\frac{1}{i\pi f_Y};\ f_X, f_Y \in \mathcal{R}\}$
$\{\mathrm{comb}(x)\,\mathrm{comb}(y);\ x, y \in \mathcal{R}\}$	$\mathrm{comb}(f_X)\,\mathrm{comb}(f_Y);\ f_X, f_Y \in \mathcal{R}\}$
$\{\exp[i\pi(x^2 + y^2)];\ x, y \in \mathcal{R}\}$	$i\exp\left[-i\pi\left(f_X^2 + f_Y^2\right)\right];\ f_X, f_Y \in \mathcal{R}\}$
$\{\exp[-(|x| + |y|)];\ x, y \in \mathcal{R}\}$	$\{\frac{2}{1+(2\pi f_X)^2}\,\frac{2}{1+(2\pi f_Y)^2};\ f_X, f_Y \in \mathcal{R}\}$
$\mathrm{circ}(\sqrt{x^2 + y^2};\ x, y \in \mathcal{R}\}$	$\frac{J_1(2\pi\sqrt{f_X^2 + f_Y^2})}{\sqrt{f_X^2 + f_Y^2}};\ f_X, f_Y \in \mathcal{R}\}$
$\delta(\sqrt{x^2 + y + 2};\ x, y \in \mathcal{R}\}$	$2\pi r_0 J_0(2\pi r_0 \sqrt{f_X^2 + f_Y^2});\ f_X, f_Y \in \mathcal{R}\}$

Table A.6: Two-dimensional Fourier transform pairs.

Bibliography

[1] B. S. Atal and S. L. Hanauer. Speech analysis and synthesis by linear prediction of the speech wave. *J. Acoust. Soc. Am.*, 50:637–655, 1971.

[2] W. R. Bennett. Spectra of quantized signals. *Bell Systems Technical Journal*, 27:446–472, July 1948.

[3] S. Bochner. *Lectures on Fourier Integrals*. Princeton University Press, 1959.

[4] H. Bohr. *Almost Periodic Functions*. Chelsea, New York, 1947. Translation by Harvey Cohn.

[5] F. Bowman. *Introduction to Bessel Functions*. Dover, New York, 1958.

[6] R. Bracewell. *The Fourier Transform and Its Applications*. McGraw-Hill, New York, 1965.

[7] R. Bracewell. *The Hartley Transform*. Oxford Press, New York, 1987.

[8] R.N. Bracewell. *Two-Dimensional Imaging*. Prentice Hall, Englewood Cliffs, New Jersey, 1995.

[9] E. O. Brigham. *The fast Fourier transform*. Prentice-Hall, Englewood Cliffs, New Jersey, 1974.

[10] H. S. Carslaw. *An Introduction to the Theory of Fourier's Series and Integrals*. Dover, New York, 1950.

[11] A. G. Clavier, P. F. Panter, and D. D. Grieg. Distortion in a pulse count modulation system. *AIEE Transactions*, 66:989–1005, 1947.

[12] A. G. Clavier, P. F. Panter, and D. D. Grieg. PCM distortion analysis. *Electrical Engineering*, pages 1110–1122, November 1947.

[13] J. W. Cooley and J. W. Tukey. An algorithm for the machine calculation of complex Fourier series. *Math. Computation*, 19:297–301, 1965.

[14] I. Daubechies. *Ten Lectures on Wavelets*. Society for Industrial and Applied Mathematics, Philadelphia, 1992.

[15] W. B. Davenport and W. L Root. *An Introduction to the Theory of Random Signals and Noise*. McGraw-Hill, New York, 1958.

[16] J. B. J. Fourier. *Théorie analytiqe de la chaleur*. 1922. Available as a Dover Reprint.

[17] R. C. Gonzalez and P. Wintz. *Digital Image Processing*. Addison-Wesley, Reading, Mass., second edition, 1987.

[18] J.W. Goodman. *Introduction to Fourier Optics*. McGraw-Hill, New York, 1988.

[19] U. Grenander and G. Szego. *Toeplitz Forms and Their Applications*. University of California Press, Berkeley and Los Angeles, 1958.

[20] F. Itakura and S. Saito. A statistical method for estimation of speech spectral density and formant frequencies. *Electron. Commun. Japan*, 53-A:36–43, 1970.

[21] I. Katznelson. *An Introduction to Harmonic Analysis*. Dover, New York, 1976.

[22] J. D. Markel and A. H. Gray, Jr. *Linear Prediction of Speech*. Springer-Verlag, New York, 1976.

[23] A. V. Oppenheim and R. W. Schafer. *Digital Signal Processing*. Prentice Hall, Englewood Cliffs,New Jersey, 1975.

[24] A. Papoulis. *The Fourier Integral and Its Applications*. McGraw-Hill, New York, 1962.

[25] J. R. Pierce. *The Science of Musical Sound*. Scientific American Books, New York, 1983.

[26] M. Rabbani and P. W. Jones. *Digital Image Compression Techniques*, volume TT7 of *Tutorial Texts in Optical Engineering*. SPIE Optical Engineering Press, Bellingham, Washington, 1991.

[27] D.R. Rao and P. Yip. *Discrete Cosine Transform*. Academic Press, San Diego, California, 1990.

[28] S. O. Rice. Mathematical analysis of random noise. In N. Wax and N. Wax, editors, *Selected Papers on Noise and Stochastic Processes*, pages 133–294. Dover, New York, NY, 1954. Reprinted from Bell Systems Technical Journal,Vol. 23:282–332 (1944) and Vol. 24: 46–156 (1945).

[29] O. Rioul and M. Vetterli. Wavelets and signal processing. *IEEE Signal Processing Magazine*, 8(4):14–38, October 1991.

[30] W. McC. Siebert. *Circuits,Signals,and Systems*. M.I.T. Press, Cambridge, 1986.

[31] Gilbert Strang. Wavelet transforms versus fourier transforms. *Bulletin (New Series) of the AMS*, 28(2):288–305, April 1993.

[32] E. C. Titchmarsh. *Introduction to the Theory of Fourier Integrals*. Oxford University Press, Oxford, 1937.

[33] James S. Walker. *Fourier Analysis*. Oxford University Press, New York, 1988.

[34] G.K. Wallace. The JPEG still picture compression standard. *Communications of the ACM*, 34(4):30–44, April 1991.

[35] G. N. Watson. *A Treatise on the Theory of Bessel Functions*. Cambridge University Press, Cambridge, 1980. Second Edition.

[36] N. Wiener. *The Fourier Integral and Certain of Its Applications*. Cambridge University Press, New York, 1933.

[37] John W. Woods, editor. *Subband Image Coding*. Kluwer Academic Publishers, Boston, 1991.

Index